Probability and Its Applications

Published in association with the Applied Probability Trust

Editors: S. Asmussen, J. Gani, P. Jagers, T.G. Kurtz

Probability and Its Applications

The *Probability and Its Applications* series publishes research monographs, with the expository quality to make them useful and accessible to advanced students, in probability and stochastic processes, with a particular focus on:

- Foundations of probability including stochastic analysis and Markov and other stochastic processes
- Applications of probability in analysis
- Point processes, random sets, and other spatial models
- Branching processes and other models of population growth
- Genetics and other stochastic models in biology
- Information theory and signal processing
- Communication networks
- Stochastic models in operations research

For further volumes:
www.springer.com/series/1560

Anatoliy Malyarenko

Invariant Random Fields on Spaces with a Group Action

 Springer

Anatoliy Malyarenko
School of Education, Culture,
   and Communication
Mälardalen University
Västerås, Sweden

Series Editors

Søren Asmussen
Department of Mathematical Sciences
Aarhus University
Aarhus, Denmark

Joe Gani
Centre for Mathematics and its Applications
Mathematical Sciences Institute
Australian National University
Canberra, Australia

Peter Jagers
Mathematical Statistics
Chalmers University of Technology
   and University of Gothenburg
Gothenburg, Sweden

Thomas G. Kurtz
Department of Mathematics
University of Wisconsin–Madison
Madison, WI, USA

ISSN 1431-7028   Probability and Its Applications
ISBN 978-3-642-44450-0   ISBN 978-3-642-33406-1 (eBook)
DOI 10.1007/978-3-642-33406-1
Springer Heidelberg New York Dordrecht London

Mathematics Subject Classification (2010): 60G60, 60G17, 60G22, 85A40

To my soulmate, supporter, and love —
my wife Iryna

Foreword

Random functions of several variables, or random fields, appear in the statistical theory of turbulence, meteorology, quantum physics, computer graphics, pattern recognition, system identification, etc. The most important case is when the finite-dimensional distributions of the field under consideration are invariant under an action of some transformation group. The precise links between the spectral theory of invariant random fields, the theory of invariant positive-definite functions, and the theory of group representations have been known since the middle of the last century.

New links have been discovered in recent applications. In particular, Differential Geometry plays a significant role in the spectral theory of invariant random sections of vector and tensor bundles. This link was described in several physical papers where the statistical properties of the Cosmic Microwave Background were studied. Rigourous mathematical results appeared very recently. This theory was poorly represented in monograph form. I am aware of only one book by D. Marinucci and G. Peccati (2011), *Random fields on the sphere*, vol. 389 of London Math. Soc. Lect. Note Ser., Cambridge University Press, Cambridge. However, this book concentrates more on spherical random fields rather than invariant random fields on general spaces with a group action.

The present monograph fills the above gap in the literature. This monograph describes the spectral theory of invariant random fields in vector bundles in a general form. Previously known classical results appear as special cases when the vector bundle is a trivial one.

Apart from the spectral theory, the author considers inversion formulae, differential equations with random fields, linear functionals of random fields, sample path properties of Gaussian random fields. Most results were proved by the author; some results are new and appear here for the first time.

Two special features of the book extend its audience. First, a separate chapter is devoted to applications in Approximation Theory, Earthquake Engineering, and Cosmology. Second, preliminary knowledge of Differential Geometry, Lie Groups and Algebras, Abstract Harmonic Analysis and some other parts of mathematics outside the scope of Probability and Statistics are not required. The necessary results are formulated in the Appendix (Chap. 6), mostly without proofs. Extensive bibliographical remarks are included in each chapter.

Malyarenko's monograph on this highly complex topic is written in a way that makes material very accessible. This is a book which is both technically interesting and a pleasure to read. The presentation is clear and the book should be useful for almost anyone who uses stochastic processes and random fields as tools for describing the real world.

The book may be interesting for all mathematicians, graduate and postgraduate students as a self-contained source of recent results in the theory of invariant random fields. Both specialists in the theory of random fields and scientists working in related applied areas that utilise this theory will no doubt find a great deal of new and useful information in this book.

Cardiff, United Kingdom   Nikolai Leonenko

Preface

Random fields appeared for the very first time in the 1920s in classical research papers about turbulence written by A.A. Friedmann, J. Kampé de Fériet, L.P. Keller, A.N. Kolmogorov, T. von Kármán, A.M. Obukhov, among others. Later on, the mathematical theory of random fields was developed and successfully applied in computer graphics, earthquake engineering, medicine, quantum physics, statistical mechanics, etc. The above theory uses such mathematical tools as Abstract Harmonic Analysis, Special Functions, Abelian and Tauberian theorems, etc.

At the end of the last millennium, new applications appeared. At that time, cosmology transformed into a predictive science whose predictions were tested against precise observations. In particular, tiny fluctuations of the cosmic microwave background were discovered. In order to build the rigourous mathematical model of the above fluctuations, one has to construct an isotropic random section of a special tensor bundle over the two-dimensional sphere. Therefore, the set of actively used mathematical tools for a specialist in random fields has to be extended with Differential Geometry, Lie Groups and Lie Algebras, just to mention a few.

New applications generated new challenges. In particular, currently there exists no single book written for specialists in probability that describes the current state of the spectral theory of invariant random fields and includes all necessary material from the above-mentioned non-probabilistic parts of mathematics. Our book originated from an attempt to fill this gap in the literature. The contents of the book are described in detail in Chap. 1. Here, we just present the book's scope.

Most random fields that appear in applied areas are invariant under an action of some group G. The group G acts either on the parametric space T of a random field or in some space of sections of a vector or tensor bundle over T. In Chap. 2 we describe a unified approach to spectral expansions of such fields based on the theory of induced group representations.

We divide the remaining part of the theory of random fields into two areas. In the first area, the so-called L^2 theory, a random field is considered as a function on T with values in the Hilbert space of square integrable random vectors. Some aspects of this theory are presented in Chap. 3. In the second area, some restrictions on the finite-dimensional distributions of the random field under consideration are imposed. In Chap. 4 we consider some questions of the theory of Gaussian random fields.

In Chap. 5 we consider applications of the above-described material to approximation theory, cosmology, and earthquake engineering. Finally, in the Appendix (Chap. 6) we consider mathematical tools outside the scope of probability and statistics, that are necessary for a specialist in random fields. Bibliographical remarks conclude each chapter and the Appendix (Chap. 6). Throughout the book, we use Halmos' abbreviation iff for "if and only if".

The book is intended for specialists in the theory of random fields, for mathematicians who would like to study the above theory, as well as for specialists in applied areas. It may be useful for graduate and postgraduate students in probability, statistics, functional analysis, cosmology, earthquake engineering, etc.

Västerås, Sweden   Anatoliy Malyarenko

Acknowledgements

This book would never have been written without the influence of two people. My teacher, Professor M.Ĭ. Yadrenko, showed me the beauty of Probability. He was both the very best teacher and the most benevolent and responsive person I have ever met in my life.

Love, care, and attention of my wife and friend, Iryna Yelsukova, played a crucial role in writing this book.

I am grateful to my university teachers Professors V.V. Buldygin, N.V. Kartashov, V.S. Korolyuk, Yu.V. Kozachenko, N.N. Leonenko, M.P. Moklyachuk, O.I. Ponomarenko, D.S. Silvestrov, A.V. Skorokhod, and A.F. Turbin.

Some results described in this book were proved due to a fruitful collaboration with my co-authors Lambros Katafygiotis, Andriy Olenko, Constantinos Papadimitriou, and Aspaziya Zerva.

I would like to thank my colleagues Professors Nicholas Bingham, Anatoliy Klimyk, Domenico Marinucci, and Carl Mueller for useful discussions on Tauberian and Abelian theorems, representation theory, cosmology, and functional limit theorems.

The remarks of the three anonymous referees of this book helped to improve its structure and exposition.

Finally, I am grateful to my colleagues from the Division of Applied Mathematics of the Mälardalen University for providing such a friendly working and research environment.

Västerås, Sweden   Anatoliy Malyarenko

Contents

List of Tables

Chapter 1
Introduction

Abstract In this introductory chapter, we describe the contents of our book.

On December 11, 1944, A.N. Kolmogorov gave a lecture "Problems of probability theory" at the meeting of the Moscow Mathematical Society. A few pages containing a summary and a sketch of this lecture were found in his archives. They were published in Kolmogorov (1993). The sixth Kolmogorov problem is as follows.

6. Distributions of scalar, vector and tensor functions which are invariant with respect to different groups of transformations. This problem is of interest from the point of view of Statistical Mechanics of Continuous Media, in particular, of Statistical Turbulence Theory. (*Translated from the Russian by V.V. Sazonov.*)

Several results concerning this problem were obtained by one of Kolmogorov's pupils, A.M. Yaglom, and published in Yaglom (1957, 1960, 1961, 1963). Applications to Statistical Mechanics of Continuous Media mentioned by Kolmogorov were presented in the books by Yaglom (1987a,b) and Monin and Yaglom (2007a,b).

A new wave of interest in Kolmogorov's sixth problem occurred at the end of the last millennium. Precise measurements of the cosmic microwave background (CMB) showed its fluctuations. Cosmologists assumed that these fluctuations are generated by a homogeneous and isotropic random field. From the observations, one can define the intensity tensor P of the CMB. In fact, it is a tensor-valued function of position $\mathbf{x}$, time t and photon direction $\mathbf{n}$. Since we measure the intensity tensor here ($\mathbf{x}$ is fixed) and now (t is fixed), P becomes a tensor-valued function on the sphere S^2. More precisely, P($\mathbf{n}$) is a random section of some tensor bundle over S^2. Therefore, cosmology requires the theory of random sections of vector and tensor bundles.

The author has been interested in questions connected to the sixth Kolmogorov problem since the second half of the 1970s, when he was a graduate student of Taras Shevchenko National University of Kyiv. At that time, the paper by Yaglom (1957) was refereed at the students' seminar in probability led by Professor M.Ĭ. Yadrenko.

In Chap. 2 we consider spectral expansions of invariant random fields in vector bundles. A scalar, i.e. real-valued (resp. complex-valued) random field $X(t)$ on a parametric space T may be considered as a random section of the one-dimensional trivial bundle $T \times \mathbb{R}$ (resp. $T \times \mathbb{C}$) over T. In order to simplify the exposition, we consider the cases of a scalar random field and a vector random field separately.

A. Malyarenko, *Invariant Random Fields on Spaces with a Group Action*, Probability and Its Applications, DOI 10.1007/978-3-642-33406-1_1, © Springer-Verlag Berlin Heidelberg 2013

All definitions in Subsection 2.1.1 are either classical or directly generalise classical definitions. This is not the case for Subsection 2.1.2, where we consider random sections of vector bundles. Indeed, let $\xi = (E, \pi, T)$ be a vector bundle, and let $\mathbf{X}(t)$ be a random section of the bundle ξ. Let s and t be two points in the base T with $s \neq t$. Random vectors $\mathbf{X}(s)$ and $\mathbf{X}(t)$ lie in different spaces. The following problem arises: how does one define a mean-square continuous random field in this situation?

To overcome this difficulty, we extend an idea of Kolmogorov formulated by him for the case of a finite-dimensional trivial vector bundle and published by his pupils Rozanov (1958) and Yaglom (1961). We define a scalar random field on the total space E, which we call the field *associated* to the vector random field $\mathbf{X}(t)$. Then, we call $\mathbf{X}(t)$ mean-square continuous if the associated scalar random field is mean-square continuous.

Let G be a topological group acting continuously from the left on the base T. We would like to call a vector random field $\mathbf{X}(t)$ *wide-sense left G-invariant*, if the associated scalar random field is wide-sense left G-invariant under some left-continuous action of G on the total space E. However, in general there exists no natural continuous left action of G on E. In Definition 2.12, we define an action of G on E *associated* to its action on the base space T. Then, we call a vector random field $\mathbf{X}(t)$ wide-sense left G-invariant, if the associated scalar random field is wide-sense left G-invariant under the associated action. For the case of the trivial vector bundle, our definition becomes a classical one.

In Example 2.16, we consider an important example of an associated action: the so-called *homogeneous*, or *equivariant* vector bundles. They are interesting for us for several reasons.

On the one hand, they have a natural associated action of some topological group G. Moreover, the above action identifies the vector space fibres over all the points of the base space. Therefore, all random vectors of a random field $\mathbf{X}(t)$ in a homogeneous vector bundle lie in the same space. We prove that for homogeneous vector bundles, our definitions of mean-square continuous field and invariant field are equivalent to the classical ones.

On the other hand, the Hilbert space of the square-integrable sections of a homogeneous vector bundle carries the so-called *induced representation* of the group G. Later on, we use the well-developed theory of induced representations to obtain spectral expansions of invariant random fields in homogeneous vector bundles.

In Example 2.17 we consider two one-dimensional vector bundles over the circle S^1: a trivial bundle (cylinder), and a nontrivial one (Möbius bundle). While there exist a lot of centred mean-square continuous wide-sense left SO(2)-invariant random fields in the trivial bundle, the one and only such field in the Möbius bundle is 0.

We begin our study with the classical case, when the parametric space T is a topological group acting on itself by left translations. The spectral expansion of invariant random fields on a wide class of topological groups was obtained by Yaglom (1961). We formulate his result in our Theorem 2.18.

Examples of using Theorem 2.18 include both classical and new results. In particular, we consider spectral expansions of wide-sense stationary sequences and processes, homogeneous random fields on the lattice $\mathbb{Z}^n$ and on the space $\mathbb{R}^n$, left-homogeneous random fields on low-dimensional Lorentz groups and their covers. The new result here is the description of homogeneous random fields on the group $\mathbb{R}_0^\infty$, which is not locally compact.

Vector invariant random fields in the above situation were also considered by Yaglom (1961). His result is formulated in our Theorem 2.22. We have also corrected some misprints in the original proof. Examples include classical cases of vector homogeneous random fields on commutative locally compact groups with countable base.

The next case is that when the parametric space T is a homogeneous space $T = G/K$, and the group G acts on T by left translations. For simplicity, we consider only the case when K is a massive compact subgroup of the group G. The spectral expansion of the left-invariant random field on T is obtained in Theorem 2.23. This result was also proved by Yaglom (1961).

In Example 2.24 we begin the study of the main objects of this book. We consider the case when T is a two-point homogeneous space. The above case is interesting for the following reasons. On the one hand, it includes classical cases of homogeneous and isotropic random fields on the space $\mathbb{R}^n$ and isotropic random fields on the sphere S^n studied in detail by Yadrenko (1983). On the other hand, the covariance function $R(x, y)$ of such a field depends only on the distance between the points x and y. In other words, R is a function of *one real variable*. This simplifies further studies of various questions connected to such fields.

In Example 2.25 we propose yet another proof of the famous theorem by Schoenberg about the general form of a positive-definite kernel $R(\mathbf{x}, \mathbf{y})$ on the infinite-dimensional separable real Hilbert space $(H, \| \cdot \|)$ that depends only on $\|\mathbf{x} - \mathbf{y}\|$. The spectral expansion of a homogeneous and isotropic field on H is obtained.

In Subsection 2.3.2 we consider an invariant random section $\mathbf{X}(t)$ of a homogeneous vector bundle. Let G be a compact topological group, let K be a compact subgroup, and let W be a unitary representation of K in a finite-dimensional complex Hilbert space H. The unitary representation U of the group G induced by the representation W is realised in the Hilbert space $L^2(G, H)$ of all square-integrable sections of the vector bundle over the homogeneous space $T = G/K$. We introduce a basis in the above space, and find the spectral expansion of the random field $\mathbf{X}(t)$ with respect to the above basis. Note that in the particular case of $G = SO(3)$, $K = SO(2)$, $H = \mathbb{C}$, and $W(\varphi) = e^{in\varphi}$, where $\varphi \in SO(2)$ and $n \in \mathbb{Z}$, the elements of the above basis are well known under the name *spin-weighted spherical harmonics*.

In Sect. 2.4 we consider the general case, when G is a compact group acting continuously from the left on a Hausdorff topological space T, and $X(t)$ is a wide-sense mean-square continuous G-invariant random field on T. Under some technical conditions, we obtain the spectral expansion of the field $X(t)$. The above expansion contains the set of uncorrelated centred mean-square continuous random fields $Z_{m\lambda}(s)$ defined on a measurable section S for the action of the group G. In the classical case, when $G = SO(n)$ acts on $T = \mathbb{R}^n$ by matrix-vector multiplication, we have $S = [0, \infty)$, the corresponding invariant random field is called *isotropic*, and we recover the well-known spectral expansion proved by Yadrenko (1963). In particular, the multiparameter fractional Brownian motion is isotropic. In Example 2.31 we calculate the autocovariance functions of the random processes $Z_{m\ell}(s)$ involved in the spectral expansion of multiparameter fractional Brownian motion.

The vector variant of the previous considerations is represented in Subsection 2.4.2. Let a connected compact Lie group G act continuously from the left on a Hausdorff topological space T. Let λ be an irreducible unitary representation of G. Let $\xi = (E, \pi, T)$ be a vector bundle over T that satisfies the following technical condition. The restriction of the vector bundle ξ to any G-orbit with stationary subgroup K is a homogeneous vector bundle $\eta = (F, \pi, O)$. Moreover, the Hilbert space of square-integrable sections of the bundle η carries the unitary representation of the group G induced by the restriction of λ to K. When the representation λ is trivial, we return to the situation described in the previous paragraph. Under some additional technical conditions we obtain the spectral expansion of an invariant random section $\mathbf{X}(t)$ of the vector bundle ξ. This is the final step in our chain of generalisations.

In the two remaining sections of Chap. 2 we consider two interesting problems. First, we solve the problem formulated by Yaglom (1957) about the spectral expansion of a vector homogeneous and isotropic random field on the space $\mathbb{R}^n$. Our solution involves special numbers, the so-called *Clebsch–Gordan coefficients* of the group $SO(n)$ (or, more generally, the Clebsch–Gordan coefficients of a compact subgroup G of the group of automorphisms of a commutative locally compact topological group T with countable base). Second, we introduce the so-called *Volterra random fields*. These are isotropic random fields $X(\mathbf{t})$ on the space R^n such that the stochastic processes $Z_{m\ell}(s)$ involved in the spectral expansion of the field $X(\mathbf{t})$ are Volterra processes. For any positive real number R, we find a family of spectral expansions of any Volterra random field. Every expansion of the above family converges in mean-square in the centred ball of radius R in the space $\mathbb{R}^n$. As an example, we calculate the spectral expansion of multiparameter fractional Brownian motion.

We divide all the theory of random fields except their spectral expansions into two areas. The first area does not require any restrictions on the finite-dimensional distributions of the random field under consideration. In other words, we consider the random field as a function with values in the Hilbert space of all centred random vectors with finite expectation of the square of the norm. Such a theory is called L^2 theory and is considered in Chap. 3.

First, we consider the following question. Assume that the random field $X(t)$ is expanded in terms of some random measures Z. How does one restore the measures Z if X is known? In other words, in Sect. 3.1 we prove *inversion formulae* for various spectral expansions obtained in Chap. 2.

The next group of questions of L^2 theory is as follows. Let G be a Lie group that acts on an analytic manifold T. Let L be a linear differential operator on T, and let $\mathscr{A}$ be the set of all centred wide-sense left G-invariant random fields on T. In Sect. 3.2 we consider the following questions.

- Describe the set $\mathscr{A}_L$ of all random fields $X(t)$ such that the field $LX(t)$ may be correctly defined as a result of mean-square differentiation.

- Find sufficient conditions on L for inclusion $L\mathscr{A}_L \subset \mathscr{A}$.
- For each $X \in \mathscr{A}_L$, calculate $\mathsf{E}[X(s)\overline{LX(t)}]$.
- Describe the image $L\mathscr{A}_L$ in terms of the spectral measure of X.

To give an idea of what is described in Sect. 3.3, consider the following situation. Let $X(t)$ be a centred G-invariant random field on a space T. Let K be a subgroup of the group G, and let C be an orbit of the group K in T. Assume that there exists a left K-invariant measure $d\mu$ on C. Let $f(t)$ be a continuous function on C with compact support. The Bochner integral

$$Y := \int_C X(t)f(t)\,d\mu(t)$$

defines a random variable Y, which is a vector of the Hilbert space $H_X^-(C)$—the closed linear span of the vectors $X(t)$, $t \in C$ in the Hilbert space of all random variables with finite variance. Is it possible to express *any* element of the space $H_X^-(C)$ in the form of the above equation?

Let t_0 be a fixed point in C. The random variable $X(t_0)$ should have the following representation:

$$X(t_0) = \int_C X(t)\delta(t - t_0)\,d\mu(t).$$

In other words, we have to be able to define integrals involving random fields and *distributions* like the Dirac delta.

In Sect. 3.3, we describe the space $L_-^2(C, d\mu)$ of distributions for which the above integral may be correctly defined as an element of the space $H_X^-(C)$. Moreover, the above integral determines an isometric isomorphism between the spaces $L_-^2(C, d\mu)$ and $H_X^-(C)$. In our proofs, we use the theory of rigged Hilbert spaces, or *Gel'fand triples*. The above spaces are negative spaces in the triples, therefore our notation includes minus as an index.

The second area lies outside L^2 theory. Here, we impose a simple condition on the finite-dimensional distributions of the random field $X(t)$. Namely, in Chap. 4 we consider *Gaussian* invariant random fields and the properties of their sample paths.

It is well known that the above properties can be studied using the Dudley semi-metric ϱ_X. Assume that the covariance function of a centred random field $X(t)$ defined on the metric space (T, ϱ), $R(t_1, t_2) = \mathsf{E}[X(t_1)X(t_2)]$, depends only on the distance $\varrho(t_1, t_2)$. For example, this is true when $T = G/K$ is a two-point homogeneous space and X is a G-invariant random field on T. Then, any two balls of the same radius in the space (T, ϱ_X) are isometric. This simplifies investigations of the properties of sample paths of such fields.

In particular, let $R(\theta)$ be the covariance between $X(s)$ and $X(t)$ when the distance between s and t is equal to θ. To estimate the Dudley integral, we have to estimate

$$\varrho_X^2(s, t) := 2\big[R(0) - R(\theta)\big]$$

in terms of the spectral measure of the random field $X(t)$. Therefore, we need to prove Abelian and/or Tauberian theorems. In Sect. 4.2, we find the upper estimates of the function $R(\theta)$ in the neighbourhood of 0 for invariant random fields on two-point homogeneous spaces. These estimates are used to find the uniform moduli of continuity of the above fields.

More advanced Abelian and Tauberian theorems in terms of regularly varying functions are considered in Sect. 4.3. The above theorems help us to find more precise moduli of continuity.

In Sect. 4.4 we further investigate one of the series expansions of the multiparameter fractional Brownian motion previously obtained in Example 2.39. We prove that the above expansion is rate optimal.

Finally, in Sect. 4.5 we prove a general functional limit theorem for the multiparameter fractional Brownian motion. The functional law of the iterated logarithm, functional Lévy modulus of continuity and many other results are particular cases.

Chapter 5 is devoted to applications. The idea of an application to approximation theory is as follows. Let $X(t)$ be a centred Gaussian random field on a compact metric space T with continuous sample paths, and let P be the corresponding Gaussian measure in the Banach space $C(T)$ of all continuous functions on T equipped with the sup-norm. Let $\mathscr{L}$ be a measurable linear subset of the linear space $C(T)$. The 0–1 law states that either $\mathsf{P}(\mathscr{L}) = 0$ or $\mathsf{P}(\mathscr{L}) = 1$. Moreover, in the latter case the reproducing kernel Hilbert space $\mathscr{H}$ of the measure P is the subset of $\mathscr{L}$.

In many cases, the space $\mathscr{H}$ consists of all functions which have some *constructive* property. For example, $\mathscr{H}$ may consist of functions that can be expanded into Fourier series with respect to some exactly determined basis. On the other hand, the linear subset $\mathscr{L}$ consists of all functions which have some *descriptive* property. For example, $\mathscr{L}$ may consist of functions whose modulus of continuity satisfies some conditions. In this situation, the 0–1 law states that some descriptive property of a continuous function on T follows from some constructive property. In other words, the above law is equivalent to a *Bernstein-type theorem* from approximation theory. Several Bernstein-type theorems are proved in Sect. 5.1.

In Sect. 5.2 we consider an application to cosmology. After reading several physical books and papers the author has found a jungle of various choices of coordinates, phase conventions etc. Therefore, Subsection 5.2.1 contains a short introduction to the *deterministic* model of the cosmic microwave background (CMB) for mathematicians. In particular, we discuss different choices of local coordinates in the tangent bundle $\xi = (TS^2, \pi, S^2)$, and fix our choice. We explain both the mathematical and physical sense of the *Stokes parameters* I, Q, U, and V.

The *probabilistic* model of the CMB is discussed in Subsection 5.2.2. We define the set of vector bundles $\xi_s = (E_s, \pi, S^2)$, $s \in \mathbb{Z}$, where the representation of the rotation group $G = \mathrm{SO}(3)$ induced by the representation $W(g_\alpha) = e^{\mathrm{i}s\alpha}$ of the subgroup $K = \mathrm{SO}(2)$ is realised. In particular, the absolute temperature of the CMB, $T(\mathbf{n})$, is a single realisation of a mean-square continuous strict-sense isotropic (i.e., $\mathrm{SO}(3)$-invariant) random field in ξ_0, while the complex polarisation, $(Q \pm \mathrm{i}U)(\mathbf{n})$, is a single realisation of a mean-square continuous strict-sense isotropic random field in $\xi_{\pm 2}$. Because any second-order strict-sense isotropic random field is automatically wide-sense isotropic, Theorem 2.27 immediately gives the spectral expansion of the above random fields. In the case of the absolute temperature, the functions involved in the formulation of the above theorem become familiar *spherical harmonics*, $Y_{\ell m}$, while in the case of the complex polarisation they become *spin-weighted spherical harmonics*, $\pm 2 Y_{\ell m}$. The expansion coefficients are uncorrelated random variables with finite variance, which do not depend on the index m. In physical terms, the variance as a function of the parameter ℓ is the *power spectrum*.

Following Zaldarriaga and Seljak (1997), we construct the random fields $E(\mathbf{n})$ and $B(\mathbf{n})$. The advantage of these fields is that they are scalar (i.e., live in ξ_0), real-valued, and isotropic. Moreover, only $T(\mathbf{n})$ and $E(\mathbf{n})$ may be correlated, while the two remaining pairs are always uncorrelated.

Our new result is Theorem 5.5. It states that the standard assumption of cosmological theories (the random fields $T(\mathbf{n})$, $E(\mathbf{n})$, and $B(\mathbf{n})$ are jointly isotropic) is equivalent to the assumption that $((Q - iU)(\mathbf{n}), T(\mathbf{n}), (Q + iU)(\mathbf{n}))$ is an isotropic random field in $\xi_{-2} \oplus \xi_0 \oplus \xi_2$.

In Sect. 5.3 we consider an application to earthquake engineering. We present an efficient approach for the simulation of homogeneous and partially isotropic random fields based on spectral expansion. It is shown that, by incorporating the partial isotropy of the field into the simulation algorithm, the computational effort required for the simulation is significantly reduced as compared with the case when only the homogeneity of the field is taken into account. The above approach has been applied by Katafygiotis et al. (1999) for simulation of a Gaussian homogeneous and spatially isotropic random field representing the ground motion during a strong earthquake with Clough–Penzien spectrum of the acceleration of apparent propagation of ground motions and Luco–Wong coherency model.

The theory of invariant random fields on spaces with a group action requires good knowledge of various part of mathematics other than Probability and Statistics. In the Appendix (Chap. 6) we discuss differentiable manifolds, vector bundles, Lie groups and Lie algebras, group actions and group representations, special functions, rigged Hilbert spaces, Abelian and Tauberian theorems.

We conclude each chapter and the Appendix (Chap. 6) with bibliographical remarks.

References

L.S. Katafygiotis, A. Zerva, and A. Malyarenko. Simulation of homogeneous and partially isotropic random fields. *J. Eng. Mech.*, 125:1180–1189, 1999.

A.N. Kolmogorov. Problems of probability theory. *Theory Probab. Appl.*, 38(2):177–178, 1993. doi:10.1137/1138020. URL http://link.aip.org/link/?TPR/38/177/1.

A.S. Monin and A.M. Yaglom. *Statistical fluid mechanics: mechanics of turbulence*, volume I. Dover, New York, 2007a.

A.S. Monin and A.M. Yaglom. *Statistical fluid mechanics: mechanics of turbulence*, volume II. Dover, New York, 2007b.

Yu.A. Rozanov. Spectral theory of n-dimensional stationary stochastic processes with discrete time. *Usp. Mat. Nauk*, 13(2 (80)):93–142, 1958. In Russian.

M.Ĭ. Yadrenko. Isotropic random fields of Markov type in Euclidean space. *Dopov. Akad. Nauk Ukr. RSR*, 1963(3):304–306, 1963. In Ukrainian.

M.Ĭ. Yadrenko. *Spectral theory of random fields*. Translat. Ser. Math. Eng. Optimization Software, Publications Division, New York, 1983.

A.M. Yaglom. Certain types of random fields in n-dimensional space similar to stationary stochastic processes. *Teor. Veroâtn. Ee Primen.*, 2:292–338, 1957. In Russian

A.M. Yaglom. Positive-definite functions and homogeneous random fields on groups and homogeneous spaces. *Sov. Math. Dokl.*, 1:1402–1405, 1960.

A.M. Yaglom. Second-order homogeneous random fields. In *Proc. 4th Berkeley Sympos. Math. Statist. and Probab.*, volume II, pages 593–622. University of California Press, Berkeley, 1961.

A.M. Yaglom. Spectral representations for various classes of random functions. In *Proc. 4th All-Union Math. Congr. (Leningrad 1961)*, volume I, pages 250–273. Izdat. Akad. Nauk SSSR, Moscow, 1963. In Russian.

A.M. Yaglom. *Correlation theory of stationary and related random functions. Vol. I. Basic results*. Springer Ser. Stat. Springer, New York, 1987a.

A.M. Yaglom. *Correlation theory of stationary and related random functions. Vol. II. Supplementary notes and references*. Springer Ser. Stat. Springer, New York, 1987b.

M. Zaldarriaga and U. Seljak. An all-sky analysis of polarisation in the microwave background. *Phys. Rev. D*, 55(4):1830–1840, 1997.

Chapter 2
Spectral Expansions

Abstract Let $\mathbb{K}$ be either the field of real numbers $\mathbb{R}$ or the field of complex numbers $\mathbb{C}$. Let $\xi = (E, \pi, T)$ be a finite-dimensional $\mathbb{K}$-vector bundle over a Hausdorff topological space T, and let G be a topological group acting on T from the left. We introduce G-invariant random sections of the bundle ξ and study their spectral expansions. We start from the simplest case, when $T = G$, then consider the more complicated case when T is a homogeneous space of the group G, and finally the general case. In each case, we consider the trivial bundle $E = T \times \mathbb{K}$ first. Finally, we consider the spectral expansions of vector homogeneous and isotropic random fields and Volterra isotropic random fields on the space $\mathbb{R}^n$.

2.1 Random Fields

In this section, we recall the classical definition of a scalar random field and introduce random sections of vector bundles.

2.1.1 The Scalar Case

Let $\mathbb{K}$ be either the field of real numbers $\mathbb{R}$ or the field of complex numbers $\mathbb{C}$. Let T be a set.

Definition 2.1 A *scalar random field* on T is a collection $\{X(t): t \in T\}$ of $\mathbb{K}$-valued random variables, defined on a probability space $(\Omega, \mathfrak{F}, \mathsf{P})$.

In the case of $T \subseteq \mathbb{R}$, it is customary to use the term *stochastic process* instead.

Definition 2.2 A random field $X(t)$ is *second-order* if $\mathsf{E}|X(t)|^2 < \infty$, $t \in T$.

In what follows, we consider only second-order random fields. We also suppose that the *parametric space* T carries a structure of a Hausdorff topological space.

Let $L_{\mathbb{K}}^2(\Omega, \mathfrak{F}, \mathsf{P})$ be the Hilbert space of all $\mathbb{K}$-valued random variables defined on the space $(\Omega, \mathfrak{F}, \mathsf{P})$ and having finite second moment. The inner product on

A. Malyarenko, *Invariant Random Fields on Spaces with a Group Action*,
Probability and Its Applications, DOI 10.1007/978-3-642-33406-1_2,
© Springer-Verlag Berlin Heidelberg 2013

$L^2_{\mathbb{K}}(\Omega, \mathfrak{F}, \mathsf{P})$ is defined as

$$(X, Y) := \mathsf{E}[X\overline{Y}], \quad X, Y \in L^2_{\mathbb{K}}(\Omega, \mathfrak{F}, \mathsf{P}).$$

Definition 2.3 A random field $X(t)$ is called *mean-square continuous* if the map

$$T \to L^2_{\mathbb{K}}(\Omega, \mathfrak{F}, \mathsf{P}), \quad t \mapsto X(t)$$

is continuous.

It follows that the *mean value*

$$E(t) := \mathsf{E}\big[X(t)\big]$$

maps T to $\mathbb{K}$ continuously. The *covariance function*

$$R(s, t) := \mathsf{E}\big[X(s)\overline{X(t)}\big]$$

maps $T \times T$ to $\mathbb{K}$ continuously. In what follows, we consider only mean-square continuous random fields.

Let G be a topological group acting continuously from the left on the parameter space T.

Definition 2.4 A random field $X(t)$ is called *strictly left G-invariant* (or just strictly left-invariant) if all finite-dimensional distributions of the field $X(t)$ are invariant under the action of G, i.e., for any non-empty finite subset $\{t_1, \ldots, t_n\}$ of T, and for any $g \in G$ the two random vectors $(X(t_1), \ldots, X(t_n))$ and $(X(gt_1), \ldots, X(gt_n))$ are identically distributed.

On the other hand, let G act continuously from the right on the parameter space T.

Definition 2.5 A random field $X(t)$ is called *strictly right G-invariant* (or just strictly right-invariant) if one has invariance as above but with each $X(gt_i)$ replaced by $X(t_i g)$.

In particular cases, it is customary to use more specialised terms. For example, if $T = \mathbb{R}$ and $G = \mathbb{R}$ acts on T by $gt = t + g$, then the corresponding strictly invariant stochastic process is called *strictly stationary*. If $T = \mathbb{R}^n$ with $n \geq 2$ and $G = \mathbb{R}^n$ acts on T by $gt = t + g$, then the corresponding strictly invariant random field is called *strictly homogeneous*. If $G = \mathrm{SO}(n)$ acts on $T = \mathbb{R}^n$ by matrix-by-vector multiplication, then the corresponding strictly invariant random field is called *strictly isotropic*. If G is the group of all orientation-preserving isometries of $T = \mathbb{R}^n$, then the corresponding strictly invariant random field is called *strictly homogeneous and isotropic*.

Let $X(t)$ be a second-order strictly left G-invariant random field on a parametric space T. It follows that

$$E(gt) = E(t), \quad g \in G,\ t \in T, \tag{2.1}$$

and

$$R(gs, gt) = R(s, t), \quad g \in G,\ s, t \in T. \tag{2.2}$$

Definition 2.6 A random field $X(t)$ is called *wide-sense left G-invariant* (or just wide-sense left-invariant) if it satisfies (2.1) and (2.2).

Let G act continuously from the right on the parameter space T.

Definition 2.7 A random field $X(t)$ is called *wide-sense right G-invariant* (or just wide-sense right-invariant) if one has invariance as above but with gs and gt replaced by sg and tg.

In what follows, "invariant random field" always means "wide-sense left-invariant random field".

There is an important class of random fields which are invariant iff they are strictly invariant (because they are completely determined by their one-dimensional and two-dimensional distributions).

Definition 2.8 A random field $X(t)$ is called *Gaussian* if all finite-dimensional distributions of the field $X(t)$ are normal (i.e., Gaussian) random vectors.

2.1.2 The Vector Case

Let $\xi = (E, \pi, T)$ be a $\mathbb{K}$-vector bundle (see Subsection 6.1.2).

Definition 2.9 A *vector random field* on T is a collection of random vectors $\{\mathbf{X}(t) : t \in T\}$ satisfying $\mathbf{X}(t) \in \pi^{-1}(t), t \in T$.

In other words, a vector random field on the base T of a vector bundle ξ is a random section of ξ.

To define a second-order vector random field, assume that every space $\pi^{-1}(t)$ carries an inner product and the function $Q \colon E \to \mathbb{R}$, $Q(\mathbf{x}) = \|\mathbf{x}\|^2_{\pi^{-1}(t)}, \mathbf{x} \in \pi^{-1}(t)$ is continuous.

Definition 2.10 A vector random field $\mathbf{X}(t)$ is *second-order* if $\mathsf{E}\|\mathbf{X}(t)\|^2_{\pi^{-1}(t)} < \infty$, $t \in T$.

Let $\mathbf{X}(t)$ be a vector random field on the base T of the vector bundle ξ. Let $X(t, \mathbf{x})$ be the scalar random field on the total space E defined by

$$X(t, \mathbf{x}) := \big(\mathbf{x}, \mathbf{X}(t)\big)_{\pi^{-1}(t)}, \quad t \in T, \ \mathbf{x} \in \pi^{-1}(t).$$

We will call $X(t, \mathbf{x})$ the scalar random field *associated* to the vector random field $\mathbf{X}(t)$.

The field $X(t, \mathbf{x})$ has the following property: its restriction to $\pi^{-1}(t)$ is linear, i.e., for any $\mathbf{x}, \mathbf{y} \in \pi^{-1}(t)$, and for any $\alpha, \beta \in \mathbb{K}$,

$$X(t, \alpha\mathbf{x} + \beta\mathbf{y}) = \alpha X(t, \mathbf{x}) + \beta X(t, \mathbf{y}). \tag{2.3}$$

Definition 2.11 A vector random field $\mathbf{X}(t)$ is *mean-square continuous* if the associated scalar random field $X(t, \mathbf{x})$ is mean-square continuous, i.e., if the map

$$E \to L^2_{\mathbb{K}}(\Omega, \mathfrak{F}, \mathsf{P}), \quad (t, \mathbf{x}) \mapsto X(t, \mathbf{x})$$

is continuous.

The *mean value* of the random field $\mathbf{X}(t)$ is

$$\mathbf{M}(t) := \mathsf{E}\big[\mathbf{X}(t)\big],$$

while its *covariance operator* is

$$R(s, t) := \mathsf{E}\big[\mathbf{X}(s) \otimes \mathbf{X}(t)^*\big].$$

Let G be a topological group acting continuously from the left on the base T of the vector bundle $\xi = (E, \pi, T)$. We would like to call a vector random field $\mathbf{X}(t)$ invariant if the associated scalar random field $X(t, \mathbf{x})$, $(t, \mathbf{x}) \in E$ is invariant. The question is: under which action? In general, there exists no natural action of G on E. First we have to define an action of G on the total space E *associated* to its action on the base space T.

Because the scalar random field $X(\mathbf{x})$ has property (2.3), it can be left-invariant under the associated action, only if the restriction of the associated action to any fibre $E_t = \pi^{-1}(t)$ is a linear invertible operator acting between the fibres. Moreover, the associated action must map the fibre E_t on the fibre E_{gt}. We will also require the above restriction to preserve inner products in the fibres.

Definition 2.12 Let $\xi = (E, \pi, T)$ be a vector bundle, and let $G \times T \to T$ be a continuous left action of a topological group G on the base space T. A left action $G \times E \to E$ of G on the total space E is called *associated* with the action $G \times T \to T$ if its restriction to any fibre E_t is a linear isometry between E_t and E_{gt}.

We are ready to formulate the main definition of this subsection.

Definition 2.13 Let $\xi = (E, \pi, T)$ be a vector bundle, $G \times T \to T$ be a continuous left action of a topological group G on the base space T, and $G \times E \to E$ be an associated action of G on the total space E. A vector random field $\mathbf{X}(t)$ on ξ is called *left-invariant and transforming according to the action $G \times T \to T$* if the associated random field $X(t, \mathbf{x})$ is left-invariant under the associated action $G \times E \to E$.

Example 2.14 Let $\xi = (T \times \mathbb{K}^n, \pi, T)$, where $\pi(t, \mathbf{x}) = t$, be a trivial vector bundle. Let $(g, t) \mapsto gt$ be a continuous left action of a topological group G on the base space T. Let $\mathbf{X}(t)$ be a vector random field on ξ. Define the associated action of G on E by

$$g(t, \mathbf{x}) := (gt, \mathbf{x}), \quad g \in G, \, t \in T, \, \mathbf{x} \in \mathbb{K}^n. \tag{2.4}$$

With this action, a vector random field $\mathbf{X}(t)$ is strictly left G-invariant if all finite-dimensional distributions of $\mathbf{X}(t)$ are invariant under action (2.4). This definition is correct, because all random vectors $\mathbf{X}(t)$ belong to the same vector space $\mathbb{K}^n$. The field $\mathbf{X}(t)$ is wide-sense left G-invariant iff

$$\mathbf{E}(gt) = \mathbf{E}(t),$$

$$R(gs, gt) = R(s, t)$$

for all $s, t \in T$, $g \in G$.

Example 2.15 Under the conditions of Example 2.14, let $\mathbb{K} = \mathbb{C}$, and let V be a representation of G in $\mathbb{C}^n$. Define the associated action of G on E by

$$g(t, \mathbf{x}) := (gt, V(g)\mathbf{x}). \tag{2.5}$$

Let $\mathbf{X}(t)$ be a vector random field on ξ, and $X(t, \mathbf{x})$ be the associated random field. By Definition 2.6, $X(\mathbf{x})$ is wide-sense left-invariant under the introduced action iff for all $g \in G$, $s, t \in T$, $\mathbf{x} \in \pi^{-1}(s)$, and $\mathbf{y} \in \pi^{-1}(t)$,

$$\mathsf{E}\big[X\big(gs, V(g)\mathbf{x}\big)\big] = \mathsf{E}\big[X(s, \mathbf{x})\big],$$

$$\mathsf{E}\big[X\big(gs, V(g)\mathbf{x}\big)\overline{X\big(gt, V(g)\mathbf{y}\big)}\big] = \mathsf{E}\big[X(s, \mathbf{x})\overline{X(t, \mathbf{y})}\big].$$

By definition of the associated scalar random field, the above equations may be rewritten as

$$\mathsf{E}\big[(V(g)\mathbf{x}, \mathbf{X}(gs))\big] = \mathsf{E}\big[(\mathbf{x}, \mathbf{X}(s))\big],$$

$$\mathsf{E}\big[(V(g)\mathbf{x}, \mathbf{X}(gs))\overline{(V(g)\mathbf{y}, \mathbf{X}(gt))}\big] = \mathsf{E}\big[(\mathbf{x}, \mathbf{X}(s))\overline{(\mathbf{y}, \mathbf{X}(t))}\big].$$

Move the operator $V(g)$ to the second term of the inner product:

$$\mathsf{E}\big[(\mathbf{x}, V^*(g)\mathbf{X}(gs))\big] = \mathsf{E}\big[(\mathbf{x}, \mathbf{X}(s))\big],$$

$$\mathsf{E}\big[(\mathbf{x}, V^*(g)\mathbf{X}(gs))\overline{(\mathbf{y}, V^*(g)\mathbf{X}(gt))}\big] = \mathsf{E}\big[(\mathbf{x}, \mathbf{X}(s))\overline{(\mathbf{x}, \mathbf{X}(t))}\big].$$

Because these equations hold true for all $\mathbf{x} \in E_s$ and $\mathbf{y} \in E_t$ we obtain: a vector random field $\mathbf{X}(t)$ is wide-sense left-invariant and transforms according to the action (2.5) if

$$E\big[V^*(g)\mathbf{X}(gs)\big] = E\big[\mathbf{X}(s)\big],$$

$$E\big[\big(V^*(g)\mathbf{X}(gs)\big) \otimes \big(V^*(g)\mathbf{X}(gt)\big)^*\big] = E\big[\big(\mathbf{X}(s)\big) \otimes \big(\mathbf{X}(t)\big)^*\big]$$

for all $g \in G$, $s, t \in T$. In other words,

$$
\begin{aligned}
\mathbf{E}(gs) &= \tilde{V}(g)\mathbf{E}(s), \\
R(gs, gt) &= \tilde{V}(g)R(s, t)V^{-1}(g),
\end{aligned}
\tag{2.6}
$$

where $\tilde{V}$ is the representation conjugate-dual to V.

Example 2.16 Let G be a locally compact separable topological group, and let K be a compact subgroup. Let W be a unitary representation of the group K in a finite-dimensional Hilbert space H. Let $\xi = (E_W, \pi, T)$ be the corresponding homogeneous vector bundle over the homogeneous space $T = G/K$. A vector random field $\mathbf{X}(t)$ in ξ is called *left-invariant* if the associated scalar random field

$$X(\mathbf{x}) = \big(\mathbf{x}, \mathbf{X}(t)\big)_H, \quad \mathbf{x} \in \pi^{-1}(t) \subset E_W, \ t \in T$$

is left-invariant under the associated action $(g_0, \mathbf{x}) \mapsto (gg_0, \mathbf{x})$.

Moreover, all random vectors $\mathbf{X}(t)$ lie in the same space H. By definition, the associated scalar random field $X(t, \mathbf{x}) = (\mathbf{x}, \mathbf{X}(t))$ is mean-square continuous iff

$$\lim_{(s,\mathbf{y}) \to (t,\mathbf{x})} E\big|X(s, \mathbf{y}) - X(t, \mathbf{x})\big|^2 = 0.$$

Let $\{\mathbf{e}_1, \mathbf{e}_2, \ldots, \mathbf{e}_{\dim H}\}$ be a basis in H. Put $\mathbf{y} = \mathbf{x} := \mathbf{e}_j$. Then we have

$$\lim_{s \to t} E\big|\overline{X_j(s)} - \overline{X_j(t)}\big|^2 = 0,$$

which is equivalent to

$$\lim_{s \to t} E\big|X_j(s) - X_j(t)\big|^2 = 0.$$

It follows that

$$\lim_{s \to t} E\big\|\mathbf{X}(s) - \mathbf{X}(t)\big\|^2 = \lim_{s \to t} E \sum_{j=1}^{\dim H} \big|X_j(s) - X_j(t)\big|^2$$

$$= \sum_{j=1}^{\dim H} \lim_{s \to t} E\big|X_j(s) - X_j(t)\big|^2 = 0.$$

Conversely, let

$$\lim_{s \to t} \mathsf{E} \|\mathbf{X}(s) - \mathbf{X}(t)\|^2 = 0. \tag{2.7}$$

Then, for any $j = 1, 2, \ldots, \dim H$,

$$0 \le \limsup_{s \to t} \mathsf{E} |X_j(s) - X_j(t)|^2$$

$$\le \limsup_{s \to t} \sum_{j=1}^{\dim H} \mathsf{E} |X_j(s) - X_j(t)|^2 = 0,$$

thus, $\lim_{s \to t} \mathsf{E} |X_j(s) - X_j(t)|^2 = 0$. It follows that

$$\lim_{(s,\mathbf{y}) \to (t,\mathbf{x})} \mathsf{E} |X(s,\mathbf{y}) - X(t,\mathbf{x})|^2 = \lim_{(s,\mathbf{y}) \to (t,\mathbf{x})} \mathsf{E} \left| \sum_{j=1}^{\dim H} \left(y_j \overline{X_j(s)} - x_j \overline{X_j(t)} \right) \right|^2$$

$$\le 2 \sum_{j=1}^{\dim H} \lim_{(s,\mathbf{y}) \to (t,\mathbf{x})} \mathsf{E} |y_j X_j(s) - x_j X_j(t)|^2 = 0.$$

We proved that in the particular case of a vector random field in a homogeneous vector bundle our definition of mean-square continuity is equivalent to the usual definition (2.7). In the same way one can easily prove that our definition of a wide-sense G-invariant field is equivalent to the following equalities: for all $s, t \in T$, and for all $g \in G$ we have

$$\mathsf{E}[\mathbf{X}(gs)] = \mathsf{E}[\mathbf{X}(s)],$$
$$\mathsf{E}[\mathbf{X}(gs) \otimes \mathbf{X}^*(gt)] = \mathsf{E}[\mathbf{X}(s) \otimes \mathbf{X}^*(t)]. \tag{2.8}$$

The first equation is equivalent to the equality

$$\mathsf{E}[\mathbf{X}(s)] = \mathsf{E}[\mathbf{X}(t)], \quad s, t \in T,$$

because G acts transitively on T. Thus, the mean value of a wide-sense G-invariant random field on T is constant.

Example 2.17 Let $T = S^1$ be the unit circle, E be the cylinder $T \times \mathbb{R}$, and let $\pi: E \to T$ with $\pi(t, x) = t$. The bundle $\xi = (E, \pi, T)$ is trivial. It is easy to see that the action of the group $G = SO(2)$ on E by cylinder rotations is associated with the action of G on T by circle rotations. Let $X_0, X_1, \ldots, Y_1, Y_2, \ldots$ be a set of uncorrelated random variables with zero mean and unit variance. Let $a_0, a_1, \ldots, a_n, \ldots$ be a sequence of real numbers with

$$\sum_{n=0}^{\infty} a_n^2 < \infty.$$

The series

$$a_0 X_0 + \sum_{n=1}^{\infty} a_n \big[X_n \cos(nt) + Y_n \sin(nt) \big]$$

converges in mean square and its sum determines a centred mean-square continuous wide-sense left G-invariant random field on ξ.

On the other hand, define an action of the additive group $\mathbb{Z}$ of integers on $\mathbb{R}^2$ by

$$n \cdot (x, y) := \big(x + n, (-1)^n y \big).$$

Let $E := \mathbb{R}^2/\mathbb{Z}$ denote the quotient manifold. The projection on the first coordinate $\pi_1 \colon \mathbb{R}^2 \to \mathbb{R}$ descends to a map $\pi \colon E \to S^1$. The triple $\xi = (E, \pi, S^1)$ is a one-dimensional vector bundle over $T = S^1$, which is called the *Möbius bundle*. It is known that any continuous section of the Möbius bundle has at least one zero.

Consider an arbitrary action of the group $G = \mathrm{SO}(2)$ on E associated to the action of G on T by circle rotations. Let $X(t)$ be a centred wide-sense mean-square continuous left G-invariant random field on ξ. On the one hand, G acts transitively on T, therefore the function $f(t) = \mathsf{E}[X^2(t)]$ is a constant. On the other hand, $f(t)$ must be a continuous section of ξ. It follows that $f(t) = 0$. So, the only centred mean-square continuous wide-sense left G-invariant random field in ξ is 0.

2.2 Invariant Random Fields on Groups

We start to study spectral expansions of invariant random fields from the classical case, when the parametric space T is a topological group acting on itself by left translations.

2.2.1 The Scalar Case

Let T be a topological group G acting on itself by left translations $g \mapsto g_1 g$. Let $X(g)$ be a zero-mean second-order random field on G, invariant under the above-described action. We will call $X(g)$ *left-homogeneous*.

It follows from (2.1) that $E(g) = E(e)$, where e is the identity element of the group G. In other words, the mean value of the random field $X(g)$ is constant. Without loss of generality we can subtract this constant from $X(g)$ and consider only the *centred* random fields with $\mathsf{E}[X(g)] = 0$.

It is well known that the class of covariance functions of random fields coincides with the class of positive-definite functions on the underlying space (see, for example, Doob, 1990, Chap. 2, Theorem 3.1). The set of all positive-definite functions of a topological group G of type I (Subsection 6.3.2) with countable base is described by Eq. (6.10). We have the following theorem.

Theorem 2.18 (Yaglom (1961)) *Let $X(g)$ be a centred mean-square continuous second-order random field on a topological group G of type I with countable base. The field $X(g)$ is left-homogeneous iff its covariance function has the form*

$$R(g_1, g_2) = \int_{\hat{G}} \text{tr}\big[\lambda\big(g_2^{-1}g_1\big)\,dF(\lambda)\big],$$

where F is a measure on $(\hat{G}, \mathfrak{B}(\hat{G}))$ whose restriction to

$$\hat{G}_n = \{\lambda \in \hat{G}: \dim\lambda = n\}$$

takes values in the set of Hermitian nonnegative-definite trace-class operators on the standard space H_n and satisfying condition

$$\sum_{n \geq 1} \text{tr}\big[F(\hat{G}_n)\big] < \infty.$$

The spectral expansion of the random field $X(g)$ has the form

$$X(g) = \int_{\hat{G}} \text{tr}\big[\lambda(g)\,dZ(\lambda)\big], \tag{2.9}$$

where Z is a random measure on $\hat{G}$ whose restriction to $\hat{G}_n$ takes values in the set of bounded linear operators on the standard space H_n satisfying the following condition: for any measurable sets Λ_1, $\Lambda_2 \subseteq \hat{G}_n$ and for any vectors $\mathbf{x}_1$, $\mathbf{x}_2$, $\mathbf{x}_3$, $\mathbf{x}_4 \in H_n$,

$$\mathsf{E}\big[\big(Z(\Lambda_1)\mathbf{x}_1, \mathbf{x}_2\big)\overline{\big(Z(\Lambda_2)\mathbf{x}_3, \mathbf{x}_4\big)}\big] = \big(F(\Lambda_1 \cap \Lambda_2)\mathbf{x}_1, \mathbf{x}_3\big)\overline{(\mathbf{x}_2, \mathbf{x}_4)}. \tag{2.10}$$

Proof (sketch) The expansion of the covariance function follows from (6.10). To change back and forth from the expansion of the covariance function to the spectral expansion of the field, we need to use the operator version of Karhunen's theorem from Karhunen (1947).

Introduce the Hilbert space $L^2(F)$ of operator-valued functions $T(\lambda)$, $\lambda \in \hat{G}$. The restriction of any function $T(\lambda) \in L^2(F)$ to $\hat{G}_n$ takes values in the set of all bounded linear operators on the standard space H_n. The inner product is defined by

$$\big(T_1(\lambda), T_2(\lambda)\big) := \sum_{n \geq 1} \int_{\hat{G}_n} \text{tr}\big[T_1(\lambda)\,dF(\lambda)T_2^*(\lambda)\big].$$

For a fixed $g \in G$, the operator-valued function $\lambda(g)$ belongs to $L^2(F)$. Indeed,

$$\|\lambda(g)\|^2 = \sum_{n \geq 1} \int_{\hat{G}_n} \mathrm{tr}\big[\lambda(g)\,dF(\lambda)\lambda^*(g)\big]$$

$$= \sum_{n \geq 1} \int_{\hat{G}_n} \mathrm{tr}\big[\lambda^*(g)\lambda(g)\,dF(\lambda)\big]$$

$$= \sum_{n \geq 1} \int_{\hat{G}_n} \mathrm{tr}\big[dF(\lambda)\big]$$

$$= \sum_{n \geq 1} \mathrm{tr}\big[F(\hat{G}_n)\big] < \infty.$$

The mapping $\lambda(g) \mapsto X(g)$ extends by linearity and continuity to an isometric isomorphism between the spaces $L^2(F)$ and H_X, the closed linear span of the set $\{X(g) : g \in G\}$ in the Hilbert space $L^2_{\mathbb{C}}(\Omega, \mathfrak{F}, P)$.

Let $\Lambda_n \in \mathfrak{B}(\hat{G}_n)$, and let $\mathbf{x}_1$ and $\mathbf{x}_2$ be two vectors in H_n. Let $T_{\Lambda_n;\mathbf{x}_1,\mathbf{x}_2}(\lambda)$ be the following element of $L^2(F)$:

$$T_{\Lambda_n;\mathbf{x}_1,\mathbf{x}_2}(\lambda) = \begin{cases} \mathbf{x}_2^* \otimes \mathbf{x}_1, & \lambda \in \Lambda_n, \\ 0, & \lambda \notin \Lambda_n. \end{cases}$$

The function under consideration is a measure on $\mathfrak{B}(\hat{G}_n)$; it depends linearly on $\mathbf{x}_1$ and semi-linearly on $\mathbf{x}_2$. Let $Z_{\Lambda_n;\mathbf{x}_1,\mathbf{x}_2}$ be the image of the function $T_{\Lambda_n;\mathbf{x}_1,\mathbf{x}_2}(\lambda)$ under the above isomorphism. The random measure $Z_{\Lambda_n;\mathbf{x}_1,\mathbf{x}_2}$ depends linearly on $\mathbf{x}_1$ and semi-linearly on $\mathbf{x}_2$. It follows that there exists a random measure $Z_n(\Lambda_n)$ with values in the space of bounded linear operators in H_n such that

$$\big(Z_n(\Lambda_n)\mathbf{x}_1, \mathbf{x}_2\big) = Z_{\Lambda_n;\mathbf{x}_1,\mathbf{x}_2}.$$

Let $Z(\Lambda)$ be the operator random measure on $\mathfrak{B}(\hat{G})$ whose restriction to $\mathfrak{B}(\hat{G}_n)$ coincides with $Z_n(\Lambda_n)$, where $\Lambda_n = \Lambda \cap \hat{G}_n$.

Let $\mathbf{e}_1, \mathbf{e}_2, \ldots, \mathbf{e}_j, \ldots$ be an orthonormal basis in H_n. Then

$$\sum_{n \geq 1} \int_{\hat{G}_n} \sum_{j,k=1}^{n} (\mu_n(g)\mathbf{e}_j, \mathbf{e}_k)\,dT_{\mu_n;\mathbf{e}_j,\mathbf{e}_k}(\lambda) = \lambda(g).$$

Since, under the inverse to the above isomorphism, $\lambda(g)$ maps to $X(g)$, we have

$$X(g) = \sum_{n \geq 1} \int_{\hat{G}_n} \mathrm{tr}\big[\lambda(g)\,dZ_n(\lambda)\big],$$

which is equivalent to (2.9).

Let $\Lambda_1, \Lambda_2 \in \hat{G}$, and let $\Lambda_{1,n} = \Lambda_1 \cap \hat{G}_n$ and $\Lambda_{2,n} = \Lambda_2 \cap \hat{G}_n$. Let $x_{j,k} = (\mathbf{x}_j, \mathbf{e}_k)$ for $1 \leq j \leq 4$. Because the isomorphism preserves inner products, we have

$$
\begin{aligned}
\mathsf{E}\big[(Z(\Lambda_{1,n})\mathbf{x}_1, \mathbf{x}_2)\overline{(Z(\Lambda_{2,n})\mathbf{x}_3, \mathbf{x}_4)}\big] &= \int_{\hat{G}_n} \mathrm{tr}\big[T_{\Lambda_{1,n};\mathbf{x}_1,\mathbf{x}_2}(\lambda)\, \mathrm{d}F(\lambda) T^*_{\Lambda_{2,n};\mathbf{x}_3,\mathbf{x}_4}(\lambda)\big] \\
&= \mathrm{tr}\big[(\mathbf{x}_2^* \otimes \mathbf{x}_1) F(\Lambda_{1,n} \cap \Lambda_{2,n})(\mathbf{x}_4^* \otimes \mathbf{x}_3)^*\big] \\
&= \sum_{j,k,\ell=1}^n \overline{x_{2,j}}\, x_{1,k} F_{k\ell}(\Lambda_{1,n} \cap \Lambda_{2,n}) x_{4,j}\overline{x_{3,\ell}} \\
&= \big(F(\Lambda_{1,n} \cap \Lambda_{2,n})\mathbf{x}_1, \mathbf{x}_3\big)\overline{(\mathbf{x}_2, \mathbf{x}_4)}.
\end{aligned}
$$

Summing up these equalities with respect to all $n \geq 1$, we obtain (2.10).

Conversely, let the random field $X(g)$ have the form (2.9). The covariance function of $X(g)$ is

$$
\begin{aligned}
R(gg_1, g_1) &= \mathsf{E}\big[X(gg_1)\overline{X(g_1)}\big] \\
&= \mathsf{E}\bigg[\int_{\hat{G}} \mathrm{tr}\big[\lambda(gg_1)\,\mathrm{d}Z(\lambda)\big]\overline{\int_{\hat{G}} \mathrm{tr}\big[\lambda(g_1)\,\mathrm{d}Z(\lambda)\big]}\bigg] \\
&= \int_{\hat{G}} \mathrm{tr}\big[\lambda(gg_1)\lambda(g_1^{-1})\,\mathrm{d}F(\lambda)\big] \\
&= \int_{\hat{G}} \mathrm{tr}\big[\lambda(g)\,\mathrm{d}F(\lambda)\big],
\end{aligned}
$$

that is, $X(g)$ is left-homogeneous. $\qquad\square$

For *right-homogeneous* fields, that is, for fields invariant under the action $g \mapsto gg_1$, condition (2.10) is replaced by

$$
\mathsf{E}\big[(Z(\Lambda)\mathbf{x}_1, \mathbf{x}_2)\overline{(Z(\Lambda)\mathbf{x}_3, \mathbf{x}_4)}\big] = \big(F(\Lambda_1 \cap \Lambda_2)\mathbf{x}_2, \mathbf{x}_4\big)\overline{(\mathbf{x}_1, \mathbf{x}_3)}.
$$

Example 2.19 (Commutative groups) Assume that G is a *commutative* locally compact topological group with countable base. Let $X(g)$ be a centred left-invariant random field on G. Such a field is called *homogeneous*. It follows from the theory of unitary representations of commutative locally compact groups (Subsection 6.3.3) and from Theorem 2.18 that

$$
R(g_1, g_2) = \int_{\hat{G}} \hat{g}(g_1 - g_2)\, \mathrm{d}\mu(\hat{g}),
$$

where μ is a finite measure on $(\hat{G}, \mathfrak{B}(\hat{G}))$, and

$$
X(g) = \int_{\hat{G}} \hat{g}(g)\, \mathrm{d}Z(\hat{g}), \tag{2.11}
$$

where Z is a complex-valued centred random measure on $\hat{G}$ with

$$\mathsf{E}\big[Z(A)\overline{Z(B)}\big] = \mu(A \cap B), \quad A, B \in \mathfrak{B}(\hat{G}).$$

In particular, when $G = \mathbb{R}$, we have the following expansion of the covariance function of the centred stationary random process $X(t)$:

$$R(s, t) = \int_{-\infty}^{\infty} e^{ip(s-t)} \, d\mu(p), \tag{2.12}$$

as well as the following expansion of the process $X(t)$ itself:

$$X(t) = \int_{-\infty}^{\infty} e^{ipt} \, dZ(p). \tag{2.13}$$

It is very important to understand that these formulae contain integration over the *character group* $\hat{\mathbb{R}}$ rather than over the group $\mathbb{R}$ itself. In applied areas like physics or time series analysis, the set $\mathbb{R}$ is usually called the *time domain*, while the set $\hat{\mathbb{R}}$ is called the *frequency domain*.

When $G = \mathbb{R}^n$, the above-described expansions take the following form:

$$R(\mathbf{x}, \mathbf{y}) = \int_{\mathbb{R}^n} e^{i(\mathbf{p}, \mathbf{x}-\mathbf{y})} \, d\mu(\mathbf{p}), \tag{2.14a}$$

$$X(\mathbf{x}) = \int_{\mathbb{R}^n} e^{i(\mathbf{p}, \mathbf{x})} \, dZ(\mathbf{p}). \tag{2.14b}$$

Again, the domain of the homogeneous random field $X(\mathbf{x})$ (*space domain*) differs from the domain of integration (*wavenumber domain*).

When $G = \mathbb{Z}$, the covariance function of the stationary random sequence has the form

$$R(m - n) = \int_{\mathbb{T}} e^{i(m-n)\varphi} \, d\mu(\varphi), \tag{2.15}$$

and the sequence itself has the form

$$X(m) = \int_{\mathbb{T}} e^{im\varphi} \, dZ(\varphi), \tag{2.16}$$

where $\mathbb{T}$ is the torus (Subsection 6.3.3).

The covariance function of the stationary random field on the lattice $\mathbb{Z}^n$ has the form

$$R(\mathbf{m}, \mathbf{n}) = \int_{\mathbb{T}^n} e^{i(\mathbf{m}-\mathbf{n}, \varphi)} \, dF(\varphi),$$

while the field itself has the form

$$X(\mathbf{n}) = \int_{\mathbb{T}^n} e^{i(\mathbf{n}, \varphi)} \, dZ(\varphi).$$

Consider the group $\mathbb{R}_0^\infty$ of finite sequences of real numbers with inductive topology (Subsection 6.3.3). Using the results of the above subsection, we obtain the following. The covariance function of the homogeneous random field on $\mathbb{R}_0^\infty$ has the form

$$R(\mathbf{x}, \mathbf{y}) = \int_{\mathbb{R}^\infty} e^{i(\mathbf{p}, \mathbf{x} - \mathbf{y})} \, d\mu(\mathbf{p}), \tag{2.17}$$

where $\mathbb{R}^\infty$ is the topological product of countably many copies of the real line $\mathbb{R}$. The field itself has the form

$$X(\mathbf{x}) = \int_{\mathbb{R}^\infty} e^{i(\mathbf{p}, \mathbf{x})} \, dZ(\mathbf{p})$$

with

$$(\mathbf{p}, \mathbf{x}) = \sum_{j=1}^\infty p_j x_j.$$

The right-hand side of the last formula contains only finitely many terms, because the sequence $\mathbf{x}$ contains only finitely many nonzero terms.

Example 2.20 (Compact groups) Let G be a compact topological group. In this case, the set $\hat{G}$ is discrete, all irreducible unitary representations $\lambda \in \hat{G}$ are finite-dimensional, and the covariance function of the left-homogeneous random field $X(g)$ has the form

$$R(g_2^{-1} g_1) = \sum_{\lambda \in \hat{G}} \mathrm{tr}\big[\lambda(g_2^{-1} g_1) F^{(\lambda)}\big], \tag{2.18}$$

where $F^{(\lambda)}$ is a Hermitian nonnegative-definite linear operator in the space H_λ of the representation λ, and

$$\sum_{\lambda \in \hat{G}} \mathrm{tr}\big[F^{(\lambda)}\big] < \infty.$$

The field $X(g)$ has the form

$$X(g) = \sum_{\lambda \in \hat{G}} \mathrm{tr}\big[\lambda(g) Z^{(\lambda)}\big], \tag{2.19}$$

where $Z^{(\lambda)}$ are random centred linear operators in H_λ satisfying the following condition: for any $\lambda, \mu \in \hat{G}$, and for any vectors $\mathbf{x}_1, \mathbf{x}_2 \in H_\lambda$, $\mathbf{x}_3, \mathbf{x}_4 \in H_\mu$,

$$\mathsf{E}\big[(Z^{(\lambda)} \mathbf{x}_1, \mathbf{x}_2) \overline{(Z^{(\mu)} \mathbf{x}_3, \mathbf{x}_4)}\big] = \delta_{\lambda\mu} (F^{(\lambda)} \mathbf{x}_1, \mathbf{x}_3) \overline{(\mathbf{x}_2, \mathbf{x}_4)},$$

where $\delta_{\lambda\mu}$ is the Kronecker delta. In particular, let $\mathbf{e}_1^{(\lambda)}, \mathbf{e}_2^{(\lambda)}, \ldots, \mathbf{e}_{\dim \lambda}^{(\lambda)}$ be a basis in H_λ, and put $\mathbf{x}_1 := \mathbf{e}_j^{(\lambda)}$, $\mathbf{x}_2 := \mathbf{e}_k^{(\lambda)}$, $\mathbf{x}_3 := \mathbf{e}_m^{(\mu)}$, and $\mathbf{x}_4 := \mathbf{e}_n^{(\mu)}$. We obtain the coordinate form of the above condition as

$$E\big[Z_{jk}^{(\lambda)}\overline{Z_{mn}^{(\mu)}}\big] = \delta_{\lambda\mu}\delta_{kn}F_{jm}^{(\lambda)}. \tag{2.20}$$

For right-homogeneous fields, condition (2.20) becomes

$$E\big[Z_{jk}^{(\lambda)}\overline{Z_{mn}^{(\mu)}}\big] = \delta_{\lambda\mu}\delta_{jm}F_{kn}^{(\lambda)}. \tag{2.21}$$

Finally, for *two-way homogeneous* fields, that is, both left-homogeneous and right-homogeneous, the two conditions (2.20) and (2.21) must hold. Therefore, the operators $Z^{(\lambda)}$ satisfy the following condition:

$$E\big[Z_{jk}^{(\lambda)}\overline{Z_{mn}^{(\mu)}}\big] = \delta_{\lambda\mu}\delta_{jm}\delta_{kn}F^{(\lambda)}, \tag{2.22}$$

with $F^{(\lambda)}$ nonnegative numbers satisfying

$$\sum_{\lambda\in\hat{G}} \dim\lambda F^{(\lambda)} < \infty.$$

In particular, let $G = \mathrm{SU}(2)$ be the group of all unitary unimodular 2×2 matrices. This case is interesting because $\mathrm{SU}(2)$ is a compact semisimple simply connected Lie group of the minimal possible dimension. The left-homogeneous random field on G has the form

$$X(\varphi,\theta,\psi) = \sum_{\ell\in\{0,1/2,1,\dots\}} \sum_{j,k=1}^{2\ell+1} D_{jk}^{\ell}(\varphi,\theta,\psi) Z_{kj}^{(\ell)},$$

where $D_{jk}^{\ell}(\varphi,\theta,\psi)$ are Wigner D-functions (6.16). We denote the corresponding irreducible unitary representation of the group G by π^{ℓ}.

Example 2.21 (Lorentz groups and their covers) Let $\mathrm{SO}_0(1,n)$ be the connected component of unity of the group of linear transformations of the space $\mathbb{R}^{n+1}$ that respect the *Minkowski quadratic form*

$$[\mathbf{x},\mathbf{x}] = x_0^2 - x_1^2 - \cdots - x_n^2.$$

The case $n = 2$ is interesting for the following reason. The group $\mathrm{SU}(1,1)$ of complex matrices

$$\begin{pmatrix} \alpha & \beta \\ \overline{\beta} & \overline{\alpha} \end{pmatrix}$$

with $|\alpha|^2 - |\beta|^2 = 1$ is a double cover of the group $\mathrm{SO}_0(1,2)$. The covering homomorphism is

$$\begin{pmatrix} \alpha & \beta \\ \overline{\beta} & \overline{\alpha} \end{pmatrix} \mapsto \begin{pmatrix} |\alpha|^2 + |\beta|^2 & 2\,\mathrm{Re}(\alpha\overline{\beta}) & -2\,\mathrm{Im}(\alpha\overline{\beta}) \\ 2\,\mathrm{Re}(\alpha\beta) & \mathrm{Re}(\alpha^2 + \beta^2) & -\,\mathrm{Im}(\alpha^2 - \beta^2) \\ 2\,\mathrm{Im}(\alpha\beta) & \mathrm{Im}(\alpha^2 + \beta^2) & \mathrm{Re}(\alpha^2 - \beta^2) \end{pmatrix}.$$

On the other hand, the group $SU(1, 1)$ is isomorphic to the group $SL(2, \mathbb{R})$ of all 2×2 real unimodular matrices. The isomorphism is

$$\begin{pmatrix} \alpha & \beta \\ \overline{\beta} & \overline{\alpha} \end{pmatrix} \mapsto \begin{pmatrix} \mathrm{Re}(\alpha - \beta) & \mathrm{Im}(\alpha + \beta) \\ -\mathrm{Im}(\alpha + \beta) & \mathrm{Re}(\alpha + \beta) \end{pmatrix}.$$

The group $G = SL(2, \mathbb{R})$ is the noncompact real semisimple Lie group of minimal possible dimension.

Using the results of Example 6.4, we obtain the following result. The centred left-homogeneous random field on the group $SL(2, \mathbb{R})$ has the form

$$\begin{aligned}
X(\varphi, \tau, \psi) = & \sum_{m,n=-\infty}^{\infty} \int_0^{\infty} \lambda_{mn}^{-1/2+i\varrho,0}(\varphi, \tau, \psi) \, dZ_{nm}^{-1/2+i\varrho,0} \\
& + \sum_{m,n=-\infty}^{\infty} \int_{0+}^{\infty} \lambda_{mn}^{-1/2+i\varrho,1/2}(\varphi, \tau, \psi) \, dZ_{nm}^{-1/2+i\varrho,1/2} \\
& + \sum_{m,n=-\infty}^{\infty} \int_{-1+}^{-1/2-} \lambda_{mn}^{\varrho,0}(\varphi, \tau, \psi) \, dZ_{nm}^{\varrho,0} \\
& + \sum_{\sigma-\varepsilon \leq -1} \sum_{m,n=-\infty}^{\sigma-\varepsilon} \lambda_{mn}^{\sigma,\varepsilon,+}(\varphi, \tau, \psi) Z_{nm}^{\sigma,\varepsilon,+} \\
& + \sum_{\sigma-\varepsilon \leq -1} \sum_{m,n=-\sigma-\varepsilon}^{\infty} \lambda_{mn}^{\sigma,\varepsilon,-}(\varphi, \tau, \psi) Z_{nm}^{\sigma,\varepsilon,-} \\
& + Z^{-1,0},
\end{aligned}$$

where (φ, τ, ψ) are the Euler angles in the group G. The first (resp. second) term in the right-hand side corresponds to the first (resp. second) principal series. The third term corresponds to the supplementary series. The fourth (resp. fifth) term apart from the summand with $(\varrho, \varepsilon) = (-1/2, 1/2)$ corresponds to the first (resp. second) discrete series. The above summands correspond to the limiting points of the first (resp. second) discrete series. Finally, the sixth term corresponds to the trivial representation. The matrix elements $\lambda_{mn}^{\sigma,\varepsilon}(\varphi, \tau, \psi)$ of various series are given by Eqs. (6.24)–(6.27).

The case $n = 3$ is interesting for the following reason. The group $SO_0(1, 3)$ has the simply connected double cover $G = SL(2, \mathbb{C})$, i.e., the group of all 2×2 complex unimodular matrices. This is the complex semisimple Lie group of minimal possible dimension. On the other hand, the space $\mathbb{R}^4$ with Minkowski quadratic form is the *physical space–time*.

Using the results of Example 6.5, we obtain the following result. The centred left-homogeneous random field on the group $SL(2, \mathbb{C})$ has the form

$$X(g) = \sum_{n=-\infty}^{\infty} \sum_{\ell_1,\ell_2=|n/2|}^{\infty} \sum_{m_1=-\ell_1}^{\ell_1} \sum_{m_2=-\ell_2}^{\ell_2} \int_0^{\infty} \lambda_{\ell_1 m_1, \ell_2 m_2}^{i\varrho-2,n}(g) \, dZ_{\ell_1 m_1, \ell_2 m_2}^{i\varrho-2,n}(\varrho)$$

$$+ \sum_{\ell_1,\ell_2=-\infty}^{\infty} \sum_{m_1=-\ell_1}^{\ell_1} \sum_{m_2=-\ell_2}^{\ell_2} \int_{-2+0}^{-0} \lambda_{\ell_1 m_1, \ell_2 m_2}^{-2-\varrho,0}(g) \, dZ_{\ell_1 m_1, \ell_2 m_2}^{-2-\varrho,0}(\varrho)$$

$$+ Z^{0,0}.$$

The first term in the right-hand side corresponds to the principal series. The second term corresponds to the supplementary series. Finally, the third term corresponds to the trivial representation. The matrix elements $\lambda^{\sigma,n}(g)$ are given by (6.28) and (6.29).

2.2.2 The Vector Case

Under the conditions of Example 2.15, let $T = G$ be a locally compact topological group of type I with countable base, acting on itself by left translations. Let $\mathbf{X}(g)$ be a wide-sense left-invariant vector random field on G transforming according to the action

$$g(g_1, \mathbf{x}) = (gg_1, V(g)\mathbf{x}),$$

where $V(g)$ is a representation of G in a finite-dimensional Hilbert space H.

In the first formula of (2.6), let s be the identity element e of the group G. Then

$$\mathbf{E}(g) = \tilde{V}(g)\mathbf{E}_0, \tag{2.23}$$

where $\mathbf{E}_0 = \mathsf{E}[\mathbf{X}(e)]$. If the representation V is not trivial, than the mean value of the corresponding random field is not constant.

For any $\mathbf{x} \in H$, consider the scalar random field

$$Y_{\mathbf{x}}(g) := (V(g)\mathbf{x}, \mathbf{X}(g)).$$

Then we have, by formulae of Example 2.15,

$$\mathsf{E}[Y_{\mathbf{x}}(gg_1)] = \mathsf{E}[(V(gg_1)\mathbf{x}, \mathbf{X}(gg_1))]$$
$$= \mathsf{E}[(V(g_1)\mathbf{x}, V^*(g)\mathbf{X}(gg_1))]$$
$$= \mathsf{E}[(V(g_1)\mathbf{x}, \mathbf{X}(g_1))]$$
$$= \mathsf{E}[Y_{\mathbf{x}}(g_1)].$$

So $\mathsf{E}[Y_{\mathbf{x}}(g_1)]$ does not depend on $g_1 \in G$.

Next,

$$
\begin{aligned}
\mathsf{E}\big[Y_{\mathbf{x}}(gg_1)\overline{Y_{\mathbf{x}}(gg_2)}\big] &= \mathsf{E}\big[\big(V(gg_1)\mathbf{x},\mathbf{X}(gg_1)\big)\overline{\big(V(gg_2)\mathbf{x},\mathbf{X}(gg_2)\big)}\big] \\
&= \mathsf{E}\big[\big(V(g_1)\mathbf{x}, V^*(g)\mathbf{X}(gg_1)\big)\overline{\big(V(g_2)\mathbf{x}, V^*(g)\mathbf{X}(gg_2)\big)}\big] \\
&= \mathsf{E}\big[\big(V(g_1)\mathbf{x}, \mathbf{X}(g_1)\big)\overline{\big(V(g_2)\mathbf{x}, \mathbf{X}(g_2)\big)}\big] \\
&= \mathsf{E}\big[Y_{\mathbf{x}}(g_1)\overline{Y_{\mathbf{x}}(g_2)}\big],
\end{aligned}
$$

i.e., the scalar random field $Y_{\mathbf{x}}(g)$ is left-homogeneous. By Theorem 2.18,

$$
Y_{\mathbf{x}}(g) = \int_{\hat{G}} \mathrm{tr}\big[\lambda(g)\,\mathrm{d}Z_{\mathbf{x}}(\lambda)\big],
$$

where $Z_{\mathbf{x}}$ is a random measure on $\hat{G}$ whose restriction to $\hat{G}_n$ takes values in the set of bounded linear operators on the standard space H_n satisfying (2.10) and depending linearly on $\mathbf{x}$. It follows that there exists a vector random measure $\mathbf{Z}$ such that $Z_{\mathbf{x}} = (\mathbf{x}, \mathbf{Z})$ and the restriction of $\mathbf{Z}$ on $\hat{G}_n$ satisfies the following condition: for any measurable set $\Lambda_1, \Lambda_2 \subseteq \hat{G}_n$ and for any vectors $\mathbf{f}_1, \mathbf{f}_2, \mathbf{f}_3, \mathbf{f}_4 \in H_n$, $\mathbf{x}, \mathbf{y} \in H$

$$
\mathsf{E}\big[\big(Z_{\mathbf{x}}(\Lambda_1)\mathbf{f}_1, \mathbf{f}_2\big)\overline{\big(Z_{\mathbf{y}}(\Lambda_2)\mathbf{f}_3, \mathbf{f}_4\big)}\big] = \big(F(\Lambda_1 \cap \Lambda_2)\mathbf{x} \otimes \mathbf{f}_1, \mathbf{y} \otimes \mathbf{f}_3\big)\overline{(\mathbf{f}_2, \mathbf{f}_4)}, \quad (2.24)
$$

where F is a measure on $(\hat{G}, \mathfrak{B}(\hat{G}))$ whose restriction to $\hat{G}_n$ takes values in the set of Hermitian nonnegative-definite trace-class operators on the space $H \otimes H_n$ and satisfying

$$
\sum_{n\geq1} \mathrm{tr}\big[F(\hat{G}_n)\big] < \infty. \tag{2.25}
$$

Using the definition of $Y_{\mathbf{x}}(g)$ we can write

$$
Y_{\mathbf{x}}(g) = \big(\mathbf{x}, V^*(g)\mathbf{X}(g)\big).
$$

It follows that

$$
\mathbf{X}(g) = \tilde{V}(g)\int_{\hat{G}} \mathrm{tr}\big[\lambda(g)\,\mathrm{d}\mathbf{Z}(\lambda)\big]. \tag{2.26}
$$

Taking expectations in (2.26), we obtain

$$
\mathbf{E}(g) = \tilde{V}(g)\int_{\hat{G}} \mathrm{tr}\big\{\lambda(g)\,\mathrm{d}\mathsf{E}\big[\mathbf{Z}(\lambda)\big]\big\}.
$$

In particular, if $g = e$, then

$$
\mathbf{E}_0 = \int_{\hat{G}} \mathrm{tr}\big\{\mathrm{d}\mathsf{E}\big[\mathbf{Z}(\lambda)\big]\big\}.
$$

It follows that

$$E[\mathbf{Z}(\Lambda)] = \begin{cases} \mathbf{E}_0, & \lambda_0 \in \Lambda, \\ \mathbf{0}, & \lambda_0 \notin \Lambda, \end{cases} \tag{2.27}$$

where λ_0 is the trivial representation of the group G.

In (2.24), put $\Lambda_1 = \Lambda_2 := \{\lambda_0\}$. We obtain: for any $\mathbf{x}, \mathbf{y} \in H$, and for any complex numbers f_1, f_2, f_3, and f_4,

$$E\left[(\mathbf{x}, \mathbf{Z}(\{\lambda_0\})) f_1 \overline{f_2(\mathbf{y}, \mathbf{Z}(\{\lambda_0\}))} f_3 f_4\right] = (F(\{\lambda_0\})\mathbf{x} f_1, \mathbf{y} f_3) \overline{f_2} f_4.$$

On the other hand, the left-hand side is clearly equal to

$$(\mathbf{x}, \mathbf{E}_0) f_1 \overline{f_2(\mathbf{y}, \mathbf{E}_0) f_3} f_4.$$

It follows that

$$(\mathbf{x}, \mathbf{E}_0)\overline{(\mathbf{y}, \mathbf{E}_0)} = (F(\{\lambda_0\})\mathbf{x}, \mathbf{y}),$$

i.e.,

$$F(\{\lambda_0\}) = \mathbf{E}_0^* \otimes \mathbf{E}_0.$$

Finally, the covariance function of the random field $\mathbf{X}(g)$ is

$$R(g_1, g_2) = E\left[\tilde{V}(g_1) \int_{\hat{G}} \mathrm{tr}[\lambda(g_1)\, d\mathbf{Z}(\lambda)] \otimes \tilde{V}(g_2) \int_{\hat{G}} \mathrm{tr}[\lambda(g_2)\, d\mathbf{Z}(\lambda)]^*\right]$$

$$= \tilde{V}(g_1) \int_{\hat{G}} \mathrm{tr}\{\lambda(g_2^{-1}g_1)\, dE[\mathbf{Z}(\lambda) \otimes \mathbf{Z}^*(\lambda)]\}\tilde{V}^*(g_2)$$

$$= \tilde{V}(g_1) \int_{\hat{G}} \mathrm{tr}[\lambda(g_2^{-1}g_1)\, dF(\lambda)]V^{-1}(g_2).$$

We have proved the following theorem.

Theorem 2.22 *Let G be a locally compact topological group of type I with countable base, acting on itself by left translations. Let $V(g)$ be a representation of G in a finite-dimensional Hilbert space H. Let $\mathbf{X}(g)$ be a wide-sense left-invariant vector random field on G transforming according to the action*

$$g(g_1, \mathbf{x}) = (gg_1, V(g)\mathbf{x}). \tag{2.28}$$

The field $\mathbf{X}(g)$ has the form (2.26), where $\mathbf{Z}(\lambda)$ is a vector random measure satisfying (2.24), and F is a measure on $(\hat{G}, \mathfrak{B}(\hat{G}))$ whose restriction to $\hat{G}_n$ takes values in the set of Hermitian nonnegative-definite trace-class operators on the space $H \otimes H_n$ and satisfying (2.25). The mean value of the field $\mathbf{X}(g)$ is given by (2.23), where $\mathbf{E}_0$ is determined from (2.27). The covariance operator of the field $\mathbf{X}(g)$ is

$$R(g_1, g_2) = \tilde{V}(g_1) \int_{\hat{G}} \mathrm{tr}[\lambda(g_2^{-1}g_1)\, dF(\lambda)]V^{-1}(g_2). \tag{2.29}$$

Conversely, any operator of the form (2.29) *is a covariance operator of the wide-sense left-invariant vector random field on G transforming according to the action* (2.28) *whose mean value has the form* (2.23), *where* $\mathbf{E}_0$ *is determined from* (2.27).

Consider examples of using Theorem 2.22. Assume that G is a *commutative* locally compact topological group with countable base. Let $U(g) = I(g)$ be the trivial representation of G in a finite-dimensional Hilbert space H. Let $\mathbf{X}(g)$ be a centred wide-sense left- (and also right-) invariant random field on G transforming according to the action

$$g(g_1, \mathbf{x}) = (gg_1, \mathbf{x}).$$

Such a field is called *vector homogeneous*. It follows from the theory of irreducible unitary representations of commutative locally compact groups (Subsection 6.3.3) and from Theorem 2.22 that

$$R(g_1, g_2) = \int_{\hat{G}} \hat{g}(g_1 - g_2) \, d\mu(\hat{g}),$$

where μ is a finite measure on $(\hat{G}, \mathfrak{B}(\hat{G}))$, with values in the cone of Hermitian nonnegative-definite linear operators on H, and

$$\mathbf{X}(g) = \int_{\hat{G}} \hat{g}(g) \, d\mathbf{Z}(\hat{g}),$$

where $\mathbf{Z}$ is an H-valued centred random measure on $\hat{G}$ with

$$\mathsf{E}\big[\mathbf{Z}(A) \otimes \mathbf{Z}^*(B)\big] = \mu(A \cap B), \quad A, B \in \mathfrak{B}(\hat{G}).$$

When $G = \mathbb{R}^n$, the above-described expansions take the following form:

$$R(\mathbf{x}, \mathbf{y}) = \int_{\mathbb{R}^n} e^{i(\mathbf{p}, \mathbf{x} - \mathbf{y})} \, d\mu(\mathbf{p}) \tag{2.30}$$

and

$$\mathbf{X}(\mathbf{x}) = \int_{\mathbb{R}^n} e^{i(\mathbf{p}, \mathbf{x})} \, d\mathbf{Z}(\mathbf{p}).$$

When $G = \mathbb{Z}^n$, the above-described expansions take the following form:

$$R(\mathbf{x}, \mathbf{y}) = \int_{\mathbb{T}^n} e^{i(\mathbf{p}, \mathbf{x} - \mathbf{y})} \, d\mu(\mathbf{p}) \tag{2.31}$$

and

$$\mathbf{X}(\mathbf{x}) = \int_{\mathbb{T}^n} e^{i(\mathbf{p}, \mathbf{x})} \, d\mathbf{Z}(\mathbf{p}).$$

2.3 Invariant Random Fields on Homogeneous Spaces

In this section, we consider the situation when the parametric space T is a homogeneous space $T = G/K$, and the group G acts on T by left translations.

2.3.1 The Scalar Case

Let G be a locally compact topological group of type I with countable base. Let K be a compact subgroup of G. Let $T = G/K$ be the homogeneous space of left cosets gK with the quotient topology. Let $X(t)$ be a wide-sense scalar random field on T that is left-invariant under the standard action of the group G. The mean value $E[X(t)]$ is constant, because the standard action is transitive. Without loss of generality we can assume that $E[X(t)] = 0$.

The covariance function of the random field $X(t)$ satisfies the condition

$$R(t_1, t_2) = R(gt_1, gt_2).$$

According to (6.20), we have

$$R(t_1, t_2) = \int_{\hat{G}_K} \mathrm{tr}\big[P_K^{(\lambda)} \lambda\big(g_2^{-1}g_1\big) P_K^{(\lambda)} \, \mathrm{d}F_K(\lambda)\big].$$

In particular, if K is a *massive* subgroup of G (Subsection 6.3.2), we obtain

$$R(t_1, t_2) = \int_{\hat{G}_K} \varphi_0^{(\lambda)}(t_1, t_2) \, \mathrm{d}F(\lambda), \qquad (2.32)$$

where $\varphi_0^{(\lambda)}(t_1, t_2)$ is the zonal spherical function of the representation $\lambda \in \hat{G}_K$.

Using (6.21), we obtain

$$R(t_1, t_2) = \int_{\hat{G}_K} \sum_{n=1}^{\dim \lambda} \varphi_n^{(\lambda)}(t_1) \overline{\varphi_n^{(\lambda)}(t_2)} \, \mathrm{d}F(\lambda).$$

It follows that

$$X(t) = \int_{\hat{G}_K} \sum_{n=1}^{\dim \lambda} \varphi_n^{(\lambda)}(t) \, \mathrm{d}Z_n(\lambda), \qquad (2.33)$$

where Z_n is a sequence of centred random measures on $\hat{G}_K$ satisfying

$$E\big[Z_m(\Lambda_1) \overline{Z_n(\Lambda_2)} \big] = \delta_{mn} F(\Lambda_1 \cap \Lambda_2). \qquad (2.34)$$

We have proved the following theorem.

Table 2.1 Compact two-point homogeneous spaces

T	G	K	α	β
S^n	$SO(n+1)$	$SO(n)$	$(n-2)/2$	$(n-2)/2$
$\mathbb{R}P^n$	$SO(n+1)$	$O(n)$	$(n-2)/2$	$-1/2$
$\mathbb{C}P^n$	$SU(n+1)$	$S(U(n) \times U(1))$	$n-1$	0
$\mathbb{H}P^n$	$Sp(n+1)$	$Sp(n) \times Sp(1)$	$2n-1$	1
$\mathbb{O}P^2$	$F_{4(-52)}$	$Spin(9)$	7	3

Table 2.2 Noncompact two-point homogeneous spaces

T	G	K	M	α	β	N
$\mathbb{R}^n$	$ISO(n)$	$SO(n)$	$SO(n-1)$	$(n-2)/2$	$(n-2)/2$	n
$\mathbb{R}\Lambda^n$	$SO_0(1,n)$	$SO(n)$	$SO(n-1)$	$(n-2)/2$	$-1/2$	n
$\mathbb{C}\Lambda^n$	$SU(n,1)$	$S(U(n) \times U(1))$	$S(U(n-1) \times U(1)^*)$	$n-1$	0	$2n$
$\mathbb{H}\Lambda^n$	$Sp(n,1)$	$Sp(n) \times Sp(1)$	$Sp(n-1) \times Sp(1)^*$	$2n-1$	1	$4n$
$\mathbb{O}\Lambda^2$	$F_{4(-20)}$	$Spin(9)$	$Spin(7)$	7	3	16

Theorem 2.23 *Let G be a locally compact topological group of type I with countable base. Let K be a massive compact subgroup of G. Let $T = G/K$ be the homogeneous space of left cosets gK with the quotient topology. The centred mean-square continuous random field $X(t)$ on T is wide-sense left-invariant under the standard action of the group G iff it can be represented by (2.33) and (2.34). The covariance function of the field $X(t)$ has the form (2.32).*

Example 2.24 Consider a random field $X(t)$ on a two-point homogeneous space T, which is left-invariant under the action of the connected component G of the isometry group of T. The complete classification of two-point homogeneous spaces is presented in Table 2.1 and Table 2.2.

In Table 2.1, S^n, $n \geq 1$ denotes the n-dimensional sphere, $\mathbb{R}P^n$, $\mathbb{C}P^n$, and $\mathbb{H}P^n$, $n \geq 2$ denote the n-dimensional projective spaces over real numbers $\mathbb{R}$, complex numbers $\mathbb{C}$ or quaternions $\mathbb{H}$ respectively, $\mathbb{O}P^2$ denotes the projective plane over the octonions. Similarly, in Table 2.2, $\mathbb{R}^n$, $n \geq 1$ denotes the n-dimensional Euclidean space, $\mathbb{R}\Lambda^n$, $\mathbb{C}\Lambda^n$, and $\mathbb{H}\Lambda^n$, $n \geq 2$ denote the n-dimensional hyperbolic (or Lobachevsky) spaces over real numbers $\mathbb{R}$, complex numbers $\mathbb{C}$ or quaternions $\mathbb{H}$ respectively, $\mathbb{O}\Lambda^2$ denotes the hyperbolic plane over octonions. N denotes the topological dimension of the space X. We use standard notation for Lie groups as well as notation from Flensted-Jensen and Koornwinder (1979):

$$U(1)^* = \{(k,k) : k \in U(1)\}, \qquad Sp(1)^* = \{(k,k) : k \in Sp(1)\}.$$

We have $T = G/K$. In the case of compact two-point homogeneous spaces, norm the Riemannian metric on T in such a way that the diameter of T is equal to π. The set $\hat{G}_K$ can be enumerated by nonnegative integers (Askey and Bingham, 1976). The zonal spherical functions $\varphi_\ell(t_1, t_2)$ of the representation $\lambda_\ell \in \hat{G}_K$ are

$$\varphi_\ell(t_1, t_2) = \frac{P_\ell^{(\alpha,\beta)}(\cos(d(t_1, t_2)))}{P_\ell^{(\alpha,\beta)}(1)}, \tag{2.35}$$

where $P_\ell^{(\alpha,\beta)}(\cos\theta)$ are Jacobi polynomials, and the numbers α and β are written in the fourth and the fifth columns of Table 2.1, respectively. By Theorem 2.23, we obtain

$$R(t_1, t_2) = \sum_{\ell=0}^{\infty} \mu_\ell \frac{P_\ell^{(\alpha,\beta)}(\cos(d(t_1, t_2)))}{P_\ell^{(\alpha,\beta)}(1)}, \tag{2.36}$$

where

$$\mu_\ell \geq 0, \qquad \sum_{\ell=0}^{\infty} \mu_\ell < \infty.$$

The dimension of the representation λ_ℓ is

$$h(T, \ell) = \frac{(2\ell + \alpha + \beta + 1)\Gamma(\beta + 1)\Gamma(\ell + \alpha + \beta + 1)\Gamma(\ell + \alpha + 1)}{\Gamma(\alpha + 1)\Gamma(\alpha + \beta + 2)\ell!\Gamma(\ell + \beta + 1)}. \tag{2.37}$$

Here $(2\ell + \alpha + \beta + 1)\Gamma(\ell + \alpha + \beta + 1)$ is to be interpreted as 1 if $\ell = 0$ and $\alpha = \beta = -1/2$.

Let $\{\mathbf{e}_m : 1 \leq m \leq h(T, \ell)\}$ be a basis for the space $H^{(\ell)}$ of the representation λ_ℓ with $\mathbf{e}_1$ being a K-invariant vector. By the Peter–Weyl theorem, the functions

$$Y_{\ell m}(x) := \sqrt{h(T, \ell)}\big(\lambda_\ell(g)\mathbf{e}_m, \mathbf{e}_1\big) \tag{2.38}$$

form a basis in the space $L^2(T, d\mu)$ of square-integrable functions with respect to the G-invariant probability measure $d\mu$ on T. In the case of the sphere S^n, it is customary to divide these functions by

$$\sqrt{\frac{2\pi^{(n+1)/2}}{\Gamma((n+1)/2)}}$$

to obtain the basis in the space of square-integrable functions with respect to the Lebesgue measure on S^n. In what follows, we denote a generic point on the sphere S^n by $\mathbf{n}$. The functions

$$Y_{\ell m}(\mathbf{n}) = \sqrt{\frac{h(S^n, \ell)\Gamma((n+1)/2)}{2\pi^{(n+1)/2}}}\big(\lambda_\ell(g)\mathbf{e}_m, \mathbf{e}_1\big)$$

are spherical harmonics.

By Theorem 2.23 and matrix multiplication we obtain: in the case of the sphere S^n

$$X(\mathbf{n}) = \sqrt{\frac{2\pi^{(n+1)/2}}{\Gamma((n+1)/2)}} \sum_{\ell=0}^{\infty} \sum_{m=1}^{h(S^n,\ell)} \sqrt{\frac{\mu_\ell}{h(S^n,\ell)}} Y_{\ell m}(\mathbf{n}) X_{\ell m},$$

where $X_{\ell m}$ is the sequence of centred uncorrelated random variables with unit variance.

In all other cases from Table 2.1 we have

$$X(t) = \sum_{\ell=0}^{\infty} \sum_{m=1}^{h(T,\ell)} \sqrt{\frac{\mu_\ell}{h(T,\ell)}} Y_{\ell m}(t) X_{\ell m}.$$

In the case of noncompact two-point homogeneous spaces, the set $\hat{G}_K$ can be identified as a subset of $\mathbb{C}$. For $\mathbb{R}^n$, we have $\hat{G}_K = [0, \infty)$. The zonal spherical functions are

$$\varphi_\lambda(t_1, t_2) = \begin{cases} 1, & \lambda = 0, \\ 2^{(n-2)/2}\Gamma(n/2)\frac{J_{(n-2)/2}(\lambda d(t_1,t_2))}{(\lambda d(t_1,t_2))^{(n-2)/2}}, & \lambda > 0, \end{cases} \qquad (2.39)$$

where $J_{(n-2)/2}$ is the Bessel function. By Theorem 2.23, we have

$$R(t_1, t_2) = 2^{(n-2)/2}\Gamma(n/2) \int_0^\infty \frac{J_{(n-2)/2}(\lambda d(t_1, t_2))}{(\lambda d(t_1, t_2))^{(n-2)/2}} \, d\nu(\lambda),$$

where ν is a finite measure on $[0, \infty)$.

For the remaining spaces, according to Kostant (1975), we have $\hat{G}_K = \{i\varrho\} \cup [0, is_0] \cup (0, \infty)$, where the parameters ϱ and s_0 are calculated as

$$\varrho = \alpha + \beta + 1, \qquad s_0 = \min\{\varrho, \alpha - \beta + 1\}.$$

The zonal spherical functions $\varphi_\lambda(t_1, t_2)$, $\lambda \in \hat{G}_K$ are

$$\varphi_\lambda^{(\alpha,\beta)}(t_1, t_2) = {}_2F_1\left(\frac{\alpha+\beta+1+i\lambda}{2}, \frac{\alpha+\beta+1-i\lambda}{2}; \alpha+1; -\sinh^2(d(t_1, t_2))\right),$$

where ${}_2F_1$ is the hypergeometric function. Note that $\varphi_{i\varrho}^{(\alpha,\beta)}(t_1, t_2) = 1$. By Theorem 2.23, we have: in the case of real and complex hyperbolic spaces

$$R(t_1, t_2) = \int_0^\infty \varphi_\lambda^{(\alpha,\beta)}(t_1, t_2) \, d\nu(\lambda) + \int_{+0}^{i\varrho} \varphi_\lambda^{(\alpha,\beta)}(t_1, t_2) \, d\nu(\lambda)$$

with $\nu([0, \infty) \cup (0, i\varrho]) < \infty$. In the case of quaternion hyperbolic spaces and the hyperbolic plane over octonions

$$R(t_1, t_2) = \int_0^\infty \varphi_\lambda^{(\alpha,\beta)}(t_1, t_2)\, dv(\lambda) + \int_{+0}^{is_0} \varphi_\lambda^{(\alpha,\beta)}(t_1, t_2)\, dv(\lambda) + c$$

with $c \geq 0$ and $v([0, \infty) \cup (0, is_0]) < \infty$.

Any element $g \in \mathrm{ISO}(n)$ can be represented as $g = k_1 r k_2$, where k_1 and k_2 are elements of the group $K = \mathrm{SO}(n)$, i.e., rotations, and $r \geq 0$ is the shift in the direction of a vector $(0, 0, \ldots, 0, r) \in A$, where

$$A = \{(0, 0, \ldots, 0, a) : a \in \mathbb{R}\}.$$

All the remaining groups in the second column of Table 2.2 are semisimple Lie groups of real rank one. By Cartan decomposition, we have again $g = k_1 a k_2$. For any point $t = gK \in T$, choose a representative $g = ka$. Then

$$\varphi_\lambda\big(g_2^{-1} g_1\big) = \varphi_\lambda\big(a_2^{-1} k_2^{-1} k_1 a_1\big).$$

Let M be the centraliser of A in K, that is, the set of all elements of K commuting with every element in A (shown in the fourth column of Table 2.2). For all $\lambda \in \hat{G}_K$, the restriction of λ to K is the direct sum of certain representations in $\hat{K}_M$, each occurring with multiplicity 1. This fact, as well as the subsequent group-theoretical facts and formulae, are proved by Vilenkin (1968) for the case of $X = \mathbb{R}^n$, and by Flensted-Jensen and Koornwinder (1979) for the remaining cases.

Let $\mathbf{e}_1^{(\lambda)}$ be a K-fixed vector of length 1 in $H^{(\lambda)}$. Let $\mu \in \hat{K}_M$, and let $\mathbf{e}_\mu^{(\lambda)}$ be an M-fixed vector of length 1 in $H^{(\lambda)}$ transforming under K according to μ. Let

$$\varphi_{\lambda,\mu}(a) = \big(\lambda(a)\mathbf{e}_1^{(\lambda)}, \mathbf{e}_\mu^{(\lambda)}\big)$$

be the corresponding associated spherical function. Let $\varphi_\mu(k)$ be the zonal spherical function of the representation μ.

For any space in Table 2.2, we have the following addition formula:

$$\varphi_\lambda\big(a_2^{-1} k a_1\big) = \sum_{\mu \in \hat{K}_M} \varphi_{\lambda,\mu}(a_1)\overline{\varphi_{\lambda,\mu}(a_2)}\varphi_\mu\big(k_2^{-1} k_1\big).$$

If T is not a complex hyperbolic space, then all representations $\mu \in \hat{K}_M$ satisfy $\mu = \mu^+$. It follows that all zonal spherical functions $\varphi_\mu(k_2^{-1}k_1)$ are real-valued. Otherwise, we have

$$\varphi_{\lambda,\mu}(a_1)\overline{\varphi_{\lambda,\mu}(a_2)} = \varphi_{\lambda,\mu^+}(a_1)\overline{\varphi_{\lambda,\mu^+}(a_2)}.$$

Therefore, in our addition formula we can take together μ and μ^+ whenever $\mu \neq \mu^+$. Then, the formula takes the form

$$\varphi_\lambda\big(a_2^{-1} k_2^{-1} k_1 a_1\big) = \sum_{\mu \in \hat{K}_M'} c_T \varphi_{\lambda,\mu}(a_1)\overline{\varphi_{\lambda,\mu}(a_2)}\, \mathrm{Re}\, \varphi_\mu\big(k_2^{-1} k_1\big), \qquad (2.40)$$

Table 2.3 Enumeration of nonzero associated spherical functions

T	$\varphi_{k,\ell}(\lambda; a) \neq 0$ if
$\mathbb{R}^n, \mathbb{C}\Lambda^1$	$\ell = 0$
$\mathbb{R}\Lambda^n, n \geq 2$	$\ell = k$
$\mathbb{C}\Lambda^n, n \geq 2; \mathbb{H}\Lambda^n, \mathbb{O}\Lambda^2$	$k \geq \ell \geq 0$

where we sum over only one representative for each pair (μ, μ^+), and

$$c_T := \begin{cases} 2, & T = \mathbb{C}\Lambda^n, \\ 1, & T \neq \mathbb{C}\Lambda^n. \end{cases} \tag{2.41}$$

In what follows, we multiply the associated spherical functions by $\sqrt{c_T}$.

In the case of $T = \mathbb{R}^n$ the set $\hat{K}'_M$ can be enumerated by nonnegative integers. In the remaining cases, the set $\hat{K}'_M$ can be parameterised by pairs (k, ℓ) of nonnegative integers. For clarity, add index $\ell = 0$ in the case of $T = \mathbb{R}^n$ as well. Then, the associated spherical functions do not equal zero in the cases shown in Table 2.3.

In the case of $T = \mathbb{R}^n$ the associated spherical functions have the form

$$\varphi_{k,0}(\lambda; a) = \sqrt{h(S^{n-1}, k)} 2^{(n-2)/2} \Gamma(n/2) \frac{J_{k+(n-2)/2}(\lambda a)}{(\lambda a)^{(n-2)/2}}, \tag{2.42}$$

while in the remaining cases they have the form

$$\varphi_{k,\ell}^{(\alpha,\beta)}(\lambda; a) = \sqrt{\pi_{k,\ell}^{(\alpha,\beta)}} \frac{\Gamma((\alpha + \beta + 1 + i\lambda)/2 + k)\Gamma((\alpha - \beta + 1 + i\lambda)/2 + \ell)}{\Gamma((\alpha + \beta + 1 + i\lambda)/2)\Gamma((\alpha - \beta + 1 + i\lambda)/2)}$$
$$\times \frac{\Gamma(\alpha + 1)}{\Gamma(\alpha + k + \ell + 1)} \sinh^{k+l} a \cosh^{k-l} a \varphi_\lambda^{(\alpha+k+\ell, \beta+k-\ell)}(a), \tag{2.43}$$

where

$$\pi_{k,\ell}^{(\alpha,\beta)} := \frac{(2k - 2\ell + 2\beta + 1)(k + \ell + \alpha + \beta + 3/2)\Gamma(\ell + \alpha + 1)}{\Gamma(\alpha + 1)\Gamma(2\beta + 2)\Gamma(\alpha + \beta + 5/2)}$$
$$\times \frac{\Gamma(k - \ell + 2\beta + 1)\Gamma(k + \alpha + \beta + 3/2)\Gamma(\beta + 3/2)}{\ell!(k - \ell)!\Gamma(k + \beta + 3/2)}. \tag{2.44}$$

Note that $\pi_{k,0}^{((n-2)/2,(n-2)/2)} = h(S^{n-1}, k)$.

The representation $\mu_{k,\ell} \in \hat{K}'_M$ that corresponds to the pair (k, ℓ), has dimension

$$\dim \mu_{k,\ell} = \begin{cases} \pi_{k,\ell}^{(\alpha,\beta)}/2, & X = \mathbb{C}\Lambda^n, \\ \pi_{k,\ell}^{(\alpha,\beta)}, & X \neq \mathbb{C}\Lambda^n. \end{cases}$$

Let $\{\mathbf{e}_m : 1 \leq m \leq \dim \mu_{k,\ell}\}$ be a basis for the space $H^{\mu}_{k,\ell}$ of the representation $\mu_{k,\ell}$ with $\mathbf{e}_1$ being an M-invariant vector. Put

$$f^m_{k\ell}(\mathbf{n}) := \sqrt{\frac{\dim \mu_{k,\ell} \Gamma(N/2)}{2\pi^{N/2}}} \left(\mu_{k,\ell}(k)\mathbf{e}_m, \mathbf{e}_1\right)$$

and

$$Y^m_{k\ell}(\mathbf{n}) := \begin{cases} f^m_{k\ell}(\mathbf{n}), & T \neq \mathbb{C}\Lambda^n, \\ \sqrt{2}\,\mathrm{Re}\, f^m_{k\ell}(\mathbf{n}), & T = \mathbb{C}\Lambda^n, 1 \leq m \leq \pi^{(\alpha,\beta)}_{k,\ell}/2, \\ \sqrt{2}\,\mathrm{Im}\, f^{m-\pi^{(\alpha,\beta)}_{k,\ell}/2}_{k\ell}(\mathbf{n}), & T = \mathbb{C}\Lambda^n, \pi^{(\alpha,\beta)}_{k,\ell}/2 + 1 \leq m \leq \pi^{(\alpha,\beta)}_{k,\ell}. \end{cases}$$

In the case of $T = \mathbb{R}^n$ these functions are just real-valued spherical harmonics.

By the Peter–Weyl theorem, the functions $Y^m_{k\ell}(\mathbf{n})$ form a basis in the space $L^2(S^{N-1})$ of square-integrable functions with respect to the Lebesgue measure $d\mathbf{n}$ on the sphere S^{N-1}. Matrix multiplication gives

$$\mathrm{Re}\,\varphi_{\mu_{k,\ell}}\left(k_2^{-1}k_1\right) = \frac{2\pi^{N/2}}{\pi^{(\alpha,\beta)}_{k,\ell}\Gamma(N/2)} \sum_{m=1}^{\pi^{(\alpha,\beta)}_{k,\ell}} Y^m_{k\ell}(\mathbf{n}_1)\overline{Y^m_{k\ell}(\mathbf{n}_2)}.$$

Let $(r, \mathbf{n})$ be the geodesic polar coordinates on T. In other words, r is the length of the geodesic line connecting o with $p(a)$. It follows from the last formula and (2.40) that

$$X(r, \mathbf{n}) = \sum_{(k,\ell) \in \hat{K}'_M} \sqrt{\frac{2\pi^{N/2}}{\pi^{(\alpha,\beta)}_{k,\ell}\Gamma(N/2)}} \sum_{m=1}^{\pi^{(\alpha,\beta)}_{k,\ell}} Y^m_{k\ell}(\mathbf{n}) \int_{\hat{G}_K} \varphi_{k,\ell}(\lambda; r)\, dZ^m_{k\ell}(\lambda), \quad (2.45)$$

where $Z^m_{k\ell}$ are centred uncorrelated random measures on $\hat{G}_K$ with

$$\mathsf{E}\left|Z^m_{k\ell}(B)\right|^2 = \mu(B)$$

for any $B \in \mathfrak{B}(\hat{G}_K)$.

Consider the restriction of the random field $X(r, \mathbf{n})$ to the sphere $\{r = 1\} = S^{N-1}$. This restriction is invariant under the action of the group K. By Theorem 2.23, it has the form

$$X(\mathbf{n}) = \sum_{(k,\ell) \in \hat{K}'_M} \sum_{m=1}^{\pi^{(\alpha,\beta)}_{k,\ell}} \sqrt{\frac{\nu_{k,\ell}}{\pi^{(\alpha,\beta)}_{k,\ell}}} Y^m_{k\ell}(\mathbf{n}) X^m_{k\ell},$$

where $X^m_{k\ell}$ is the sequence of centred uncorrelated random variables with unit variance, and

$$\nu_{k,\ell} \geq 0, \qquad \sum_{(k,\ell) \in \hat{K}'_M} \nu_{k,\ell} < \infty.$$

Let H_m be the space of restrictions of all homogeneous harmonic polynomials in N variables of a degree m to the sphere S^{N-1}. It is easy to see that the random field $X(\mathbf{n})$ is invariant under the action of the bigger group $SO(N)$ iff two representations $\mu_{k,\ell}$ and $\mu_{k',\ell'}$ whose spaces are the subspaces of the same space H_m, satisfy $\nu_{k,\ell} = \nu_{k',\ell'}$. The structure of the above subspaces is described in Subsection 6.4.3.

Example 2.25 Let ℓ_2 be the set of sequences $\mathbf{x}$ of real numbers with

$$\sum_{j=1}^{\infty} x_j^2 < \infty$$

with the Hilbert space topology induced by the inner product

$$(\mathbf{x}, \mathbf{y}) = \sum_{j=1}^{\infty} x_i y_j.$$

Let G be the group of isometries of the space ℓ_2, and let K be the stationary subgroup of the point $\mathbf{0}$. Let $X(\mathbf{x})$ be a centred G-invariant mean square continuous random field on $\ell_2 = G/K$. Consider the restriction of the random field $X(\mathbf{x})$ to the subset $\mathbb{R}_0^{\infty} \subset \ell_2$. The inductive topology of the space $\mathbb{R}_0^{\infty}$ is weaker than that induced by the topology of the Hilbert space ℓ_2. Therefore, the above restriction is mean-square continuous in the inductive topology. Moreover, ℓ_2 is a subgroup of the group G. Therefore, both the random field $X(\mathbf{x})$ and the above restriction are homogeneous. By Eq. (2.17), the covariance function of the above restriction has the form

$$R(\mathbf{x}, \mathbf{y}) = \int_{\mathbb{R}^{\infty}} e^{i(\mathbf{p}, \mathbf{x}-\mathbf{y})} \, d\mu(\mathbf{p}).$$

The random field $X(\mathbf{x})$, $\mathbf{x} \in \ell_2$, is K-invariant. It follows that the above restriction is invariant under the group $SO(\infty)$, which is the inductive limit of groups $SO(n)$ as $n \to \infty$. Therefore, for any $g \in SO(\infty)$ we have

$$R(g\mathbf{x}, g\mathbf{y}) = R(\mathbf{x}, \mathbf{y}), \quad \mathbf{x}, \mathbf{y} \in \mathbb{R}_0^{\infty}.$$

On the other hand,

$$R(g\mathbf{x}, g\mathbf{y}) = \int_{\mathbb{R}^{\infty}} e^{i(\mathbf{p}, g\mathbf{x}-g\mathbf{y})} \, d\mu(\mathbf{p})$$

$$= \int_{\mathbb{R}^{\infty}} \exp\left(i(g^{-1}\mathbf{p}, \mathbf{x}-\mathbf{y})\right) d\mu(\mathbf{p})$$

$$= \int_{\mathbb{R}^{\infty}} e^{i(\mathbf{p}, \mathbf{x}-\mathbf{y})} \, d\mu(g\mathbf{p}).$$

It follows that the measure μ is SO(∞)-invariant. By the result of Umemura (1965), μ has the following form:

$$\mu(\mathbf{p}) = \int_0^\infty \mu_\lambda(\mathbf{p}) \, d\nu(\lambda),$$

where μ_λ is the product of countably many Gaussian measures on $\mathbb{R}$ with zero mean and variance λ, while ν is a finite measure on $[0, \infty)$. Therefore, we have

$$
\begin{aligned}
R(\mathbf{x}, \mathbf{y}) &= \int_{\mathbb{R}^\infty} e^{i(\mathbf{p}, \mathbf{x} - \mathbf{y})} \int_0^\infty d\mu_\lambda(\mathbf{p}) \, d\nu(\lambda) \\
&= \int_0^\infty \int_{\mathbb{R}^\infty} e^{i(\mathbf{p}, \mathbf{x} - \mathbf{y})} \, d\mu_\lambda(\mathbf{p}) \, d\nu(\lambda) \\
&= \int_0^\infty \exp\left(-\lambda \|\mathbf{x} - \mathbf{y}\|^2/2\right) d\nu(\lambda).
\end{aligned}
$$

The rightmost side of the above display is continuous in the Hilbert space topology. It follows that the autocovariance function of the G-invariant, i.e., homogeneous and isotropic field on ℓ_2, has the form

$$R(\mathbf{x}, \mathbf{y}) = \int_0^\infty \exp\left(-\lambda \|\mathbf{x} - \mathbf{y}\|^2/2\right) d\nu(\lambda). \tag{2.46}$$

The integrand may be written as

$$\exp\left(-\lambda \|\mathbf{x} - \mathbf{y}\|^2/2\right) = \exp\left(-\lambda \|\mathbf{x}\|^2/2\right) \exp\left(-\lambda \|\mathbf{x}\|^2/2\right) \exp\left(\lambda(\mathbf{x}, \mathbf{y})\right).$$

By Taylor expansion

$$
\begin{aligned}
\exp\left(\lambda(\mathbf{x}, \mathbf{y})\right) &= \prod_{j=1}^\infty \exp(\lambda x_j y_j) \\
&= \prod_{j=1}^\infty \sum_{k=0}^\infty \frac{(\lambda x_j y_j)^k}{k!} \\
&= \sum_{m=0}^\infty \sum_{(i_1, \ldots, i_m, k_1, \ldots, k_m) \in L_m} \frac{\lambda^m}{k_1! \cdots k_m!} \prod_{\ell=1}^m (x_{i_\ell} y_{i_\ell})^{k_\ell},
\end{aligned}
$$

where $L_0 := \varnothing$, L_m, $m \geq 1$ is the set of all multi-indices $\varrho = (i_1, \ldots, i_m, k_1, \ldots, k_m)$ where $i_1, \ldots, i_m$ are positive integers with $1 \leq i_1 < \cdots < i_m$, $k_1, \ldots, k_m$ are positive integers with $k_1 + \cdots + k_m = m$. It follows that

$$R(\mathbf{x}, \mathbf{y}) = \sum_{m=0}^\infty \sum_{\varrho \in L_m} \int_0^\infty f_\varrho(\mathbf{x}, \lambda, m) f_\varrho(\mathbf{y}, \lambda, m) \, d\nu(\lambda),$$

where

$$f_\varrho(\mathbf{x}, \lambda, m) = \exp(-\lambda \|\mathbf{x}\|^2/2) \lambda^{m/2} \frac{x_{i_1}^{k_1} \cdots x_{i_m}^{k_m}}{\sqrt{k_1! \cdots k_m!}}.$$

Finally, by Karhunen's theorem,

$$X(\mathbf{x}) = \sum_{m=0}^{\infty} \sum_{\varrho \in L_m} \int_0^\infty \varphi_\varrho(\mathbf{x}, \lambda, m) \, dZ_m^\varrho(\lambda),$$

where Z_m^ϱ, $m \geq 0$, $\varrho \in L_m$ is the sequence of centred random measures on $[0, \infty)$ with

$$\mathsf{E}\big[Z_m^\varrho(A)\overline{Z_{m'}^{\varrho'}(B)}\big] = \delta_{mm'}\delta_{\varrho\varrho'}\nu(A \cap B)$$

for all Borel subsets A, B of $[0, \infty)$.

2.3.2 The Vector Case

In Example 2.16, let G be a compact topological group, and let K be a closed subgroup. Let T be the homogeneous space G/K. Let W be a unitary representation of K in a finite-dimensional complex Hilbert space H, and let $\xi = (E_W, \pi, T)$ be the corresponding homogeneous vector bundle. Let $\hat{G}$ (resp. $\hat{K}$) be the set of all equivalence classes of irreducible unitary representations of G (resp. K). For simplicity, assume that K is massive in G.

First, consider the case when $W = \lambda$ is an *irreducible* unitary representation of K. Let $\hat{G}_K(\lambda)$ be the set of all $\mu \in \hat{G}$ whose restrictions to K contain λ (necessarily once, because K is massive in G). For any $\mu \in \hat{G}_K(\lambda)$, let i_μ be the embedding of H into the space H_μ of the representation μ. Let p_μ be the orthogonal projection from H_μ on H. By the result of Camporesi (2005), any $\mathbf{f} \in L^2(G, H)$ can be represented by the series

$$\mathbf{f}(g) = \frac{1}{\dim \lambda} \sum_{\mu \in \hat{G}_K(\lambda)} \dim \mu \int_G p_\mu \mu(g^{-1}h) i_\mu \mathbf{f}(h) \, dh.$$

The above series converges in the strong topology of the Hilbert space $L^2(G, H)$, i.e., the following Parseval identity holds:

$$\|\mathbf{f}\|_{L^2(G,H)}^2 = \frac{1}{\dim \lambda} \sum_{\mu \in \hat{G}_K(\lambda)} \dim \mu \int_G \left(\int_G p_\mu \mu(g^{-1}h) i_\mu \mathbf{f}(h) \, dh, \mathbf{f}(g) \right) dg.$$

Fix a basis $\{\mathbf{e}_1, \mathbf{e}_2, \ldots, \mathbf{e}_{\dim H}\}$ of the space H. Let $\{\mathbf{e}_1^{(\mu)}, \mathbf{e}_2^{(\mu)}, \ldots, \mathbf{e}_{\dim \mu}^{(\mu)}\}$ be a basis in H_μ with

$$i_\mu \mathbf{e}_j = \mathbf{e}_{p+j}^{(\mu)}, \quad 1 \leq j \leq \dim \lambda, \tag{2.47}$$

for some $p \geq 0$. Let $f_j(g) = (\mathbf{f}(g), \mathbf{e}_j)$ be the coordinates of $\mathbf{f}(g)$. Equation (2.47) means that λ acts on the linear span of the $\dim \lambda$ basis vectors of H_μ that are enumerated without lacunas. Then we have

$$i_\mu \mathbf{f}(h) = \big(0, \ldots, 0, f_1(h), \ldots, f_{\dim \lambda}(h), 0, \ldots, 0\big).$$

Let $\mu_{m,n}(g) = (\mu(g)\mathbf{e}_m^{(\mu)}, \mathbf{e}_n^{(\mu)})$ be the matrix elements of the representation μ. Then

$$\big(\mu\big(g^{-1}h\big)i_\mu \mathbf{f}(h)\big)_{p+j} = \sum_{m=1}^{\dim \mu} \overline{\mu_{m,p+j}(g)} \sum_{n=1}^{\dim \lambda} \mu_{m,p+n}(h) f_n(h)$$

and

$$f_j(g) = \frac{1}{\dim \lambda} \sum_{\mu \in \hat{G}_K(\lambda)} \dim \mu \sum_{m=1}^{\dim \mu} \sum_{n=1}^{\dim \lambda} \int_G f_n(h)\mu_{m,p+n}(h)\, dh\, \overline{\mu_{m,p+j}(g)}.$$

From now on, let G be a connected compact Lie group, and let $p \colon G \to T$ denote the natural projection: $p(g) := gK$. Let D_G be an open dense subset in G, and let $(D_G, \mathbf{J}(g))$ with

$$\mathbf{J}(g) = \big(\theta_1(g), \ldots, \theta_{\dim G}(g)\big) \colon D_G \to \mathbb{R}^{\dim G}$$

be a chart of the atlas of the manifold G with the following property: if $k \in K$ and both g and kg lie in D_G, then $\theta_j(kg) = \theta_j(g)$ for $1 \leq j \leq \dim T$. Then, $(D_T, \mathbf{I}(t))$ with

$$\begin{aligned} D_T &= p D_G, \\ \mathbf{I}(t) &= \big(\theta_1(t), \ldots, \theta_{\dim T}(t)\big) \colon D_T \to \mathbb{R}^{\dim T} \end{aligned} \qquad (2.48)$$

is a chart of the atlas of the manifold T, and the domain D_T of this chart is dense in T. Let $t \in D_T$ have local coordinates $(\theta_1, \ldots, \theta_{\dim T})$ in the chart (2.48). Then, the representation of the section $\mathbf{s} \in L^2(E_W)$ associated to $\mathbf{f} \in L^2(G, H)$ has the form

$$s_j(t) = \frac{1}{\dim \lambda} \sum_{\mu \in \hat{G}_K(\lambda)} \dim \mu \sum_{m=1}^{\dim \mu} \sum_{n=1}^{\dim \lambda} \int_T s_n(t)\mu_{m,p+n}(t)\, dt\, \overline{\mu_{m,p+j}(t)}, \qquad (2.49)$$

where dt is the G-invariant measure on T with $\int_T dt = 1$, and

$$\mu_{m,p+n}(t) = \mu_{m,p+n}\big(\theta_1, \ldots, \theta_{\dim T}, \theta_{\dim T+1}^{(0)}, \ldots, \theta_{\dim G}^{(0)}\big).$$

Introduce the following notation:

$$_\lambda \mathbf{Y}_{\mu m}(t) := \sqrt{\frac{\dim \mu}{\dim \lambda}} \big(\overline{\mu_{m,p+1}(t)}, \overline{\mu_{m,p+2}(t)}, \ldots, \overline{\mu_{m,p+\dim \lambda}(t)}\big). \qquad (2.50)$$

Note that the correct notation must be $_\lambda \mathbf{Y}_{\mathbf{I}\mu m}(t)$, because functions (2.50) depend on the choice of a chart. In what follows, we use only chart (2.48) and suppress the symbol $\mathbf{I}$ for notational simplicity.

Equation (2.49) means that the functions $\{_\lambda \mathbf{Y}_{\mu m}(t): \mu \in \hat{G}_K(\lambda), 1 \leq m \leq \dim \mu\}$ form a basis in $L^2(E_W)$, i.e.,

$$s_j(t) = \sum_{\mu \in \hat{G}_K(\lambda)} \sum_{m=1}^{\dim \mu} \sum_{n=1}^{\dim \lambda} \int_T s_n(t) \overline{(_\lambda Y_{\mu m})_n(t)} \, dt \, (_\lambda Y_{\mu m})_j(t). \tag{2.51}$$

Let $\mathbf{X}(t)$ be a mean-square continuous random field in ξ. Consider the following random variables:

$$Z_{mn}^{(\mu)} := \int_T X_n(t) \overline{(_\lambda Y_{\mu m})_n(t)} \, dt, \tag{2.52}$$

where $\mu \in \hat{G}_K(\lambda)$, $1 \leq m \leq \dim \mu$, and $1 \leq n \leq \dim \lambda$. This integral has to be understood as a Bochner integral of a function taking values in the space $L^2_{\mathbb{K}}(\Omega, \mathfrak{F}, \mathsf{P})$. Let $H_{\mathbf{X}}$ be the closed linear span of the set $\{X_n(t): t \in T, 1 \leq n \leq \dim W\}$ in the above Hilbert space.

Theorem 2.26 *Let G be a connected compact Lie group, let K be a massive subgroup, let W be an irreducible unitary representation of the group K, and let ξ be the corresponding homogeneous vector bundle. In the chart (2.48), a mean-square continuous random field $\mathbf{X}(t)$ in ξ has the form*

$$X_j(t) = \sum_{V \in \hat{G}_K(W)} \sum_{m=1}^{\dim V} \sum_{n=1}^{\dim W} Z_{mn}^{(V)} (_W Y_{Vm})_j(t), \tag{2.53}$$

in the sense that both sides represent the same element of the space $H_{\mathbf{X}}$, where random variables $Z_{mn}^{(V)}$ have the form (2.52).

Proof It is enough to show that both sides of (2.53) have the same covariance with every element in the total subset $\{X_k(s): s \in T, 1 \leq k \leq \dim W\}$ of the space $H_{\mathbf{X}}$. In other words, we have to prove that

$$\mathsf{E}\left[\sum_{V \in \hat{G}_K(W)} \sum_{m=1}^{\dim V} \sum_{n=1}^{\dim W} Z_{mn}^{(V)} (_W Y_{Vm})_j(t) \overline{X_k(s)} \right] = \mathsf{E}\left[X_j(t) \overline{X_k(s)} \right].$$

We have

$$
\mathsf{E}\left[\sum_{V \in \hat{G}_K(W)} \sum_{m=1}^{\dim V} \sum_{n=1}^{\dim W} Z_{mn}^{(V)}({}_W Y_{Vm})_j(t)\overline{X_k(s)} \right]
$$

$$
= \sum_{V \in \hat{G}_K(W)} \sum_{m=1}^{\dim V} \sum_{n=1}^{\dim W} \mathsf{E}\left[\int_T X_n(u)\overline{({}_W Y_{Vm})_n(u)}\, du ({}_W Y_{Vm})_j(t)\overline{X_k(s)} \right]
$$

$$
= \sum_{V \in \hat{G}_K(W)} \sum_{m=1}^{\dim V} \sum_{n=1}^{\dim W} \int_T \mathsf{E}\left[X_n(u)\overline{X_k(s)} \right]\overline{({}_W Y_{Vm})_n(u)}\, du ({}_W Y_{Vm})_j(t)
$$

$$
= \mathsf{E}\left[X_j(t)\overline{X_k(s)} \right],
$$

by (2.51) with $s_j(t) = \mathsf{E}[X_j(t)\overline{X_k(s)}]$. $\square$

Denote by μ_0 the trivial irreducible representation of the group G.

Theorem 2.27 *Under the conditions of Theorem 2.26, the following statements are equivalent.*

1. $\mathbf{X}(t)$ *is a mean-square continuous wide-sense invariant random field in* ξ.
2. $\mathbf{X}(t)$ *has the form* (2.53), *where* $Z_{mn}^{(\mu)}$, $\mu \in \hat{G}_K(\lambda)$, $1 \le m \le \dim \mu$, $1 \le n \le \dim \lambda$ *are random variables satisfying the following conditions.*

- *If* $\mu \ne \mu_0$, *then* $\mathsf{E}[Z_{mn}^{(\mu)}] = 0$.
- $\mathsf{E}[Z_{mn}^{(\mu)}\overline{Z_{m'n'}^{(\mu')}}] = \delta_{\mu\mu'}\delta_{mm'} R_{nn'}^{(\mu)}$, *with*

$$
\sum_{\mu \in \hat{G}_K(\lambda)} \dim \mu \operatorname{tr}\left[R^{(\mu)} \right] < \infty. \tag{2.54}
$$

Proof Let $\mathbf{X}(t)$ be a mean-square continuous wide-sense invariant random field in ξ. By Theorem 2.26, $\mathbf{X}(t)$ has the form (2.53). Calculating the mean value of both sides of (2.53), we obtain

$$
\mathsf{E}\left[X_j(t)\right] = \sum_{V \in \hat{G}_K(W)} \sum_{m=1}^{\dim V} \sum_{n=1}^{\dim W} \mathsf{E}[Z_{mn}^{(V)}]({}_W Y_{Vm})_j(t).
$$

Let $g \in G$. Substitute gt in place of t in the last display. We obtain

$$\mathsf{E}\big[X_j(gt)\big] = \sum_{V \in \hat{G}_K(W)} \sum_{m=1}^{\dim V} \sum_{n=1}^{\dim W} \mathsf{E}\big[Z_{mn}^{(V)}\big](wY_{Vm})_j(gt)$$

$$= \sum_{V \in \hat{G}_K(W)} \sum_{m=1}^{\dim V} \sum_{n=1}^{\dim W} \mathsf{E}\big[Z_{mn}^{(V)}\big] \sum_{\ell=1}^{\dim V} \overline{V_{m\ell}(g)} (wY_{V\ell})_j(t)$$

$$= \sum_{V \in \hat{G}_K(W)} \sum_{m=1}^{\dim V} \sum_{n=1}^{\dim W} \sum_{\ell=1}^{\dim V} \overline{V_{\ell m}(g)} \mathsf{E}\big[Z_{\ell n}^{(V)}\big](wY_{Vm})_j(t).$$

The left-hand sides of the two last displays are equal. So, the coefficients of the expansions must be equal:

$$\sum_{\ell=1}^{\dim V} \overline{V_{\ell m}(g)} \mathsf{E}\big[Z_{\ell n}^{(V)}\big] = \mathsf{E}\big[Z_{mn}^{(V)}\big].$$

Denote $\mathbf{M}_n^{(V)} := (\mathsf{E}[Z_{1n}^{(V)}], \dots, \mathsf{E}[Z_{\dim V n}^{(V)}])$. Then

$$V^+(g)\mathbf{M}_n^{(V)} = \mathbf{M}_n^{(V)}, \quad g \in G,$$

where $V^+(g) = V(g^{-1})^\top$ is the representation dual to the representation V. It follows that either $\mathbf{M}_n^{(V)} = \mathbf{0}$ or the one-dimensional subspace generated by $\mathbf{M}_n^{(V)}$ is an invariant subspace of the irreducible representation V^+. In the latter case, V^+ must be one-dimensional. If V^+ is trivial, then V is also trivial, and $\mathbf{M}_n^{(V)}$ is any complex number. If V^+ is not trivial, neither is V. Then, there exist $g \in G$ with $V(g) \neq 1$. It follows that $\mathbf{M}_n^{(V)} = V(g)\mathbf{M}_n^{(V)} = \mathbf{0}$.

Calculate the covariance operator of the random field (2.53). We obtain

$$R_{jj'}(t_1, t_2)$$

$$= \sum_{V,V' \in \hat{G}_K(W)} \sum_{m=1}^{\dim V} \sum_{m'=1}^{\dim V'} \sum_{n,n'=1}^{\dim W} \mathsf{E}\big[Z_{mn}^{(V)}\overline{Z_{m'n'}^{(V')}}\big](wY_{Vm})_j(t_1)\overline{(wY_{V'm'})_{j'}(t_2)}.$$

It follows that

$$R_{jj'}(gt_1, gt_2)$$

$$= \sum_{V,V'\in\hat{G}_K(W)} \sum_{m=1}^{\dim V} \sum_{m'=1}^{\dim V'} \sum_{n,n'=1}^{\dim W} \mathsf{E}\big[Z_{mn}^{(V)}\overline{Z_{m'n'}^{(V')}}\big](wY_{Vm})_j(gt_1)\overline{(wY_{V'm'})_{j'}(gt_2)}$$

$$= \sum_{V,V'\in\hat{G}_K(W)} \sum_{m=1}^{\dim V} \sum_{m'=1}^{\dim V'} \sum_{n,n'=1}^{\dim W} \mathsf{E}\big[Z_{mn}^{(V)}\overline{Z_{m'n'}^{(V')}}\big]$$

$$\times \sum_{\ell=1}^{\dim V}\overline{V_{m\ell}(g)}(wY_{V\ell})_j(t_1)\sum_{\ell'=1}^{\dim V'}V_{m'\ell'}'(g)\overline{(wY_{V'\ell'})_{j'}(t_2)}$$

$$= \sum_{V,V'\in\hat{G}_K(W)} \sum_{m=1}^{\dim V} \sum_{m'=1}^{\dim V'} \sum_{n,n'=1}^{\dim W} \sum_{\ell=1}^{\dim V} \sum_{\ell'=1}^{\dim V'} \overline{V_{m\ell}(g)}V_{m'\ell'}'(g)$$

$$\times \mathsf{E}\big[Z_{\ell n}^{(V)}\overline{Z_{\ell'n'}^{(V')}}\big](wY_{Vm})_j(t_1)\overline{(wY_{V'm'})_{j'}(t_2)}.$$

By equating the coefficients of the two expansions, we obtain

$$\sum_{\ell=1}^{\dim V}\sum_{\ell'=1}^{\dim V'}\overline{V_{m\ell}(g)}V_{m'\ell'}'(g)\mathsf{E}\big[Z_{\ell n}^{(V)}\overline{Z_{\ell'n'}^{(V')}}\big] = \mathsf{E}\big[Z_{mn}^{(V)}\overline{Z_{m'n'}^{(V')}}\big].$$

Let $P_{nn'}^{(V,V')}$ be the matrix with elements

$$\big(P_{nn'}^{(V,V')}\big)_{mm'} = \mathsf{E}\big[Z_{mn}^{(V)}\overline{Z_{m'n'}^{(V')}}\big].$$

Then

$$\big(V^+ \otimes V'\big)(g)P_{nn'}^{(V,V')} = P_{nn'}^{(V,V')}, \quad g \in G.$$

It follows that either $P_{nn'}^{(V,V')}$ is the zero matrix or the one-dimensional subspace generated by the matrix $P_{nn'}^{(V,V')}$ is an invariant subspace of the representation $V^+ \otimes V'$. In the latter case, the representation $V^+ \otimes V'$ contains a one-dimensional irreducible component, say $\mathscr{V}$. By a well-known result from representation theory (see, for example, Naĭmark and Štern, 1982, Sect. 2.6), the trivial representation of G is an irreducible component of the representation $V^+ \otimes V'$ iff $V = V'$. Therefore, if $\mathscr{V}$ is trivial, then $V = V'$ and $\mathscr{V}$ acts on the one-dimensional subspace generated by the identity matrix. It follows that the matrix $P_{nn'}^{(V,V')}$ is a multiple of the identity matrix, say $(P_{nn'}^{(V,V')})_{mm'} = \delta_{mm'}R_{nn'}^{(V)}$. If $\mathscr{V}$ is not trivial, then there exists $g \in G$ with $\mathscr{V}(g) \neq 1$. It follows that $P_{nn'}^{(V,V')} = \mathscr{V}(g)P_{nn'}^{(V,V')}$, so $P_{nn'}^{(V,V')}$ is the zero matrix.

This gives

$$\mathsf{E}\big[Z_{mn}^{(V)}\,\overline{Z_{m'n'}^{(V')}}\big] = \delta_{VV'}\delta_{mm'}R_{nn'}^{(V)}.$$

Let $t_0 \in T$ be the left coset of the identity element of G. We may assume $t_0 \in D_T$ (otherwise use a chart $(gD_T, \mathbf{I}(g^{-1}t))$ for a suitable $g \in G$). Then

$$(wY_{Vm})_j(t_0) = \sqrt{\frac{\dim V}{\dim W}}\,\delta_{m,p+j}$$

and

$$X_j(t_0) = \sum_{V\in\hat{G}_K(W)}\sum_{m=1}^{\dim V}\sum_{n=1}^{\dim W} Z_{mn}^{(V)}(wY_{Vm})_j(t)$$

$$= \frac{1}{\sqrt{\dim W}}\sum_{V\in\hat{G}_K(W)}\sqrt{\dim V}\sum_{n=1}^{\dim W} Z_{jn}^{(V)}.$$

So

$$\mathsf{E}\big|X_j(t_0)\big|^2 = \frac{1}{\dim W}\sum_{V\in\hat{G}_K(W)}\dim V\,\dim W\,\mathsf{E}\big|Z_{j1}^{(V)}\big|^2$$

$$= \sum_{V\in\hat{G}_K(W)}\dim V\,R_{jj}^{(V)},$$

and

$$\sum_{V\in\hat{G}_K(W)}\dim V\,\mathrm{tr}\big[R^{(V)}\big] = \sum_{j=1}^{\dim W}\mathsf{E}\big|X_j(t_0)\big|^2 < \infty.$$

Conversely, let $Z_{mn}^{(V)}$, $V \in \hat{G}_K(W)$, $1 \le m \le \dim V$, $1 \le n \le \dim W$ be random variables satisfying conditions of Theorem 2.27. Consider the random field (2.53). Then its mean value is

$$E\big[X_j(t)\big] = \begin{cases}\mathsf{E}[Z_{11}^{(V_0)}], & V_0 \in \hat{G}_K(W), \\ 0, & \text{otherwise,}\end{cases}$$

which is constant. Note that $V_0 \in \hat{G}_K(W)$ iff W is trivial (by Frobenius reciprocity, Subsection 6.3.6).

The covariance operator of the random field (2.53) is

$$R_{jj'}(t_1, t_2)$$

$$= \sum_{V,V' \in \hat{G}_K(W)} \sum_{m=1}^{\dim V} \sum_{m'=1}^{\dim V'} \sum_{n,n'=1}^{\dim W} \mathsf{E}\big[Z_{mn}^{(V)} \overline{Z_{m'n'}^{(V')}}\big] (w\, Y_{Vm})_j(t_1) \overline{(w\, Y_{V'm'})_{j'}(t_2)}$$

$$= \sum_{V \in \hat{G}_K(W)} \sum_{n,n'=1}^{\dim W} R_{nn'}^{(V)} \sum_{m=1}^{\dim V} (w\, Y_{Vm})_j(t_1) \overline{(w\, Y_{Vm})_{j'}(t_2)}$$

$$= \frac{1}{\dim W} \sum_{V \in \hat{G}_K(W)} \dim V \sum_{n,n'=1}^{\dim W} R_{nn'}^{(V)} V_{p+j,p+j'}(g_1^{-1} g_2),$$

where g_1 (resp. g_2) is an arbitrary element from the left coset corresponding to t_1 (resp. t_2). The terms of this functional series are bounded by the terms of the convergent series (2.54), because $|V_{p+j,p+j'}(g_1^{-1} g_2)| \leq 1$. Therefore, the series converges uniformly, and its sum is a continuous function. This means that $\mathbf{X}(t)$ is mean-square continuous.

For any $g \in G$, we have

$$R_{jj'}(gt_1, gt_2) = \frac{1}{\dim W} \sum_{V \in \hat{G}_K(W)} \dim V \sum_{n,n'=1}^{\dim W} R_{nn'}^{(V)} V_{p+j,p+j'}\big((gg_1)^{-1} gg_2\big)$$

$$= R_{jj'}(t_1, t_2),$$

so $\mathbf{X}(t)$ is invariant. □

Assume that W is a not necessarily irreducible representation of K in a finite-dimensional complex Hilbert space H. Because K is compact, the representation W is equivalent to a direct sum $\lambda_1 \oplus \lambda_2 \oplus \cdots \oplus \lambda_N$ of irreducible unitary representations of K. The representation induced by W is a direct sum of representations induced by λ_k, $1 \leq k \leq N$. It is realised in a homogeneous vector bundle $\xi = \xi_1 \oplus \xi_2 \oplus \cdots \oplus \xi_N$, where ξ_k is the homogeneous vector bundle that carries the irreducible component λ_k.

Let $\mathbf{X}(t)$ be an invariant random field in ξ. Denote the components of $\mathbf{X}(t)$ by $X_j^{(k)}(t)$, $1 \leq k \leq N$, $1 \leq j \leq \dim \lambda_k$. Denote by P_k the orthogonal projection from H on the space H_k where the irreducible component λ_k acts.

Theorem 2.28 *Under conditions of Theorem 2.26, the following statements are equivalent.*

1. $\mathbf{X}(t)$ *is a mean-square continuous wide-sense invariant random field in* ξ.

2. $\mathbf{X}(t)$ *has the form*

$$X_j^{(k)}(t) = \sum_{\mu \in \hat{G}_K(\lambda_k)} \sum_{m=1}^{\dim \mu} \sum_{n=1}^{\dim \lambda_k} Z_{mn}^{(\mu k)}(\lambda_k Y_{\mu m})_j(t), \qquad (2.55)$$

where $Z_{mn}^{(\mu k)}$, $1 \le k \le N$, $\mu \in \hat{G}_K(\lambda_k)$, $1 \le m \le \dim \mu$, $1 \le n \le \dim \lambda_k$ *are random variables satisfying the following conditions.*

- *If* $\mu \ne \mu_0$, *then* $\mathsf{E}[Z_{mn}^{(\mu k)}] = 0$.
- $\mathsf{E}[Z_{mn}^{(\mu k)} \overline{Z_{m'n'}^{(\mu' k')}}] = \delta_{\mu\mu'} \delta_{mm'} R_{kn,k'n'}^{(\mu)}$, *with*

$$\sum_{k=1}^{N} \sum_{\mu \in \hat{G}_K(\lambda_k)} \dim \mu \operatorname{tr}\left[P_k R^{(\mu)} P_k \right] < \infty.$$

Proof Use mathematical induction. The induction base, when $N = 1$, is Theorem 2.27. Assume the induction hypothesis: Theorem 2.28 is proved up to $N - 1$.

Let $\mathbf{X}(t)$ be a mean-square continuous wide-sense invariant random field in ξ. Then the field

$$\mathbf{Y}_1(t) = \left(X_1^{(1)}(t), \dots, X_{\dim \lambda_1}^{(1)}(t), \dots, X_1^{(N-1)}(t), \dots, X_{\dim \lambda_{N-1}}^{(N-1)}(t) \right),$$

is a mean-square continuous wide-sense invariant random field in $\xi_1 \oplus \cdots \oplus \xi_{N-1}$. By the induction hypothesis,

$$X_j^{(k)}(t) = \sum_{\mu \in \hat{G}_K(\lambda_k)} \sum_{m=1}^{\dim \mu} \sum_{n=1}^{\dim \lambda_k} Z_{mn}^{(\mu k)}(\lambda_k Y_{\mu m})_j(t), \qquad 1 \le k \le N - 1,$$

where $\mathsf{E}[Z_{mn}^{(\mu k)}] = 0$ unless $\mu \ne \mu_0$ and $\mathsf{E}[Z_{mn}^{(\mu k)} \overline{Z_{m'n'}^{(\mu' k')}}] = \delta_{\mu\mu'} \delta_{mm'} R_{kn,k'n'}^{(\mu, N-1)}$, with

$$\sum_{k=1}^{N-1} \sum_{\mu \in \hat{G}_K(\lambda_k)} \dim \mu \operatorname{tr}\left[P_k R^{(\mu, N-1)} P_k \right] < \infty.$$

The field

$$\mathbf{Y}_2(t) = \left(X_1^{(N)}(t), \dots, X_{\dim \lambda_N}^{(N)}(t) \right)$$

is a mean-square continuous wide-sense invariant random field in ξ_N. By Theorem 2.27,

$$X_j^{(N)}(t) = \sum_{\mu \in \hat{G}_K(\lambda_N)} \sum_{m=1}^{\dim \mu} \sum_{n=1}^{\dim \lambda_N} Z_{mn}^{(\mu N)}(\lambda_N Y_{\mu m})_j(t),$$

where $\mathsf{E}[Z_{mn}^{(\mu N)}] = 0$ unless $\mu \neq \mu_0$ and $\mathsf{E}[Z_{mn}^{(\mu N)} \overline{Z_{m'n'}^{(\mu' N)}}] = \delta_{\mu\mu'} \delta_{mm'} R_{nn'}^{(\mu N)}$, with

$$\sum_{\mu \in \hat{G}_K(\lambda_N)} \dim \mu \, \mathrm{tr}[R^{(\mu N)}] < \infty.$$

The matrix $R_{kn,k'n'}^{(\mu)}$ with elements

$$R_{kn,k'n'}^{(\mu)} = \mathsf{E}[Z_{1n}^{(\mu k)} \overline{Z_{1n'}^{(\mu k')}}]$$

obviously satisfies the conditions of the second item of Theorem 2.28.

Conversely, let $Z_{mn}^{(\mu k)}$, $1 \leq k \leq N$, $\mu \in \hat{G}_K(\lambda_k)$, $1 \leq m \leq \dim \mu$, $1 \leq n \leq \dim \lambda_k$ be random variables satisfying conditions of Theorem 2.28. Consider the random field (2.55). Its mean value is obviously constant. Its covariance operator is

$$R_{jj'}^{(kk')}(t_1, t_2)$$

$$= \sum_{\mu \in \hat{G}_K(\lambda_k) \cap \hat{G}_K(\lambda_{k'})} \sum_{n=1}^{\dim \lambda_k} \sum_{n'=1}^{\dim \lambda_{k'}} R_{kn,k'n'}^{(\mu)} \sum_{m=1}^{\dim \mu} (\lambda_k Y_{\mu m})_j(t_1) \overline{(\lambda_{k'} Y_{\mu m})_{j'}(t_2)}$$

$$= \frac{1}{\sqrt{\dim \lambda_k \dim \lambda_{k'}}} \sum_{\mu \in \hat{G}_K(\lambda_k) \cap \hat{G}_K(\lambda_{k'})} \dim \mu \sum_{n=1}^{\dim \lambda_k} \sum_{n'=1}^{\dim \lambda_{k'}} R_{kn,k'n'}^{(\mu)}$$

$$\times \mu_{p+j,p+j'}(g_1^{-1} g_2),$$

with the same notation as in the proof of Theorem 2.27. The uniform convergence of the above series and the invariance of the field (2.55) are proved exactly in the same way as in the proof of Theorem 2.27. $\square$

An example will be considered in Sect. 5.2.

2.4 Invariant Random Fields on General Spaces

In this section, we consider the general case, when G is a compact group acting continuously from the left on a Hausdorff topological space T, and $X(t)$ is a wide-sense mean-square continuous G-invariant random field on T.

2.4.1 The Scalar Case

Let G be a compact topological group acting continuously from the left on a Hausdorff topological space T. Call two points t_1, $t_2 \in T$ *equivalent* if their stationary subgroups are isomorphic. Let $\{T_\alpha : \alpha \in \mathscr{A}\}$ be the corresponding equivalence classes. Suppose that the action of G has a *measurable section*, i.e. a Borel subset $S \subset T$ with the following properties:

- The intersection of S with any G-orbit contains exactly one point;
- For any $\alpha \in \mathscr{A}$, any two points $t_1, t_2 \in S \cap T_\alpha$ have the *same* stationary subgroup.

Let K_α be the stationary subgroup of points in $S \cap T_\alpha$. There exists a unique G-bi-invariant probability measure $d\nu_\alpha(x)$ on the set $K_\alpha \backslash G / K_\alpha$ of double cosets. For simplicity, assume that all K_α are *massive* subgroups of G. Let $X(t)$ be a centred mean-square continuous G-invariant random field on T. Let $\{Z_{m\lambda}(s) \colon \lambda \in \bigcup_{\alpha \in \mathscr{A}} \hat{G}_{K_\alpha}, 1 \le m \le \dim \lambda\}$ be the set of centred mean-square continuous random fields on S satisfying the following conditions:

$$\mathsf{E}\left[Z_{m\lambda}(s_1)\overline{Z_{n\mu}(s_2)}\right] = \delta_{\lambda\mu}\delta_{mn}r_\lambda(s_1, s_2),$$

$$r_\lambda(s, s) = 0, \quad \text{if } s \in T_\alpha \quad \text{but } \lambda \notin \hat{G}_{K_\alpha},$$

$$\sum_{\lambda \in \bigcup_{\alpha \in \mathscr{A}} \hat{G}_{K_\alpha}} r_\lambda(s, s) < \infty. \tag{2.56}$$

Theorem 2.29 *Let G be a compact topological group acting continuously from the left on a Hausdorff topological space T. Let $\{T_\alpha : \alpha \in \mathscr{A}\}$ be the equivalence classes of the following relation: two points $t_1, t_2 \in T$ are equivalent if their stationary subgroups are isomorphic. Let $S \subset T$ be a measurable section, and let K_α be the stationary subgroup of points in $S \cap T_\alpha$. Let all K_α be massive subgroups of G. Let $t \in T$ and $t = gs$ with $g \in G$ and $s \in S$. The following statements are equivalent.*

1. $X(t)$ *is a centred mean-square continuous G-invariant random field on T with covariance function $R(t_1, t_2)$.*
2. $X(t)$ *has the form*

$$X(t) = \sum_{\lambda \in \bigcup_{\alpha \in \mathscr{A}} \hat{G}_{K_\alpha}} \sum_{m=1}^{\dim \lambda} Z_{m\lambda}(s)\varphi_m^{(\lambda)}(g),$$

where $\{Z_{m\lambda}(s) \colon \lambda \in \bigcup_{\alpha \in \mathscr{A}} \hat{G}_{K_\alpha}, 1 \le m \le \dim \lambda\}$ is the set of centred mean-square continuous random fields on S satisfying (2.56). The fields $Z_{m\lambda}(s)$ have the form

$$Z_{m\lambda}(s) = \frac{1}{\dim \lambda} \int_G X(gs)\overline{\varphi_m^{(\lambda)}(g)}\,dg. \tag{2.57}$$

If $t_1 = g_1 s_1 \in T_\alpha$ and $t_2 = g_2 s_2 \in T_\beta$, then the covariance function $R(t_1, t_2)$ depends only on s_1, s_2 and a point x in the set $K_\alpha \backslash G / K_\alpha$ and has the form

$$R(s_1, s_2; x) = \sum_{\lambda \in \hat{G}_{K_\alpha} \cap \hat{G}_{K_\beta}} \varphi_1^{(\lambda)}(x)r_\lambda(s_1, s_2). \tag{2.58}$$

The covariance function $r_\lambda(s_1, s_2)$ has the form

$$r_\lambda(s_1, s_2) = \frac{1}{\dim \lambda} \int_{K_\alpha \backslash G / K_\alpha} R(s_1, s_2; x) \overline{\varphi_1^{(\lambda)}(x)} \, d\nu_\alpha(x). \tag{2.59}$$

Proof Let $X(t)$ be a centred mean-square continuous G-invariant random field on T. If $t \in T_\alpha$, then the orbit of t is the homogeneous space G/K_α, and the restriction of $X(t)$ to G/K_α is a mean-square continuous G-invariant random field on G/G_α. Write $t = gs$, where s is the unique point in S lying on the same orbit as t. By Theorem 2.23,

$$X(gs) = \sum_{\lambda \in \hat{G}_{K_\alpha}} \sum_{m=1}^{\dim \lambda} \varphi_m^{(\lambda)}(g) Z_{m\lambda}(s), \tag{2.60}$$

where $Z_{m\lambda}(s)$ satisfy the following conditions:

$$\mathsf{E}\left[Z_{m\lambda}(s_1) \overline{Z_{n\mu}(s_2)} \right] = \delta_{\lambda\mu} \delta_{mn} r_\lambda(s_1, s_2),$$

$$\sum_{\lambda \in \hat{G}_{K_\alpha}} r_\lambda(s, s) < \infty. \tag{2.61}$$

Multiply both sides of (2.60) by $\overline{\varphi_m^{(\lambda)}(g)}$ and integrate with respect to the Haar measure over G. As usual, the integral has to be understood as a Bochner integral. By the Peter–Weyl theorem, we obtain (2.57) for $s \in T_\alpha$ and $\lambda \in \hat{G}_{K_\alpha}$. Because $X(t)$ is mean-square continuous, so are $Z_{m\lambda}(s)$. The second equation of (2.56) follows. Then, the region of summation in (2.60) and in the second equation in (2.61) can be extended to $\bigcup_{\alpha \in \mathscr{A}} \hat{G}_{K_\alpha}$. This gives (2.56).

Let $t_1 = g_1 s_1 \in T_\alpha$ and $t_2 = g_2 s_2 \in T_\beta$. Then, by (2.60) and the first equation in (2.56),

$$R(g_1 s_1, g_2 s_2) = \sum_{\lambda \in \hat{G}_{K_\alpha} \cap \hat{G}_{K_\beta}} \sum_{m=1}^{\dim \lambda} \varphi_m^{(\lambda)}(g_1) \overline{\varphi_m^{(\lambda)}(g_2)} r_\lambda(s_1, s_2). \tag{2.62}$$

By matrix multiplication (6.21), we obtain (2.58). Multiply both sides by $\overline{\varphi_1^{(\lambda)}(x)}$ and integrate over $K_\alpha \backslash G / K_\alpha$ with respect to the measure $d\nu_\alpha(x)$. We obtain (2.59).

Conversely, let all conditions of item 2 of Theorem 2.29 be satisfied. Apply (6.21) to (2.58). We obtain (2.62). Apply Karhunen's theorem to (2.62). We obtain (2.60). Due to the second equation in (2.56), the region of summation in (2.60) can be extended to $\bigcup_{\alpha \in \mathscr{A}} \hat{G}_{K_\alpha}$. By Theorem 2.23, the field $X(t)$ is mean-square continuous and isotropic. Theorem 2.29 is proved. $\qquad\square$

Example 2.30 Let $G = \mathrm{SO}(N)$ act on $T = \mathbb{R}^N$ by matrix multiplication. There exist two equivalence classes:

$$T_1 = \mathbf{0},$$

$$T_2 = \mathbb{R}^N \setminus \{\mathbf{0}\}.$$

The measurable section S may be identified with the interval $[0, \infty)$ by

$$[0, \infty) \ni s \mapsto (0, 0, \dots, 0, s) \in \mathbb{R}^N.$$

The stationary subgroup of the point $\mathbf{0}$ is $K_1 = \mathrm{SO}(N)$, while the stationary subgroup of any of the remaining points is $K_2 = \mathrm{SO}(N - 1)$. Both subgroups are massive in $\mathrm{SO}(N)$. Let $X(\mathbf{t})$ be a centred mean-square continuous G-invariant random field on T. Such a field is called *isotropic* after Yadrenko (1963).

The set $\hat{G}_{K_1}$ may be identified with the set $\mathbb{Z}_+$ of nonnegative integers, while the set G_{K_2} obviously contains only the trivial representation of G, which corresponds to number $\ell = 0 \in \mathbb{Z}_+$. The dimension of the representation $\lambda_\ell \in \hat{G}_{K_2}$, $h(S^{N-1}, \ell)$, is given by (6.56). The centred mean-square continuous random processes $Z_{m\ell}(s)$, $s \in [0, \infty)$, $\ell \in \mathbb{Z}_+$, $1 \le m \le h(S^{N-1}, \ell)$ satisfy the following conditions:

$$\mathsf{E}\big[Z_{m\ell}(s_1)\overline{Z_{m'\ell'}(s_2)}\big] = \delta_{\ell\ell'}\delta_{mm'}r_\ell(s_1, s_2),$$

$$r_\ell(0, 0) = 0, \quad \text{if } \ell > 0,$$

$$\sum_{\ell=0}^{\infty} r_\ell(s, s) < \infty.$$

Let $(s, \theta_1, \dots, \theta_{N-2}, \varphi)$ be the spherical coordinates in $\mathbb{R}^N$. The spherical functions $\varphi_m^{(\ell)}$ are obtained by inverting (6.57):

$$\varphi_m^{(\ell)}(\theta_1, \dots, \theta_{n-2}, \varphi) = \sqrt{\frac{2\pi^{N/2}}{h(S^{N-1}, \ell)\Gamma(N/2)}}\, Y_{\ell m}(\theta_1, \dots, \theta_{N-2}, \varphi).$$

It follows that the random field $X(s, \theta_1, \dots, \theta_{N-2}, \varphi)$ has the form

$$X(s, \theta_1, \dots, \theta_{N-2}, \varphi) = \sum_{\ell=0}^{\infty} \sum_{m=1}^{h(S^{N-1}, \ell)} \sqrt{\frac{2\pi^{N/2}}{h(S^{N-1}, \ell)\Gamma(N/2)}}$$

$$\times Y_{\ell m}(\theta_1, \dots, \theta_{N-2}, \varphi) Z_{\ell m}(s).$$

It is customary to include the square root into $Z_{\ell m}(s)$. We obtain

$$X(s, \theta_1, \dots, \theta_{N-2}, \varphi) = \sum_{\ell=0}^{\infty} \sum_{m=1}^{h(S^{N-1}, \ell)} Z_{\ell m}(s) Y_{\ell m}(\theta_1, \dots, \theta_{N-2}, \varphi)$$

with

$$\sum_{\ell=0}^{\infty} h\big(S^{N-1}, \ell\big) r_\ell(s, s) < \infty.$$

The processes $Z_{\ell m}(s)$ have the form

$$Z_{\ell m}(s) = \int_{S^{N-1}} X(s, \theta_1, \ldots, \theta_{N-2}, \varphi) \overline{Y_{\ell m}(\theta_1, \ldots, \theta_{N-2}, \varphi)}\, dS,$$

where dS is Lebesgue measure on S^{N-1}.

Let $\mathbf{x}_1 = g_1 s_1$ and $\mathbf{x}_2 = g_2 s_2$ be two points in T_2 with $g_1, g_2 \in K_2$, $s_1, s_2 > 0$. The set of double cosets $K_2 \backslash G / K_2$ may be identified with the interval $[-1, 1]$ by $[-1, 1] \ni \cos \theta \mapsto \theta$, where θ is the angle between $\mathbf{x}_1$ and $\mathbf{x}_2$. The zonal spherical function $\varphi_1^{(\ell)}(\theta)$ is

$$\varphi_1^{(\ell)}(\theta) = \frac{C_\ell^{(N-2)/2}(\cos \theta)}{C_\ell^{(N-2)/2}(1)},$$

where $C_\ell^{(N-2)/2}(\cos \theta)$ are Gegenbauer polynomials. The covariance function of the random field $X(\mathbf{x})$ has the form

$$R(s_1, s_2, \cos \theta) = \frac{\Gamma(N/2)}{2\pi^{N/2}} \sum_{\ell=0}^{\infty} h\big(S^{N-1}, \ell\big) \frac{C_\ell^{(N-2)/2}(\cos \theta)}{C_\ell^{(N-2)/2}(1)} r_\ell(s_1, s_2).$$

Finally, the G-bi-invariant probability measure on $[-1, 1]$ has the form

$$d\nu(x) = 2^{(N-2)/2} \sqrt{(N-2)!} \big(1 - x^2\big)^{(N-3)/2}\, dx,$$

and the covariance function $r_\ell(s_1, s_2)$ has the form

$$r_\ell(s_1, s_2) = \frac{2\pi^{(N-1)/2}}{\Gamma((N-1)/2) C_\ell^{(n-2)/2}(1)}$$

$$\times \int_{-1}^{1} R(s_1, s_2, x) C_\ell^{(N-2)/2}(x)\big(1 - x^2\big)^{(N-3)/2}\, dx.$$

Substitute the value of the Gegenbauer polynomial (6.44) in the last display and use the duplication formula (6.33) with $z = (N-2)/2$. We obtain

$$r_\ell(s_1, s_2) = c \int_{-1}^{1} R(s_1, s_2, x) C_\ell^{(N-2)/2}(x)\big(1 - x^2\big)^{(N-3)/2}\, dx$$

with

$$c = \frac{2^{N-2} \pi^{(N-2)/2} \ell! \, \Gamma((N-2)/2)}{\Gamma(\ell + N - 2)}.$$

Example 2.31 The *multiparameter fractional Brownian motion* with parameter $H \in (0, 1)$ is the centred Gaussian random field $X(\mathbf{t})$ on the space $\mathbb{R}^N$ with the covariance function

$$R(\mathbf{t}_1, \mathbf{t}_2) := \frac{1}{2}\left(\|\mathbf{t}_1\|^{2H} + \|\mathbf{t}_2\|^{2H} - \|\mathbf{t}_1 - \mathbf{t}_2\|^{2H}\right), \tag{2.63}$$

where $\|\cdot\|$ denote the Euclidean norm in $\mathbb{R}^N$.

It is easy to see that $X(\mathbf{t})$ is isotropic. Its covariance function may be written as

$$R(s, t, u) = \frac{1}{2}\left[s^{2H} + t^{2H} - \left(s^2 - 2stu + t^2\right)^H\right],$$

while the covariance function $r_\ell(s, t)$ has the form

$$r_\ell(s, t) = \frac{c}{2}\int_{-1}^{1}\left[s^{2H} + t^{2H} - \left(s^2 - 2stu + t^2\right)^H\right]C_\ell^{(N-2)/2}(x)\left(1 - x^2\right)^{(N-3)/2}\,dx$$

which may be rewritten as

$$r_\ell(s, t) = I_1 - \lim_{\alpha \to (N-1)/2} I_2(\alpha),$$

where

$$I_1 = \frac{c}{2}\left(s^{2H} + t^{2H}\right)\int_{-1}^{1}C_m^{(N-2)/2}(u)\left(1 - u^2\right)^{(N-3)/2}\,du,$$

$$I_2(\alpha) = \frac{c}{2}(2st)^H\int_{-1}^{1}(u+1)^{\alpha-1}(1-u)^{(N-3)/2}\left(\frac{s^2 + t^2}{2st} - u\right)^H C_m^{(N-2)/2}(u)\,du.$$

To calculate I_1, we use formula 2.21.2.17 from Prudnikov et al. (1988):

$$\int_{-a}^{a}\left(a^2 - x^2\right)^{\lambda - 1/2}C_\ell^\lambda(x/a)\,dx = \delta_{0\ell}\sqrt{\pi}\,a^{2\lambda}\frac{\Gamma(\lambda + 1/2)}{\Gamma(\lambda + 1)}$$

with $a = 1$, and $\lambda = (N - 2)/2$. After simplification, we obtain

$$I_1 = \frac{2^{N-3}\pi^{(N-1)/2}\Gamma((N-2)/2)\Gamma((N-1)/2)}{\Gamma(N-2)\Gamma(N/2)}\left(s^{2H} + t^{2H}\right)\delta_{0\ell}.$$

This expression can be further simplified using the duplication formula (6.33) with $z = (N - 2)/2$. We get

$$I_1 = \frac{\pi^{N/2}}{\Gamma(N/2)}\left(s^{2H} + t^{2H}\right)\delta_{0\ell}.$$

To calculate $I_2(\alpha)$, we use formula 2.21.4.15 from Prudnikov et al. (1988):

$$\int_{-a}^{a} (x+a)^{\alpha-1}(a-x)^{\lambda-1/2}(z-x)^{-\vartheta} C_\ell^\lambda(x/a)\,dx$$

$$= \frac{(-1)^\ell}{\ell!}(1/2+\lambda-\alpha)_n(2\lambda)_\ell \frac{\Gamma(\alpha)\Gamma(\lambda+1/2)}{\Gamma(\alpha+\lambda+n+1/2)}(2a)^{\alpha+\lambda-1/2}(z+a)^{-\vartheta}$$

$$\times {}_3F_2\big(\alpha,\theta,1/2+\alpha-\lambda;1/2+\alpha-\lambda-\ell,1/2+\alpha+\lambda+\ell;2a/(a+z)\big)$$

with $a=1$, $\lambda=(N-2)/2$, $z=(s^2+t^2)/(2st)$, and $\vartheta=-H$. After simplification, we obtain

$$I_2(\alpha) = (-1)^\ell 2^{\alpha+3(N-3)/2}\pi^{(N-2)/2}$$

$$\times \frac{\Gamma((N-2)/2)\Gamma((N-1)/2-\alpha+\ell)\Gamma(\alpha)\Gamma((N-1)/2)}{\Gamma((N-1)/2-\alpha)\Gamma(N-2)\Gamma(\alpha+(N-1)/2+\ell)}(s+t)^{2H}$$

$$\times {}_3F_2\left(\alpha,-H,\alpha-\frac{N-3}{2};\alpha-\frac{N-3}{2}-\ell,\alpha+\frac{N-1}{2}+\ell;\frac{4st}{(s+t)^2}\right).$$

Using the duplication formula (6.33) with $z=(N-2)/2$, we get

$$I_2(\alpha) = (-1)^\ell 2^{\alpha+(N-3)/2}\pi^{(N-1)/2}\frac{\Gamma((N-1)/2-\alpha+\ell)\Gamma(\alpha)(s+t)^{2H}}{\Gamma((N-1)/2-\alpha)\Gamma(\alpha+(N-1)/2+\ell)}$$

$$\times {}_3F_2\left(\alpha,-H,\alpha-\frac{N-3}{2};\alpha-\frac{N-3}{2}-\ell,\alpha+\frac{N-1}{2}+\ell;\frac{4st}{(s+t)^2}\right).$$

$$(2.64)$$

In the case of $\ell=0$, (2.64) simplifies as follows.

$$I_2(\alpha) = 2^{\alpha+(N-3)/2}\pi^{(N-1)/2}\frac{\Gamma(\alpha)(s+t)^{2H}}{\Gamma(\alpha+(N-1)/2)}$$

$$\times {}_3F_2\left(\alpha,-H,\alpha-\frac{N-3}{2};\alpha-\frac{N-3}{2},\alpha+\frac{N-1}{2};\frac{4st}{(s+t)^2}\right).$$

According to paragraph 7.2.3.2 from Prudnikov et al. (1990), the value of the generalised hypergeometric function ${}_pF_q(a_1,\ldots,a_p;b_1,\ldots,b_q;z)$ is independent of the order of *upper parameters* $a_1,\ldots,a_p$ and *lower parameters* $b_1,\ldots,b_q$. Moreover, formula 7.2.3.7 from Prudnikov et al. (1990) states that

$$ {}_pF_q(a_1,\ldots,a_{p-r},c_1,\ldots,c_r;b_1,\ldots,b_{q-r},c_1,\ldots,c_r;z)$$

$$= {}_{p-r}F_{q-r}(a_1,\ldots,a_{p-r};b_1,\ldots,b_{q-r};z).$$

$$(2.65)$$

Using these properties, we get

$$I_2(\alpha) = 2^{\alpha+(N-3)/2}\pi^{(N-1)/2}\frac{\Gamma(\alpha)(s+t)^{2H}}{\Gamma(\alpha+(N-1)/2)}{}_2F_1$$
$$\times\left(\alpha,-H;\alpha+\frac{N-1}{2};\frac{4st}{(s+t)^2}\right).$$

In particular,

$$\lim_{\alpha\to(N-1)/2}I_2(\alpha) = 2^{N-2}\pi^{(N-1)/2}\frac{\Gamma((N-1)/2)(s+t)^{2H}}{\Gamma(N-1)}$$
$$\times{}_2F_1\left(\frac{N-1}{2},-H;N-1;\frac{4st}{(s+t)^2}\right).$$

Application of the duplication formula (6.33) with $z=(N-1)/2$ yields

$$\lim_{\alpha\to(N-1)/2}I_2(\alpha) = \frac{\pi^{N/2}}{\Gamma(N/2)}(s+t)^{2H}{}_2F_1\left(\frac{N-1}{2},-H;N-1;\frac{4st}{(s+t)^2}\right)$$

and, finally

$$r_0(s,t) = \frac{\pi^{N/2}}{\Gamma(N/2)}\left[s^{2H}+t^{2H}-(s+t)^{2H}{}_2F_1\left(\frac{N-1}{2},-H;N-1;\frac{4st}{(s+t)^2}\right)\right].$$

$$(2.66)$$

In the case of $\ell\geq 1$, (2.64) can be rewritten as

$$\lim_{\alpha\to(N-1)/2}I_2(\alpha)$$

$$= (-1)^\ell 2^{N-2}\pi^{(N-1)/2}\frac{(\ell-1)!\Gamma((N-1)/2)(s+t)^{2H}}{\Gamma(N-1+\ell)}$$

$$\times\lim_{\alpha\to(N-1)/2}\frac{{}_3F_2(\alpha,-H,\alpha-\frac{N-3}{2};\alpha-\frac{N-3}{2}-\ell,N-1+\ell;\frac{4st}{(s+t)^2})}{\Gamma(\alpha-(N-3)/2-\ell)}$$

$$\times\lim_{\alpha\to(N-1)/2}\frac{\Gamma(\alpha-(N-3)/2-\ell)}{\Gamma((N-1)/2-\alpha)}.$$

To calculate the first limit, we use the following formula:

$$\lim_{b_1\to-m}\frac{{}_pF_q(a_1,\ldots,a_p;b_1,\ldots,b_q;z)}{\Gamma(b_1)} = \frac{z^{m+1}}{(m+1)!}\frac{\prod_{j=1}^p(a_j)_{n+1}}{\prod_{j=2}^q(b_j)_{n+1}}$$

$$\times{}_pF_q(a_1+m+1,\ldots,a_p+m+1;b_2+m+1,\ldots,b_q+m+1,m+2;z),$$

with $m = \ell - 1$, $p = 3$, $q = 2$, $a_1 = (N - 1)/2$, $a_2 = -H$, $a_3 = 1$, $b_1 = 1 - \ell$, $b_2 = N - 1 + \ell$, and $z = 4st/(s + t)^2$. We get

$$\lim_{\alpha \to (N-1)/2} \frac{{}_3F_2(\alpha, -H, \alpha - \frac{N-3}{2}; \alpha - \frac{N-3}{2} - \ell, N - 1 + \ell; \frac{4st}{(s+t)^2})}{\Gamma(\alpha - (N - 3)/2 - \ell)}$$

$$= \frac{2^{2\ell}(st)^\ell \Gamma((N - 1)/2 + \ell)\Gamma(\ell - H)\Gamma(N - 1 + \ell)}{(s + t)^{2\ell}\Gamma((N - 1)/2)\Gamma(-H)\Gamma(N - 1 + 2\ell)}$$

$$\times {}_3F_2\big((N - 1)/2 + \ell, \ell - H, \ell + 1; N - 1 + 2\ell, \ell + 1; 4st/(s + t)^2\big).$$

Using the duplication formula (6.33) with $z = (N - 1)/2 + \ell$ and (2.65), we get

$$\lim_{\alpha \to (N-1)/2} \frac{{}_3F_2(\alpha, -H, \alpha - \frac{N-3}{2}; \alpha - \frac{N-3}{2} - \ell, N - 1 + \ell; \frac{4st}{(s+t)^2})}{\Gamma(\alpha - (N - 3)/2 - \ell)}$$

$$= \frac{(st)^\ell \Gamma(\ell - H)\Gamma(N - 1 + \ell)\sqrt{\pi}}{(s + t)^{2\ell}\Gamma((N - 1)/2)\Gamma(-H)2^{N-2}\Gamma(N/2 + \ell)}$$

$$\times {}_2F_1\big((N - 1)/2 + \ell, \ell - H; N - 1 + 2\ell; 4st/(s + t)^2\big).$$

Rewrite the second limit as

$$\lim_{\alpha \to (N-1)/2} \frac{\Gamma(\alpha - (N - 3)/2 - \ell)}{\Gamma((N - 1)/2 - \alpha)} = \lim_{\alpha \to (N-1)/2} \frac{\Gamma(\alpha - (N - 3)/2 - \ell)}{\Gamma(\alpha - (N - 1)/2)}$$

$$\times \lim_{\alpha \to (N-1)/2} \frac{\Gamma(\alpha - (N - 1)/2)}{\Gamma((N - 1)/2 - \alpha)}.$$

For the first part, use (6.32) with $z = \alpha - (N - 1)/2$ and $m = \ell - 1$. For the second part, use (6.31) with $z = \alpha - (N - 3)/2$. We get

$$\lim_{\alpha \to (N-1)/2} \frac{\Gamma(\alpha - (N - 3)/2 - \ell)}{\Gamma((N - 1)/2 - \alpha)} = \frac{(-1)^{\ell-1}}{(\ell - 1)!}(-1) = \frac{(-1)^\ell}{(\ell - 1)!}.$$

Combining everything together, we obtain

$$r_\ell(s, t) = -\frac{\pi^{N/2}(s + t)^{2(H-\ell)}(st)^\ell \Gamma(\ell - H)}{\Gamma(-H)\Gamma(N/2 + \ell)}$$

$$\times {}_2F_1\big((N - 1)/2 + \ell, \ell - H; N - 1 + 2\ell; 4st/(s + t)^2\big), \quad \ell \geq 1.$$

The last display and (2.66) may be written as

$$r_\ell(s, t) = \frac{\pi^{N/2}}{\Gamma(N/2 + \ell)}\Big[(s^{2H} + t^{2H})\delta_{0\ell} - \frac{\Gamma(\ell - H)(s + t)^{2(H-\ell)}(st)^\ell}{\Gamma(-H)}$$

$$\times {}_2F_1\Big(\frac{N - 1}{2} + \ell, \ell - H; N - 1 + 2\ell; \frac{4st}{(s + t)^2}\Big)\Big]. \tag{2.67}$$

2.4.2 The Vector Case

Let G be a connected compact Lie group acting continuously from the left on a Hausdorff topological space T. Assume that there are only finitely many equivalence classes $T_1, T_2, \ldots, T_M$. Let S be a measurable section for the action of G. Let K_j be the stationary subgroup of points in $S \cap T_j$, $1 \le j \le M$. For simplicity, assume that all K_j are *massive* subgroups of G.

Let λ be an irreducible unitary representation of the group G. Because all K_j are compact groups, the restriction of the representation λ to K_j is equivalent to the direct sum $\lambda_{j1} \oplus \lambda_{j2} \oplus \cdots \oplus \lambda_{jN_j}$ of irreducible unitary representations of K_j. Let $\xi = (E, \pi, T)$ be such a vector bundle over T, that for each $j = 1, 2, \ldots, M$ its restriction to any orbit $O \subseteq T_j$ is a homogeneous vector bundle where the representation of G induced by the representation $\lambda_{j1} \oplus \lambda_{j2} \oplus \cdots \oplus \lambda_{jN_j}$ of K_j is realised. Let $\mathbf{X}(t)$ be a centred mean-square continuous G-invariant random field in ξ. Suppose that the stationary subgroups K_j can be ordered by inclusion in the following way:

$$G \supseteq K_1 \supset K_2 \supset \cdots \supset K_M \supseteq \{e\}.$$

Moreover, suppose that the above chain may be extended to the chain (6.12), and let k_j, $1 \le j \le M$ be the number of the group K_j in the chain (6.12), i.e., $K_j = G_{k_j}$. Let $\lambda = (\lambda_1, \ldots, \lambda_{N-1})$ be a Gel'fand–Tsetlin scheme (Subsection 6.3.4) with $\lambda_1 = \lambda$. By definition of the Gel'fand–Tsetlin scheme, for each j with $1 \le j \le M$, there is a one-to-one correspondence between the irreducible components λ_{k_j} of the restriction of λ to K_j and the schemes

$$\boldsymbol{\lambda}_{k_j} := (\lambda_1, \ldots, \lambda_{k_j}), \tag{2.68}$$

satisfying (6.14) and (6.15). The Gel'fand–Tsetlin basis in the space of each of the above representations is described by the schemes

$$\boldsymbol{\lambda}^{k_j} := (\lambda_{k_j}, \ldots, \lambda_{N-1}). \tag{2.69}$$

We have $\lambda = \boldsymbol{\lambda}_{k_j} \cup \boldsymbol{\lambda}^{k_j}$. Let Λ_{k_j} be the set of all schemes (2.68), and let Λ^{k_j} be the set of all schemes (2.69) satisfying (6.14) and (6.15).

Note that if $1 \le j < \ell \le M$, then $\hat{G}_{K_j}(\boldsymbol{\lambda}_{k_j}) \subseteq \hat{G}_{K_\ell}(\boldsymbol{\lambda}_{k_\ell})$. Indeed, let $\mu \in \hat{G}_{K_j}(\boldsymbol{\lambda}_{k_j})$. Then the restriction of μ to K_j contains λ_j, while the restriction of λ_j to K_ℓ contains λ_{k_ℓ} by definition of λ.

Consider the set of centred mean-square continuous random fields $Z_{m\boldsymbol{\lambda}^{k_M}}^{(\mu\boldsymbol{\lambda}_{k_M})}(s)$, where $\boldsymbol{\lambda}_{k_M} \in \Lambda_{k_M}$, $\mu \in \hat{G}_{K_M}(\boldsymbol{\lambda}_{k_M})$, $1 \le m \le \dim \mu$, $\boldsymbol{\lambda}^{k_M} \in \Lambda^{k_M}$, satisfying the following conditions:

$$\mathsf{E}\big[Z_{m\lambda^{k_M}}^{(\mu\lambda_{k_M})}(s_1)\overline{Z_{m'\lambda'^{k_M}}^{(\mu'\lambda'_{k_M})}(s_2)}\big] = \delta_{\mu\mu'}\delta_{mm'}R_{\lambda_{k_M}\cup\lambda^{k_M},\lambda'_{k_M}\cup\lambda'^{k_M}}^{(\mu)}(s_1,s_2),$$

$$R^{(\mu)}(s,s) = 0, \quad \text{if } s \in T_j, \quad \text{but } \mu \notin \hat{G}_{K_j}(\lambda_{k_j}), \tag{2.70}$$

$$\sum_{\lambda_{k_M}\in\Lambda_{k_M}}\sum_{\mu\in\hat{G}_{K_M}(\lambda_{k_M})} \dim\mu \, \mathrm{tr}\big[P_{\lambda_{k_M}}R^{(\mu)}(s,s)P_{\lambda_{k_M}}\big] < \infty,$$

where $P_{\lambda_{k_M}}$ is the orthogonal projection from the space of the representation λ to the space of the representation which corresponds to the scheme λ_{k_M}.

Theorem 2.32 *Under the above assumptions, let $t \in T$ and $t = gs$ with $g \in G$ and $s \in S$. The following statements are equivalent.*

1. $\mathbf{X}(t)$ *is a centred mean-square continuous G-invariant random field in ξ.*
2. $\mathbf{X}(t)$ *has the form*

$$X_\lambda(gs) = \sum_{\mu\in\hat{G}_{K_M}(\lambda_{k_M})} \sum_{m=1}^{\dim\mu} \sum_{\lambda_{k_M}\in\Lambda^{k_M}} Z_{m\lambda^{k_M}}^{(\mu\lambda_{k_M})}(s)(_{\lambda_{k_M}}\mathbf{Y}_{\mu m})_{\lambda^{k_M}}(g),$$

where $Z_{m\lambda^{k_M}}^{(\mu\lambda_{k_M})}(s)$ is the set of centred mean-square continuous random fields on S satisfying (2.70). The fields $Z_{m\lambda^{k_M}}^{(\mu\lambda_{k_M})}(s)$ have the form

$$Z_{m\lambda^{k_M}}^{(\mu\lambda_{k_M})}(s) = \int_G X_{\lambda_{k_M}\cup\lambda^{k_M}}(gs)\overline{(_{\lambda_{k_M}}\mathbf{Y}_{\mu m})_{\lambda^{k_M}}(g)}\,dg. \tag{2.71}$$

If $t_1 = g_1s_1 \in T_j$ and $t_2 = g_2s_2 \in T_\ell$, then the covariance function $R_{\lambda\lambda'}(t_1,t_2)$ has the form

$$R_{\lambda\lambda'}\big(s_1,s_2;g_2^{-1}g_1\big) = \sqrt{\dim\lambda_{k_j}\dim\lambda'_{k_\ell}} \sum_{\mu\in\hat{G}_{K_j}(\lambda_{k_j})\cap\hat{G}_{K_\ell}(\lambda'_{k_\ell})} \sum_{\lambda^{k_j}\in\Lambda^{k_j}} \sum_{\lambda'^{k_\ell}\in\Lambda^{k_\ell}}$$

$$R_{\lambda_{k_j}\cup\lambda^{k_M},\lambda'_{k_\ell}\cup\lambda'^{k_M}}^{(\mu)}(s_1,s_2)\mu_{\lambda_{k_M}\lambda'_{k_M}}\big(g_2^{-1}g_1\big). \tag{2.72}$$

The covariance function $R_{\lambda_{k_j}\cup\lambda^{k_j},\lambda'_{k_\ell}\cup\lambda'^{k_\ell}}^{(\mu)}(s_1,s_2)$ has the form

$$R_{\lambda_{k_j}\cup\lambda^{k_j},\lambda'_{k_\ell}\cup\lambda'^{k_\ell}}^{(\mu)}(s_1,s_2) = \frac{\sqrt{\dim\lambda_{k_j}\dim\lambda'_{k_\ell}}}{\dim\mu}\int_G R_{\lambda\lambda'}(s_1,s_2;g)\overline{\mu_{\lambda^{k_j}\lambda'^{k_\ell}}(g)}\,dg. \tag{2.73}$$

Proof Let $\mathbf{X}(t)$ be a centred mean-square continuous G-invariant random field in ξ. Let $t \in T_M$. Then the orbit of t is the homogeneous space G/K_M, and the restriction of $\mathbf{X}(t)$ to G/K_M is a mean-square continuous G-invariant random field in the homogeneous vector bundle where the representation of G induced by the representation $\lambda_{M1} \oplus \lambda_{M2} \oplus \cdots \oplus \lambda_{MN_M}$ of K_M is realised. By Theorem 2.28,

$$X_{\lambda^{k_M} \cup \lambda^{k_M}}(gs) = \sum_{\mu \in \hat{G}_{K_M}(\lambda_{k_M})} \sum_{m=1}^{\dim \mu} \sum_{\lambda^{k_M} \in \Lambda^{k_M}} Z_{m\lambda^{k_M}}^{(\mu \lambda_{k_M})}(s)(\lambda_{k_M} \mathbf{Y}_{\mu m})_{\lambda^{k_M}}(g), \quad (2.74)$$

where $Z_{m\lambda^{k_M}}^{(\mu \lambda_{k_M})}(s)$ satisfy the following conditions:

$$\mathsf{E}\big[Z_{m\lambda^{k_M}}^{(\mu \lambda_{k_M})}(s_1)\overline{Z_{m'\lambda'^{k_M}}^{(\mu' \lambda'_{k_M})}(s_2)}\big] = \delta_{\mu\mu'}\delta_{mm'} R_{\lambda_{k_M} \cup \lambda^{k_M}, \lambda'_{k_M} \cup \lambda'^{k_M}}^{(\mu)}(s_1, s_2),$$

$$\sum_{\lambda_{k_M} \in \Lambda_{k_M}} \sum_{\mu \in \hat{G}_{K_M}(\lambda_{k_M})} \dim \mu \, \mathrm{tr}\big[P_{\lambda_{k_M}} R^{(\mu)}(s, s) P_{\lambda_{k_M}}\big] < \infty. \quad (2.75)$$

Multiply both sides of (2.74) by $\overline{(\lambda_{k_M} \mathbf{Y}_{\mu m})_{\lambda^{k_M}}(g)}$ and integrate with respect to the Haar measure over G. By the Peter–Weyl theorem, we obtain (2.71) for $s \in S \cap T_M$ and $\mu \in \hat{G}_{k_M}(\lambda_{k_M})$. Because $\mathbf{X}(t)$ is mean-square continuous, so are $Z_{m\nu}^{(\mu \lambda_{k_M})}(s)$. The second equation of (2.70) follows.

Let $t_1 = g_1 s_1 \in T_j$ and $t_2 = g_2 s_2 \in T_\ell$. Then by (2.74) and the first equation in (2.70),

$$R_{\lambda\lambda'}(s_1, s_2; g_2^{-1}g_1) = \sum_{\mu \in \hat{G}_{K_j}(\lambda_{k_j}) \cap \hat{G}_{K_\ell}(\lambda'_{k_\ell})} \sum_{\lambda^{k_j} \in \Lambda^{k_j}} \sum_{\lambda'^{k_\ell} \in \Lambda^{k_\ell}}$$

$$R_{\lambda_{k_j} \cup \lambda^{k_j}, \lambda'_{k_\ell} \cup \lambda'^{k_\ell}}^{(\mu)}(s_1, s_2) \sum_{m=1}^{\dim \mu} (\lambda_{k_j} \mathbf{Y}_{\mu m})_{\lambda^{k_j}}(g_1)\overline{(\lambda_{k_\ell} \mathbf{Y}_{\mu m})_{\lambda'^{k_\ell}}(g_2)}. \quad (2.76)$$

By definition (2.50) and matrix multiplication, we obtain (2.72). Multiply both sides by $\overline{\mu_{\lambda^{k_j}\lambda'^{k_\ell}}(g)}$ and integrate over G with respect to Haar measure dg. We obtain (2.73).

Conversely, let all conditions of item 2 of Theorem 2.32 be satisfied. Apply matrix multiplication to (2.72). We obtain (2.76). Apply Karhunen's theorem to (2.76). We obtain (2.74). By Theorem 2.28, the field $\mathbf{X}(t)$ is mean-square continuous and isotropic. Theorem 2.32 is proved. □

Example 2.33 Under the conditions of Example 2.30, let λ be an irreducible unitary representation of the group $G = SO(N)$, and let $(p_1, p_2, \ldots, p_{[N/2]})$ be its signature. We have $M = 2$, $k_1 = 1$, $k_2 = 2$. The restriction of λ to $K_1 = SO(N)$ is just $\lambda_{11} = \lambda$ with $N_1 = 1$ and $\hat{G}_{K_1}(\lambda_{11}) = \{\lambda\}$, while the restriction of λ to

$K_2 = SO(N-1)$ is equivalent to the direct sum $\lambda_{21} \oplus \cdots \oplus \lambda_{2N_2}$, where the signature $(q_{1\ell}, \ldots, q_{[(N-1)/2]\ell})$ of the irreducible component $\lambda_{2\ell}$ satisfies either (6.14) for odd N or (6.15) for even N. It is easy to check that

$$
N_2 = \begin{cases} (2p_j + 1) \prod_{\ell=1}^{j-1} (p_\ell - p_{\ell+1} + 1), & N = 2j+1, \\ \prod_{\ell=1}^{j-1} (p_\ell - |p_{\ell+1}| + 1), & N = 2j. \end{cases}
$$

The set $\hat{G}_{K_2}(\lambda_{2\ell})$ contains the elements $\mu \in \hat{G}$ with signatures $(r_1, r_2, \ldots, r_j)$ satisfying conditions

$$r_1 \geq q_{1\ell},$$

$$q_{k-1,\ell} \geq r_k \geq |q_{k\ell}|, \quad 2 \leq k \leq j$$

for $N = 2j+1$, and

$$r_1 \geq q_{1\ell},$$

$$q_{k\ell} \geq r_{k+1} \geq q_{k+1,\ell}, \quad 2 \leq k \leq j-2$$

$$|r_j| \leq q_{j-1,\ell}$$

for $N = 2j$. In particular, $\hat{G}_{K_1}(\lambda_{11}) \subset \hat{G}_{K_2}(\lambda_{2\ell})$ for all ℓ, $1 \leq \ell \leq N_2$.

Let $\lambda = (\lambda_1, \lambda_2, \ldots, \lambda_{N-1})$ with $\lambda_1 = \lambda$ be a vector of the Gel'fand–Tsetlin basis, and let $(s, \theta_1, \ldots, \theta_{N-2}, \varphi)$ be the spherical coordinates in $\mathbb{R}^N$. A centred mean-square continuous G-invariant random field in ξ has the form

$$X_\lambda(s, \theta_1, \ldots, \theta_{N-2}, \varphi)$$

$$= \sum_{\mu \in \hat{G}_{K_2}(\lambda_2)} \sum_{m=1}^{\dim \mu} \sum_{\lambda^2 \in \Lambda^2} Z_{m\lambda^2}^{(\mu\lambda_2)}(s) (_{\lambda_2} \mathbf{Y}_{\mu m})_{\lambda^2}(\theta_1, \ldots, \theta_{N-2}, \varphi),$$

where $Z_{m\lambda^2}^{(\mu\lambda_2)}(s)$ is the set of centred mean-square continuous random processes on $[0, \infty)$ satisfying

$$\mathsf{E}\big[Z_{m\lambda^2}^{(\mu\lambda_2)}(s_1) \overline{Z_{m'\lambda'^2}^{(\mu'\lambda_2')}(s_2)} \big] = \delta_{\mu\mu'} \delta_{mm'} R_{\lambda_2 \cup \lambda^2, \lambda_2' \cup \lambda^2}^{(\mu)}(s_1, s_2),$$

$$R^{(\mu)}(0,0) = 0, \quad \text{if } \mu \neq \lambda,$$

$$\sum_{\lambda^2 \in \Lambda^2} \sum_{\mu \in \hat{G}_{K_2}(\lambda_2)} \dim \mu \, \mathrm{tr}\big[P_{\lambda_2} R^{(\mu)} P_{\lambda_2} \big] < \infty.$$

The fields $Z_{m\nu}^{(\mu\lambda_2)}(s)$ have the form

$$Z_{m\nu}^{(\mu\lambda_2)}(s) = \int_{S^{N-1}} X_{\lambda_2 \cup \lambda^2}(s, \theta_1, \ldots, \theta_{N-2}, \varphi) \overline{(_{\lambda_2} \mathbf{Y}_{\mu m})_{\lambda^2}(\theta_1, \ldots, \theta_{N-2}, \varphi)} \, \mathrm{d}S,$$

where dS is Lebesgue measure on S^{N-1}. The covariance function $R_{\lambda\lambda'}(t_1, t_2)$ has the form

$$R_{\lambda\lambda'}\left(s_1, s_2; g_2^{-1}g_1\right) = \sqrt{\dim \lambda_2 \dim \lambda_2'} \sum_{\mu \in \hat{G}_{K_2}(\lambda_2) \cap \hat{G}_{K_2}(\lambda_2')} \sum_{\lambda^2 \in \Lambda^2} \sum_{\lambda'^2 \in \Lambda^2}$$

$$R^{(\mu)}_{\lambda_2 \cup \lambda^2, \lambda_2' \cup \lambda'^2}(s_1, s_2) \mu_{\lambda_2 \lambda_2'}\left(g_2^{-1}g_1\right).$$

The covariance function $R^{(\mu)}_{\lambda_2 \cup \lambda^2, \lambda_2' \cup \lambda'^2}(s_1, s_2)$ has the form

$$R^{(\mu)}_{\lambda_2 \cup \lambda^2, \lambda_2' \cup \lambda'^2}(s_1, s_2) = \frac{\sqrt{\dim \lambda_2 \dim \lambda_2'}}{\dim \mu} \int_G R_{\lambda\lambda'}(s_1, s_2; g) \overline{\mu_{\lambda^2\lambda'^2}(g)} \, dg.$$

Let $N = 3$. Denote $\nu := \lambda_2$. Then

$$X_\nu(s, \varphi, \theta) = \sum_{\mu=|\nu|}^{\infty} \sum_{m=-\mu}^{\mu} Z_m^{(\mu\nu)}(s)_\nu Y_{\mu m}(\varphi, \theta),$$

$$Z_m^{(\mu\nu)}(s) = \int_{S^2} X_\nu(s, \varphi, \theta)_\nu \overline{Y_{\mu m}(\varphi, \theta)} \, dS,$$

$$R_{\nu\nu'}\left(s_1, s_2, g_2^{-1}g_1\right) = \sum_{\mu=\max\{|\nu|,|\nu'|\}}^{\infty} R_{\nu\nu'}^{(\mu)}(s_1, s_2) D_{\nu\nu'}^{(\mu)}\left(g_2^{-1}g_1\right),$$

$$R_{\nu\nu'}^{(\mu)}(s_1, s_2) = \int_{SO(3)} R_{\nu\nu'}(s_1, s_2, g) \overline{D_{\nu\nu'}^{(\mu)}(g)} \, dg,$$

where $D_{\nu\nu'}^{(\mu)}(g)$ are Wigner D-functions, i.e., the matrix elements of the irreducible unitary representation μ in Gel'fand–Tsetlin basis.

2.5 Isotropic Random Fields on Commutative Groups

Yaglom (1957) formulated the following problem. Let n be a positive integer, and let λ be an irreducible unitary representation of the group $SO(n)$ in a finite-dimensional complex Hilbert space H. Let $\mathbf{X}(\mathbf{t})$ be a homogeneous H-valued vector random field on $\mathbb{R}^n$. Let $\mathbf{E}(\mathbf{t})$ be the mean value of the field $\mathbf{X}(\mathbf{t})$, and let $R(\mathbf{t}_1, \mathbf{t}_2)$ be its covariance operator. The field $\mathbf{X}(\mathbf{t})$ is called an *isotropic vector field* if for all $g \in SO(n)$,

$$\mathbf{E}(g\mathbf{t}) = \lambda(g)\mathbf{E}(\mathbf{t}),$$

$$R(g\mathbf{t}_1, g\mathbf{t}_2) = \lambda(g)R(\mathbf{t}_1, \mathbf{t}_2)\lambda^{-1}(g).$$

The problem is to find the spectral expansion of a homogeneous and isotropic vector field.

We will solve this problem in a more general setting. Let T be a commutative locally compact group with countable base, let G be a compact group of automorphisms of T, and let λ be an irreducible unitary representation of the group G in a finite-dimensional complex Hilbert space H. Let $\mathbf{X}(t)$ be a wide-sense homogeneous vector random field on T taking values in H. Let $\mathbf{E}(t)$ be the mean value of the field $\mathbf{X}(t)$, and let $R(t_1, t_2)$ be its covariance operator. We will find the spectral expansion of the field $\mathbf{X}(t)$ that satisfies the following conditions: for any $g \in G$,

$$\mathbf{E}(gt) = \lambda(g)\mathbf{E}(t),$$
$$R(gt_1, gt_2) = \lambda(g)R(t_1, t_2)\lambda^{-1}(g). \tag{2.77}$$

Assume that the following conditions hold true.

1. There are only finitely many equivalence classes $\hat{T}_1, \hat{T}_2, \ldots, \hat{T}_N$ for the action of G on $\hat{T}$.
2. The action $g^{-1}\hat{t}(t) = \hat{t}(gt)$ of G on $\hat{T}$ has a measurable section $\hat{S}$.
3. The stationary subgroups K_j of points in $\hat{S}_j := \hat{S} \cap \hat{T}_j$, $1 \le j \le N$, are massive subgroups of G.
4. The stationary subgroups K_j can be ordered by inclusion in the following way:

$$G \supseteq K_1 \supset K_2 \supset \cdots \supset K_N \supseteq \{e\}.$$

Moreover, the above chain may be extended to the chain (6.12).

5. There are only finitely many equivalence classes $T_1, T_2, \ldots, T_M$ for the action of G on T.
6. The action G on T has a measurable section S.
7. The stationary subgroups $\tilde{K}_p$ of points in $S_p := S \cap T_p$, $1 \le p \le M$, are massive subgroups of G.

In what follows we will see that these conditions hold true for the classical case considered by Yaglom (1957), see Example 2.35.

By Theorem 2.22, the mean value of the field $\mathbf{X}(t)$ is a constant, say $\mathbf{E}_0$. It follows from the first equation in (2.77) that $\mathbf{E}_0 = \lambda(g)\mathbf{E}_0$, $g \in G$. If λ is trivial, $\mathbf{E}_0$ can be any constant vector. Otherwise, $\mathbf{E}_0 = \mathbf{0}$ by Lemma 6.3.

By Theorem 2.22, the covariance operator of the field $\mathbf{X}(t)$ has the form

$$R(t_1, t_2) = \int_{\hat{T}} \hat{t}(t_1 - t_2)\,dF(\hat{t}), \tag{2.78}$$

where F is a measure on $\hat{T}$ with values in the set of Hermitian nonnegative-definite operators in the space H. Let σ be the following measure:

$$\sigma(B) := \operatorname{tr}[F(B)], \qquad B \in \mathfrak{B}(\hat{T}).$$

By Berezansky (1968), the measure F is absolutely continuous with respect to σ, and the density

$$A(\hat{t}) := \frac{dF(\hat{t})}{d\sigma(\hat{t})}$$

is a measurable function on $\hat{T}$ taking values in the set of Hermitian nonnegative-definite operators in the space H with unit trace. Thus, (2.78) may be written as

$$R(t_1, t_2) = \int_{\hat{T}} \hat{t}(t_1 - t_2) A(\hat{t}) \, d\sigma(\hat{t}). \tag{2.79}$$

We calculate the expression $R(gt_1, gt_2)$ by two different methods. On the one hand, by the second equation in (2.77),

$$R(gt_1, gt_2) = \lambda(g) R(t_1, t_2) \lambda^{-1}(g)$$

$$= \lambda(g) \int_{\hat{T}} \hat{t}(t_1 - t_2) A(\hat{t}) \, d\sigma(\hat{t}) \lambda^{-1}(g)$$

$$= \int_{\hat{T}} \hat{t}(t_1 - t_2) \lambda(g) A(\hat{t}) \lambda^{-1}(g) \, d\sigma(\hat{t}),$$

because integration commutes with continuous linear operators. On the other hand, G also acts on $\hat{T}$ by duality: $g^{-1}\hat{t}(t) = \hat{t}(gt)$. Then, by (2.79),

$$R(gt_1, gt_2) = \int_{\hat{T}} \hat{t}\big(g(t_1 - t_2)\big) A(\hat{t}) \, d\sigma(\hat{t})$$

$$= \int_{\hat{T}} (g^{-1}\hat{t})(t_1 - t_2) A(\hat{t}) \, d\sigma(\hat{t})$$

$$= \int_{\hat{T}} \hat{t}(t_1 - t_2) A(g\hat{t}) \, d\sigma(g\hat{t}).$$

In the last display we denote $g^{-1}\hat{t}$ again by $\hat{t}$. Because the expansion (2.79) is unique, we have, for each $g \in G$ and for each $B \in \mathfrak{B}(\hat{T})$,

$$A(g\hat{t}) = \lambda(g) A(\hat{t}) \lambda^{-1}(g),$$
$$\sigma(gB) = \sigma(B). \tag{2.80}$$

The first equation may be written as

$$A(g\hat{t}) = (\lambda^+ \otimes \lambda)(g) A(\hat{t}). \tag{2.81}$$

Indeed, let $\mathbf{x} \in H$ and $\mathbf{x}^* \in H^*$. The linear operator $\mathbf{x}^* \otimes \mathbf{x}$ acts on H by (6.6):

$$(\mathbf{x}^* \otimes \mathbf{x})\mathbf{y} = \mathbf{x}^*(\mathbf{y})\mathbf{x}.$$

For this operator,

$$
\begin{aligned}
\left(\lambda(g)\mathbf{x}^* \otimes \mathbf{x}\lambda^{-1}(g)\right)\mathbf{y} &= \lambda(g)\mathbf{x}^*\left(\lambda^{-1}(g)\mathbf{y}\right)\mathbf{x} \\
&= \left(\lambda^+(g)\mathbf{x}^*\right)(\mathbf{y})\lambda(g)\mathbf{x} \\
&= \left(\lambda^+(g)\mathbf{x}^*\right) \otimes \left(\lambda(g)\mathbf{x}\right)\mathbf{y} \\
&= \left(\lambda^+ \otimes \lambda\right)(g)\left(\mathbf{x}^* \otimes \mathbf{x}\right)\mathbf{y}.
\end{aligned}
$$

By linearity, this equality follows for any linear operator $A(\hat{t})$.

The representation $\lambda^+ \otimes \lambda$ is equivalent to the direct sum of the irreducible unitary representations $\mu \in \hat{G}$. Let $n(\lambda^+ \otimes \lambda, \mu)$ be the multiplicity of the representation μ in the representation $\lambda^+ \otimes \lambda$. Then, the representations $\lambda^+ \otimes \lambda$ and $\sum_{\mu \in \hat{G}} \sum_{m=1}^{n(\lambda^+ \otimes \lambda, \mu)} \oplus \mu_m$ are equivalent. Let C be the unitary intertwining operator between the above representations, i.e.,

$$
\left(\lambda^+ \otimes \lambda\right)(g) = C^{-1} \sum_{\mu \in \hat{G}} \sum_{n=1}^{n(\lambda^+ \otimes \lambda, \mu)} \oplus \mu_n(g)C.
$$

Substitute this equation into (2.81). We obtain

$$
A(g\hat{t}) = C^{-1} \sum_{\mu \in \hat{G}} \sum_{n=1}^{n(\lambda^+ \otimes \lambda, \mu)} \oplus \mu_n(g)CA(\hat{t}).
$$

Multiply both sides by C from the left. We have

$$
CA(g\hat{t}) = \sum_{\mu \in \hat{G}} \sum_{n=1}^{n(\lambda^+ \otimes \lambda, \mu)} \oplus \mu_n(g)CA(\hat{t}).
$$

Let k_j, $1 \le j \le N$ be the number of the group K_j in the chain (6.12), i.e., $K_j = G_{k_j}$. Let $\hat{t}_0 \in \hat{S}_j$ and $g \in K_j$. Then $g\hat{t}_0 = \hat{t}_0$, and we obtain

$$
CA(\hat{t}_0) = \sum_{\mu \in \hat{G}} \sum_{n=1}^{n(\lambda^+ \otimes \lambda, \mu)} \oplus \mu_n(g)CA(\hat{t}_0), \quad \hat{t}_0 \in \hat{S}_j, \ g \in K_j. \tag{2.82}
$$

Let P_{μ_n} be the orthogonal projector from $H^* \otimes H$ to the space of the representation μ_n. Then $P_{\mu_n}CA(\hat{t}_0)$ is a common eigenvector of all operators $\mu_n(g)$, $g \in K_j$. Let $\nu_1, \ldots, \nu_p$ be the irreducible components of the restriction of μ_n to the massive subgroup K_j, and let P_{ν_p} be the orthogonal projector from the space of the representation μ_n on the space of the representation ν_p. Then, $P_{\nu_p}P_{\mu_n}CA(\hat{t}_0)$ is a common eigenvector of all operators $\nu_p(g)$, $g \in K_j$. If ν_p is trivial, then $P_{\nu_p}P_{\mu_n}CA(\hat{t})$ lies in the one-dimensional subspace. The intersection of this subspace with the cone of nonnegative-definite operators is the half-line of nonnegative numbers, and we have $P_{\nu_p} P_{\mu_n} CA(\hat{t}_0) = r_{\mu_n}(\hat{t}_0) \geq 0$. If ν_p is not trivial, then $P_{\nu_p} P_{\mu_n} CA(\hat{t}_0) = 0$ by Lemma 6.3. In other words, if $\boldsymbol{\mu}_n = (\mu_1, \ldots, \mu_L)$, then

$$
[CA(\hat{t}_0)]_{\mu_n} = \begin{cases} 0, & \mu_{k_j} \text{ is not trivial,} \\ r_{\mu_n}(\hat{t}_0), & \mu_{k_j} \text{ is trivial.} \end{cases}
$$

Substitute this expression into (2.82) and multiply both sides by $C^{-1} = C^*$ from the left. We obtain

$$
A_{\lambda\lambda^+}(\hat{t}_0) = \sum_{\mu \in \hat{G}_{K_j}} \sum_{n=1}^{n(\lambda^+ \otimes \lambda, \mu)} \overline{C(\lambda, \lambda^+; \boldsymbol{\mu}_{nj})} r_{\mu_n}(\hat{t}_0), \quad \hat{t}_0 \in \hat{S}_j,
$$

where $\boldsymbol{\mu}_{nj} \in M_n$ is the unique vector of the Gel'fand–Tsetlin basis M_n of the representation μ_n with trivial μ_{k_j}, while $\overline{C(\lambda, \lambda^+; \boldsymbol{\mu}_{nj})}$ are the Clebsch–Gordan coefficients. By the first equation from (2.80),

$$
A_{\lambda\lambda^+}(g\hat{t}_0) = \sum_{\nu_1 \in \Lambda} \sum_{\nu_2 \in \Lambda^+} \sum_{\mu \in \hat{G}_{K_j}} \sum_{n=1}^{n(\lambda^+ \otimes \lambda, \mu)} D_{\lambda\nu_1}^{(\lambda)}(g) \overline{C(\nu_1, \nu_2; \boldsymbol{\mu}_{nj})} r_{\mu_n}(\hat{t}_0) D_{\lambda^+\nu_2}^{(\lambda^+)}(g),
$$

where Λ (resp. Λ^+) is the Gel'fand–Tsetlin basis of the representation λ (resp. λ^+).

Bourbaki (2004) proved that the measures σ satisfying the second equation from (2.80) have the form

$$
\sigma = \int_{\hat{S}_j} \sigma_{\hat{t}} \, d\tau_j(\hat{t}),
$$

where $\sigma_{\hat{t}}$ is the G-invariant probability measure on the G-orbit of the point $\hat{t}$, and $d\tau_j$ is a finite measure on $\hat{S}_j$.

By (2.79),

$$
R_{\lambda\lambda^+}(t_1, t_2) = \sum_{j=1}^{N} \sum_{\nu_1 \in \Lambda} \sum_{\nu_2 \in \Lambda^+} \sum_{\mu \in \hat{G}_{K_j}} \sum_{n=1}^{n(\lambda^+ \otimes \lambda, \mu)} \int_{\hat{S}_j} \int_G (g\hat{t}_0)(t_1 - t_2) D_{\lambda\nu_1}^{(\lambda)}(g)
$$
$$
\times \overline{C(\nu_1, \nu_2; \boldsymbol{\mu}_{nj})} r_{\mu_n}(\hat{t}_0) D_{\lambda^+\nu_2}^{(\lambda^+)}(g) \, dg \, d\tau_j(\hat{t}_0),
$$

$$(2.83)$$

while by (6.18)

$$
D_{\lambda\nu_1}^{(\lambda)}(g) D_{\lambda^+\nu_2}^{(\lambda^+)}(g) = \sum_{\varrho \in \hat{G}} \sum_{p=1}^{n(\lambda^+ \otimes \lambda, \varrho)} \sum_{\varrho_p, \varrho'_p \in R_p} \overline{C(\lambda, \lambda^+; \varrho_p)} C(\nu_1, \nu_2; \varrho'_p) D_{\varrho\varrho'}^{(\varrho)}(g),
$$

where R_p is the Gel'fand–Tsetlin basis of the representation ϱ_p. Substitute this formula into (2.83). We obtain

$$R_{\lambda\lambda^+}(t_1, t_2) = \sum_{j=1}^{N} \sum_{\nu_1 \in \Lambda} \sum_{\nu_2 \in \Lambda^+} \sum_{\mu \in \hat{G}_{K_j}} \sum_{n=1}^{n(\lambda^+ \otimes \lambda, \mu)} \int_{\hat{S}_j} \int_G (g\hat{t}_0)(t_1 - t_2)$$

$$\times \sum_{\varrho \in \hat{G}} \sum_{p=1}^{n(\lambda^+ \otimes \lambda, \varrho)} \sum_{\varrho_p, \varrho'_p \in R_p} \overline{C(\lambda, \lambda^+; \varrho_p)} C(\nu_1, \nu_2; \varrho'_p) D_{\varrho\varrho'}^{(\varrho)}(g)$$

$$\times \overline{C(\nu_1, \nu_2; \mu_{nj})} \, dg \, d\tau_{\mu_n j}(\hat{t}_0),$$

where

$$d\tau_{\mu_n j}(\hat{t}_0) := r_{\mu_n}(\hat{t}_0) \, d\tau_j(\hat{t}_0).$$

The Clebsch–Gordan coefficients form a unitary matrix. Therefore,

$$\sum_{\nu_1 \in \Lambda} \sum_{\nu_2 \in \Lambda^+} C(\nu_1, \nu_2; \varrho'_p) \overline{C(\nu_1, \nu_2; \mu_{nj})} = \delta_{\varrho'_p}^{\mu_{nj}}.$$

It follows that $\varrho'_p = \mu_{nj}$ and $\varrho_p = \mu_n$. Equation (2.83) takes the form

$$R_{\lambda\lambda^+}(t_1, t_2) = \sum_{j=1}^{N} \sum_{\mu \in \hat{G}_{K_j}} \sum_{n=1}^{n(\lambda^+ \otimes \lambda, \mu)} \sum_{\mu_n \in M_n} \overline{C(\lambda, \lambda^+; \mu_n)}$$

$$\times \int_{\hat{S}_j} \int_G (g\hat{t}_0)(t_1 - t_2) D_{\mu\mu_j}^{(\mu)}(g) \, dg \, d\tau_{\mu_n j}(\hat{t}_0). \qquad (2.84)$$

Let $t_1 = g_1 t_1^{(0)}$ with $g_1 \in G$ and $t_1^{(0)} \in S_p$. The function $g \mapsto (g\hat{t}_0)(t_1^{(0)})$ is both left- and right-invariant under the action of $\tilde{K}_p$. Indeed, let $k \in \tilde{K}_p$. Then $k^{-1}(t_1^{(0)}) = t_1^{(0)}$, and

$$kg\hat{t}_0(t_1^{(0)}) = g\hat{t}_0(k^{-1} t_1^{(0)}) = g\hat{t}_0(t_1^{(0)}),$$

$$gk\hat{t}_0(t_1^{(0)}) = g\hat{t}_0(t_1^{(0)}).$$

By the generalised Peter–Weyl theorem

$$(g\hat{t}_0)(t_1^{(0)}) = \sum_{\varrho \in \hat{G}_{\tilde{K}_p}} \dim \varrho \int_G (g\hat{t}_0)(t_1^{(0)}) \overline{D_{\varrho_p \varrho_p}^{(\varrho)}(g)} \, dg \, D_{\varrho_p \varrho_p}^{(\varrho)}(g),$$

where $\varrho_p \in R$ is the unique vector of the Gel'fand–Tsetlin basis R of the representation ϱ with trivial ϱ_{k_p}.

Gross and Kunze (1976) introduced the *Bessel functions of order* ϱ by

$$J^{(\varrho)}(\hat{t}, t) := \int_G (g^{-1}\hat{t})(t) D^{(\varrho)}(g) \, dg. \qquad (2.85)$$

If $\hat{T} = \mathbb{C}$ and G is the circle group, then the above definition reduces to the usual integral representation of the integer-order Bessel functions. Because the Haar measure dg on a compact group G is invariant under the *inversion* $g \mapsto g^{-1}$, (2.85) may be rewritten as

$$J^{(\varrho)}(\hat{t}, t) = \int_G (g\hat{t})(t) D^{(\varrho)}(g^{-1}) \, dg,$$

or in coordinate form,

$$J^{(\varrho)}_{\varrho\varrho'}(\hat{t}, t) = \int_G (g\hat{t})(t) \overline{D^{(\varrho)}_{\varrho'\varrho}(g)} \, dg.$$

It follows that the Peter–Weyl expansion of the function $(g\hat{t}_0)(t_1^{(0)})$ may be written as

$$(g\hat{t}_0)(t_1^{(0)}) = \sum_{\varrho \in \hat{G}_{\tilde{K}_p}} \dim \varrho \, J^{(\varrho)}_{\varrho_p \varrho_p}(\hat{t}_0, t_1^{(0)}) D^{(\varrho)}_{\varrho_p \varrho_p}(g).$$

For the point $t_1 = g_1 t_1^{(0)}$ we obtain

$$(g\hat{t}_0)(g_1 t_1^{(0)}) = (g_1^{-1} g \hat{t}_0)(t_1^{(0)})$$

$$= \sum_{\varrho \in \hat{G}_{\tilde{K}_p}} \dim \varrho \, J^{(\varrho)}_{\varrho_p \varrho_p}(\hat{t}_0, t_1^{(0)}) D^{(\varrho)}_{\varrho_p \varrho_p}(g_1^{-1} g)$$

$$= \sum_{\varrho \in \hat{G}_{\tilde{K}_p}} \dim \varrho \sum_{\sigma \in R} J^{(\varrho)}_{\varrho_p \varrho_p}(\hat{t}_0, t_1^{(0)}) \overline{D^{(\varrho)}_{\sigma \varrho_p}(g_1)} D^{(\varrho)}_{\sigma \varrho_p}(g).$$

If $t_2 = g_2 t_2^{(0)}$ with $t_2^{(0)} \in S_{p'}$, then

$$\overline{(g\hat{t}_0)(g_1 t_1^{(0)})} = \sum_{\varrho' \in \hat{G}_{\tilde{K}_{p'}}} \dim \varrho' \sum_{\sigma' \in R'} \overline{J^{(\varrho')}_{\varrho'_{p'} \varrho'_{p'}}(\hat{t}_0, t_2^{(0)})} D^{(\varrho')}_{\sigma' \varrho'_{p'}}(g_2) \overline{D^{(\varrho')}_{\sigma' \varrho'_{p'}}(g)}.$$

Substitute the last two formulae into (2.84). We obtain

$$
R_{\lambda\lambda^+}\left(g_1 t_1^{(0)}, g_2 t_2^{(0)}\right) = \sum_{j=1}^{N} \sum_{\mu \in \hat{G}_{K_j}} \sum_{n=1}^{n(\lambda^+ \otimes \lambda, \mu)} \sum_{\mu_n \in M_n} \overline{C(\lambda, \lambda^+; \mu_n)}
$$

$$
\times \sum_{\varrho \in \hat{G}_{\tilde{K}_p}} \sum_{\sigma \in R} \sum_{\varrho' \in \hat{G}_{\tilde{K}_{p'}}} \sum_{\sigma' \in R'} \dim \varrho \dim \varrho'
$$

$$
\times \int_G D_{\mu\mu_j}^{(\mu)}(g) D_{\sigma\varrho_p}^{(\varrho)}(g) \overline{D_{\sigma'\varrho'_{p'}}^{(\varrho')}(g)}\, dg
$$

$$
\times \int_{\hat{S}_j} J_{\varrho_p\varrho_p}^{(\varrho)}\left(\hat{t}_0, t_1^{(0)}\right) \overline{J_{\varrho'_{p'}\varrho'_{p'}}^{(\varrho')}\left(\hat{t}_0, t_2^{(0)}\right)}\, d\tau_{\mu_n j}(\hat{t}_0)
$$

$$
\times \overline{D_{\sigma\varrho_p}^{(\varrho)}(g_1)} D_{\sigma'\varrho'_{p'}}^{(\varrho')}(g_2). \tag{2.86}
$$

To calculate integral over G, use (6.18):

$$
D_{\mu\mu_j}^{(\mu)}(g) D_{\sigma\varrho_p}^{(\varrho)}(g) = \sum_{\varphi \in \hat{G}} \sum_{q=1}^{n(\mu \otimes \varrho, \varphi)} \sum_{\varphi_q, \varphi'_q \in F_q} \overline{C(\mu, \sigma; \varphi_q)} C(\mu_j, \varrho_p; \varphi'_q) D_{\varphi\varphi'}^{(\varphi)}(g),
$$

where F_q is the Gel'fand–Tsetlin basis of the representation φ_q. By the Peter–Weyl theorem,

$$
\int_G D_{\mu\mu_j}^{(\mu)}(g) D_{\sigma\varrho_p}^{(\varrho)}(g) \overline{D_{\sigma'\varrho'_{p'}}^{(\varrho')}(g)}\, dg = \frac{1}{\dim \varrho'} \sum_{q=1}^{n(\mu \otimes \varrho, \varrho')} \overline{C(\mu, \sigma; \sigma'_q)} C(\mu_j, \varrho_p; \varrho'_{p'q}).
$$

Substitute this into (2.86). We obtain

$$
R_{\lambda\lambda^+}\left(g_1 t_1^{(0)}, g_2 t_2^{(0)}\right) = \sum_{j=1}^{N} \sum_{\mu \in \hat{G}_{K_j}} \sum_{n=1}^{n(\lambda^+ \otimes \lambda, \mu)} \sum_{\mu_n \in M_n} \overline{C(\lambda, \lambda^+; \mu_n)}
$$

$$
\times \sum_{\varrho \in \hat{G}_{\tilde{K}_p}} \sum_{\sigma \in R} \sum_{\varrho' \in \hat{G}_{\tilde{K}_{p'}}} \sum_{\sigma' \in R'} \dim \varrho
$$

$$
\times \sum_{q=1}^{n(\mu \otimes \varrho, \varrho')} \overline{C(\mu, \sigma; \sigma'_q)} C(\mu_j, \varrho_p; \varrho'_{p'q})
$$

$$
\times \int_{\hat{S}_j} J_{\varrho_p\varrho_p}^{(\varrho)}\left(\hat{t}_0, t_1^{(0)}\right) \overline{J_{\varrho'_{p'}\varrho'_{p'}}^{(\varrho')}\left(\hat{t}_0, t_2^{(0)}\right)}\, d\tau_{\mu_n j}(\hat{t}_0)
$$

$$
\times \overline{D_{\sigma\varrho_p}^{(\varrho)}(g_1)} D_{\sigma'\varrho'_{p'}}^{(\varrho')}(g_2).
$$

By Karhunen's theorem,

$$
X_\lambda\left(gt^{(0)}\right) = \sum_{j=1}^{N} \sum_{\mu \in \hat{G}_{K_j}} \overset{n(\lambda^+ \otimes \lambda, \mu)}{\sum_{n=1}} \sum_{\mu_n \in M_n} \sum_{\varrho \in \hat{G}_{\tilde{K}_p}} \sum_{\sigma \in R} \overline{D_{\sigma \varrho_p}^{(\varrho)}(g)}
$$

$$
\times \int_{\hat{S}_j} J_{\varrho_p \varrho_p}^{(\varrho)}\left(\hat{t}_0, t^{(0)}\right) \mathrm{d}Z_{\lambda \varrho_p j \mu_n \sigma}\left(\hat{t}^{(0)}\right), \tag{2.87}
$$

where $Z_{\lambda \varrho_p j \mu_n \sigma}$ is the sequence of centred random measures on $\hat{S}_j$ with

$$
\mathsf{E}\left[Z_{\lambda \varrho_p j \mu_n \sigma}(A_1) \overline{Z_{\lambda^+ \varrho'_{p'} j' \mu'_{n'} \sigma'}(A_2)}\right] = \delta_{\varrho'_{p'}}^{\varrho_p} \delta_{j'}^{j} \delta_{\mu'_{n'}}^{\mu_n} \delta_{\sigma'}^{\sigma} \dim \varrho \overline{C\left(\lambda, \lambda^+; \mu_n\right)}
$$

$$
\times \overset{n(\mu \otimes \varrho, \varrho')}{\sum_{q=1}} \overline{C\left(\mu, \sigma; \sigma'_q\right)} C\left(\mu_j, \varrho_p; \varrho'_{p'q}\right) \tau_{\mu_n j}(A_1 \cap A_2) \tag{2.88}
$$

for any Borel subsets A_1 and A_2 of the set $\hat{S}_j$.

Theorem 2.34 *Let T be a commutative locally compact group with countable base, let G be a compact group of automorphisms of T, and let λ be an irreducible unitary representation of the group G in a finite-dimensional complex Hilbert space H. Let $\mathbf{X}(t)$ be a wide-sense homogeneous isotropic vector random field on T taking values in H. Assume that conditions 1–7 hold true. Let k_j, $1 \le j \le N$ be the number of the group K_j in the chain (6.12), i.e., $K_j = G_{k_j}$. Then,*

- *If λ is trivial, then $\mathsf{E}[\mathbf{X}(t)]$ is any constant vector. Otherwise, $\mathsf{E}[\mathbf{X}(t)] = \mathbf{0}$.*
- *The field $\mathbf{X}(t)$ has the form (2.87), where $Z_{\varrho_p j \mu_n \sigma}$ is the sequence of centred random measures on $\hat{S}_j$ satisfying (2.88) for any Borel subsets A_1 and A_2 of the set $\hat{S}_j$.*

Example 2.35 Let $T = \mathbb{R}^n$ and $G = \mathrm{SO}(n)$. The action of G on $\hat{T} = \mathbb{R}^n$ has $N = 2$ equivalence classes: $\hat{T}_1 = \{0\}$ and $\hat{T}_2 = \mathbb{R}^n \setminus \{0\}$. The measurable section $\hat{S}$ is

$$
\hat{S} = \left\{(0, 0, \ldots, 0, x_n): x_n \ge 0\right\},
$$

while the sets $\hat{S}_1$ and $\hat{S}_2$ are

$$
\hat{S}_1 = \{0\},
$$

$$
\hat{S}_2 = \left\{(0, 0, \ldots, 0, x_n): x_n > 0\right\}.
$$

The stationary subgroups are $K_1 = \mathrm{SO}(n)$ and $K_2 = \mathrm{SO}(n-1)$. Both subgroups are massive in G, and the chain $G = K_1 \supset K_2$ may be extended to the chain (6.13) with $k_1 = 1$ and $k_2 = 2$.

The action of G on $T = \mathbb{R}^n$ has $M = 2$ equivalence classes: $T_1 = \{0\}$ and $T_2 = \mathbb{R}^n \setminus \{0\}$. The measurable section S is

$$S = \{(0, 0, \ldots, 0, x_n) : x_n \geq 0\},$$

while the sets S_1 and S_2 are

$$S_1 = \{0\},$$
$$S_2 = \{(0, 0, \ldots, 0, x_n) : x_n > 0\}.$$

The stationary subgroups are $\tilde{K}_1 = SO(n)$ and $\tilde{K}_2 = SO(n-1)$. Both subgroups are massive in G.

The group $G = SO(n)$ is *simply reducible*. This means that the tensor product of any two irreducible unitary representations decomposes into different irreducible components. In particular, $n(\lambda^+ \otimes \lambda, \mu) \leq 1$.

Let $p = 2$, i.e., $t^{(0)} \neq 0$. By (6.14) and (6.15), the representation $\varrho \in \hat{G}$ belongs to $\hat{G}_{\tilde{K}_2}$ iff its signature has the form $p_1 = k \geq 0$ and $p_j = 0, 2 \leq j \leq [n/2]$. We denote such a representation by $[k]$. The set of representations $[k]$ with $n(\lambda^+ \otimes \lambda, [k]) = 1$ is non-empty, because the trivial representation $[0]$ is always an irreducible component of $\lambda^+ \otimes \lambda$ acting in the linear span of the identity matrix. The domain of the measure $\tau_{[0]2}$ can be extended from S_2 to S by adding the only measure $\tau_{\mu_n j}$ with $j = 1$, $\tau_{[0]1}$. Thus, we may denote the above measures on $\hat{S}$ just by τ_k with restriction that only τ_0 may have an atom at zero.

Let $(r, \varphi, \theta_1, \ldots, \theta_{n-2})$ be the spherical coordinates in $\mathbb{R}^n$. By (6.57) the matrix elements $D^{(\varrho)}_{\sigma \varrho_p}(\varphi, \theta_1, \ldots, \theta_{n-2})$ are

$$D^\ell_m(\varphi, \theta_1, \ldots, \theta_{n-2}) = \sqrt{\frac{2\pi^{n/2}}{\dim[\ell]\Gamma(n/2)}} Y_{\ell m}(\varphi, \theta_1, \ldots, \theta_{n-2}).$$

It is customary to use *real* spherical harmonics $S^m_\ell(\varphi, \theta_1, \ldots, \theta_{n-2})$ instead.

The Bessel functions have the form

$$J^\ell_m(R, r) = \int_{S^{n-1}} e^{iRr} D^\ell_m(\varphi, \theta_1, \ldots, \theta_{n-2}) \, dS$$

$$= \sqrt{\frac{2\pi^{n/2}}{\dim[\ell]\Gamma(n/2)}} \int_{S^{n-1}} e^{iRr} S^m_\ell(\varphi, \theta_1, \ldots, \theta_{n-2}) \, dS,$$

where dS is the $SO(n)$-invariant probability measure on the unit sphere S^{n-1}. Vilenkin (1968) proved that the functions in the last line are not equal to zero iff $m = m(\ell, 0, \ldots, 0, +)$ and

$$\int_{S^{n-1}} e^{iRr} S^{m(\ell, 0, \ldots, 0, +)}_\ell(\varphi, \theta_1, \ldots, \theta_{n-2}) \, dS$$

$$= 2^{(n-2)/2} \Gamma(n/2) \sqrt{\dim[\ell]} \frac{J_{\ell+(n-2)/2}(Rr)}{(Rr)^{(n-2)/2}}.$$

Substituting these expressions into (2.87), we obtain

$$X_\lambda(r, \varphi, \theta_1, \ldots, \theta_{n-2}) = \sqrt{2^{n-1} \Gamma(n/2) \pi^{n/2}} \sum_{[k] \subset \lambda \otimes \lambda^+} \sum_{\ell=0}^{\infty} \sum_{m=1}^{\dim[\ell]} S_\ell^m(\varphi, \theta_1, \ldots, \theta_{n-2})$$

$$\times \int_0^\infty \frac{J_{\ell+(n-2)/2}(Rr)}{(Rr)^{(n-2)/2}} \, dZ_{\lambda k \ell m}(R).$$

In particular, when λ is a trivial representation, we have

$$X(r, \varphi, \theta_1, \ldots, \theta_{n-2}) = \sqrt{2^{n-1} \Gamma(n/2) \pi^{n/2}} \sum_{\ell=0}^{\infty} \sum_{m=1}^{\dim[\ell]} S_m^\ell(\varphi, \theta_1, \ldots, \theta_{n-2})$$

$$\times \int_0^\infty \frac{J_{\ell+(n-2)/2}(Rr)}{(Rr)^{(n-2)/2}} \, dZ_{\ell m}(R).$$

Let $n = 3$. We write θ instead of θ_1. The signature of any irreducible unitary representation of the group SO(3) is just one nonnegative integer or half-integer, p, while the Gel'fand–Tsetlin scheme is (p, q), where q is an integer, if p is an integer, and is a half-integer otherwise, and $|q| \leq p$. Fix p and denote the corresponding representation by $[p]$. The representation, dual to $[p]$, is $[p]$ itself, while the tensor product $[p] \otimes [p]$ is equivalent to the direct sum of representations $[k]$ with integer k and $0 \leq k \leq 2p$. Therefore, we obtain

$$X_q(r, \varphi, \theta) = \sqrt{\pi} \sum_{k=0}^{2p} \sum_{\ell=0}^{\infty} \sum_{m=1}^{2\ell+1} S_m^\ell(\varphi, \theta) \int_0^\infty \frac{J_{\ell+1/2}(Rr)}{\sqrt{Rr}} \, dZ_{qk\ell m}(R).$$

2.6 Volterra Isotropic Random Fields

Let $X(t)$ be a real-valued centred mean-square continuous Gaussian random field on the space $\mathbb{R}^N$. Let $(s, \theta_1, \ldots, \theta_{N-2}, \varphi)$ be the spherical coordinates on $\mathbb{R}^N$. The results of Example 2.30 may be written using the real spherical harmonics $S_\ell^m(\theta_1, \ldots, \theta_{N-2}, \varphi)$:

$$X(s, \theta_1, \ldots, \theta_{N-2}, \varphi) = \sum_{\ell=0}^{\infty} \sum_{m=1}^{h(S^{N-1}, \ell)} Z_{\ell m}(s) S_\ell^m(\theta_1, \ldots, \theta_{N-2}, \varphi)$$

with

$$\sum_{\ell=0}^{\infty} h(S^{N-1}, \ell) r_\ell(s, s) < \infty,$$

where λ_ℓ are irreducible unitary representations of the group $SO(N)$ of class 1 with respect to the subgroup $SO(N-1)$, while $Z_{\ell m}(s)$ is the sequence of independent real-valued centred Gaussian processes on $[0, \infty)$ with known covariance functions $r_\ell(s_1, s_2)$. Assume that $\{b_{\ell mn}(s): n \geq 1\}$ is a sequence of functions on the interval $[0, R]$ such that

$$Z_{\ell m}(s) = \sum_{n \geq 1} X_{\ell mn} b_{\ell mn}(s), \tag{2.89}$$

where $\{X_{\ell mn}: \ell \geq 0, 1 \leq m \leq h(S^{N-1}, \ell), n \geq 1\}$ is a sequence of independent standard normal random variables, and where $h(S^{N-1}, \ell)$ is determined by (6.56). We assume that the above series converges in mean square. Then we obtain the mean-square convergent spectral expansion of the field $X(t)$ as

$$X(s, \theta_1, \ldots, \theta_{N-2}, \varphi) = \sum_{\ell=0}^{\infty} \sum_{m=1}^{h(S^{N-1}, \ell)} \sum_{n \geq 1} X_{\ell mn} b_{\ell mn}(s) S_\ell^m(\theta_1, \ldots, \theta_{N-2}, \varphi). \tag{2.90}$$

One way to find the functions $b_{\ell mn}(s)$ is as follows. Recall that a function $a(s, u): (0, \infty) \times (0, \infty) \to \mathbb{R}$ is called a *Volterra kernel* if it is locally square-integrable, and

$$a(s, u) = 0 \quad \text{for } s < u. \tag{2.91}$$

Definition 2.36 A *Volterra process* with Volterra kernel $a(s, u)$ is a centred Gaussian stochastic process $Y(s)$ with covariance function

$$R(s_1, s_2) = \int_0^{\min\{s_1, s_2\}} a(s_1, u) a(s_2, u) \, du.$$

Note that the upper limit in the integral above may be changed to ∞ because of (2.91).

For example, the Wiener process is a Volterra process with Volterra kernel

$$a(s, u) = \begin{cases} 1, & s \geq u, \\ 0, & s < u. \end{cases}$$

Another example is the so-called *Riemann–Liouville process*, with Volterra kernel

$$a(s, u) = \frac{(\max\{0, s - u\})^{\alpha-1}}{\Gamma(\alpha)}, \quad \alpha > 1/2.$$

It is not difficult to calculate the covariance function of the Riemann–Liouville process. Indeed, by (6.41) we obtain

$$a(s, u) = u^{\alpha-1} G_{1,1}^{0,1}\left(\frac{s}{u} \middle| \begin{matrix} \alpha \\ 0 \end{matrix}\right).$$

Using the symmetry relation (6.39) yields

$$a(s, u) = u^{\alpha-1} G_{1,1}^{1,0}\left(\frac{u}{s}\left|\begin{matrix} 1 \\ 1 - \alpha \end{matrix}\right.\right).$$

The covariance function may be calculated using the classical Meijer integral (6.38):

$$R(s_1, s_2) = \int_0^\infty a(s_1, u)a(s_2, u)\,du$$

$$= s_1^{2\alpha-1} G_{2,2}^{1,1}\left(\frac{s_1}{s_2}\left|\begin{matrix} 1 - \alpha, 1 \\ 1 - \alpha, 1 - 2\alpha \end{matrix}\right.\right).$$

Assume that $Z_{\ell m}(s)$ are Volterra processes with Volterra kernels $a_\ell(s, u)$. Then

$$r_\ell(s_1, s_2) = \int_0^{\min\{s_1, s_2\}} a_\ell(s_1, u)a_\ell(s_2, u)\,du.$$

By (2.91), the last display can be rewritten as

$$r_\ell(s_1, s_2) = \int_0^R a_\ell(s_1, u)a_\ell(s_2, u)\,du, \quad s_1, s_2 \in [0, R].$$

Let $\{e_{\ell mn}(u): n \geq 1\}$ be a basis in the Hilbert space $L^2[0, R]$. We obtain

$$r_\ell(s_1, s_2) = \sum_{n=1}^\infty b_{\ell mn}(s_1)b_{\ell mn}(s_2), \quad 1 \leq m \leq h(S^{N-1}, \ell),$$

where

$$b_{\ell mn}(s) := \int_0^R a_\ell(s, u)\overline{e_{\ell mn}(u)}\,du. \tag{2.92}$$

It follows from Karhunen's theorem that the stochastic process $Z_{\ell m}(s)$ itself has the form (2.89), where the series converges in mean square for all $s \in [0, R]$.

Definition 2.37 A centred isotropic Gaussian random field $X(t)$ on the space $\mathbb{R}^n$ is called a *Volterra field* if all random processes $Z_{\ell m}(s)$ are Volterra processes.

In particular, any Volterra field has spectral expansion (2.90).

Theorem 2.38 *The multiparameter fractional Brownian motion is a Volterra field.*

Proof Put

$$a_\ell(s, u) := \begin{cases} c_{NH} u^{H-1/2} G_{2,2}^{2,0}\left(\dfrac{u^2}{s^2}\left|\begin{matrix} N/2, 1 \\ 0, 1 - H \end{matrix}\right.\right), & \ell = 0, \\[2ex] c_{NH} s^{2H-\ell} u^{\ell-H-1/2} G_{1,1}^{1,0}\left(\dfrac{u^2}{s^2}\left|\begin{matrix} N/2 + H \\ 0 \end{matrix}\right.\right), & \ell \geq 1, \end{cases} \tag{2.93}$$

where

$$c_{NH} := \sqrt{2\pi^{(N-2)/2}\Gamma(N/2+H)\Gamma(H+1)\sin(\pi H)}. \qquad (2.94)$$

The above functions are Volterra kernels by (6.37). By (6.38),

$$\int_0^\infty a_0(s_1,u)a_0(s_2,u)\,du = \frac{c_{NH}^2}{2}s_1^{2H}G_{4,4}^{2,2}\left(\frac{s_1^2}{s_2^2}\middle|\begin{array}{c}1-H,0,N/2,1\\0,1-H,1-H-N/2,-H\end{array}\right),$$

$$\int_0^\infty a_\ell(s_1,u)a_\ell(s_2,u)\,du = \frac{c_{NH}^2}{2}s_1^\ell s_2^{2H-\ell}G_{2,2}^{1,1}\left(\frac{s_1^2}{s_2^2}\middle|\begin{array}{c}H+1-\ell,N/2+H\\0,1-N/2-\ell\end{array}\right),$$

with $\ell \geq 1$ in the second formula. It remains to prove that the right-hand sides of the above equations coincide with the covariance functions (2.67) calculated in Example 2.31.

Let $\ell = 0$. By (6.45) with $m = n = 2$, $p = q = 4$, $z = s_1^2/s_2^2$, $a_1 = 1 - H$, $a_2 = b_1 = 0$, $a_3 = N/2$, $a_4 = 1$, $b_2 = 1 - H$, $b_3 = 1 - H - N/2$, and $b_4 = -H$, we obtain

$$r_0(s_1,s_2)$$

$$= \frac{c_{NH}^2}{2}s_1^{2H}\left[\frac{\Gamma(1-H)\Gamma(H)}{\Gamma(\frac{N}{2})\Gamma(\frac{N}{2}+H)\Gamma(1+H)}{}_4F_3\left(\begin{array}{c}H,1,1-\frac{N}{2},0;\\H,H+\frac{N}{2},1+H;s_1^2/s_2^2\end{array}\right)\right.$$

$$+ \frac{\Gamma(H-1)\Gamma(2-H)}{\Gamma(\frac{N}{2}+H-1)\Gamma(H)\Gamma(\frac{N}{2}+1)}\frac{s^{2(1-H)}}{t^{2(1-H)}}{}_4F_3$$

$$\left.\times\left(\begin{array}{c}1,2-H,2-H-\frac{N}{2},1-H;\\2-H,\frac{N}{2}+1,2;s_1^2/s_2^2\end{array}\right)\right].$$

The first term is simplified, using (6.48), while the second term is simplified by (2.65). We get

$$r_0(s_1,s_2)$$

$$= \frac{c_{NH}^2}{2}s_1^{2H}\left[\frac{\Gamma(1-H)\Gamma(H)}{\Gamma(\frac{N}{2})\Gamma(\frac{N}{2}+H)\Gamma(1+H)}\right.$$

$$\left.+ \frac{\Gamma(H-1)\Gamma(2-H)}{\Gamma(\frac{N}{2}+H-1)\Gamma(H)\Gamma(\frac{N}{2}+1)}\frac{s_1^{2(1-H)}}{s_2^{2(1-H)}}{}_3F_2\left(\begin{array}{c}1,2-H-\frac{N}{2},1-H;\\\frac{N}{2}+1,2;s_1^2/s_2^2\end{array}\right)\right].$$

Using (6.49) with $b = 2 - H - N/2$, $c = 1 - H$, $e = N/2 + 1$, and $z = s_1^2/s_2^2$ yields

$$r_0(s_1, s_2) = \frac{c_{NH}^2}{2} s_1^{2H} \left[\frac{\Gamma(1-H)\Gamma(H)}{\Gamma(N/2)\Gamma(N/2+H)\Gamma(1+H)} \right.$$

$$+ \frac{\Gamma(H-1)\Gamma(2-H)(N/2)}{\Gamma(N/2+H-1)\Gamma(H)\Gamma(N/2+1)(1-H-N/2)(-H)} \frac{s_2^{2H}}{s_1^{2H}}$$

$$\left. \times \left[{}_2F_1\left(1-H-N/2, -H; N/2; s_1^2/s_2^2\right) - 1 \right] \right].$$

To simplify the second line, use (6.31) in the following way:

$$\Gamma(H-1) = \Gamma(H)/(H-1),$$

$$\Gamma(2-H) = (1-H)\Gamma(1-H),$$

$$\Gamma(N/2+H-1)(1-H-N/2) = -\Gamma(N/2+H),$$

$$\Gamma(H)(-H) = -\Gamma(H+1),$$

$$\Gamma(N/2+1) = (N/2)\Gamma(N/2).$$

After simplification, we obtain

$$r_0(s_1, s_2) = \frac{c_{NH}^2 \Gamma(1-H)\Gamma(H)}{2\Gamma(N/2)\Gamma(N/2+H)\Gamma(H+1)}$$

$$+ \left[s_1^{2H} + s_2^{2H}\left(1 - {}_2F_1\left(-H, 1-H-N/2; N/2; s_1^2/s_2^2\right)\right) \right],$$

or, by (6.34) with $z = H$,

$$r_0(s_1, s_2) = \frac{c_{NH}^2 \pi}{2\Gamma(N/2)\Gamma(N/2+H)\Gamma(H+1)\sin(\pi H)}$$

$$\times \left[s_1^{2H} + s_2^{2H}\left(1 - {}_2F_1\left(-H, 1-H-N/2; N/2; s_1^2/s_2^2\right)\right) \right].$$

Substituting (2.94) to the last display, we get

$$r_0(s_1, s_2) = \frac{\pi^{N/2}}{\Gamma(N/2)} \left[s^{2H} + t^{2H}\left(1 - {}_2F_1\left(-H, 1-H-N/2; N/2; s^2/t^2\right)\right) \right],$$

or, by the argument-simplification formula (6.50),

$$r_0(s_1, s_2) = \frac{\pi^{N/2}}{\Gamma(N/2)} \left[s_1^{2H} + s_2^{2H} - (s_1+s_2)^{2H} \right.$$

$$\left. \times {}_2F_1\left(\frac{N-1}{2}, -H; N-1; \frac{4s_1 s_2}{(s_1+s_2)^2}\right) \right].$$

This completes the proof of the case of $\ell = 0$.

For the case of $\ell \geq 1$, we start from (2.66) and use (6.31) with $z = 1 - H$:

$$r_\ell(s_1, s_2) = \frac{\pi^{N/2}(s_1 + s_2)^{2(H-\ell)}(s_1 s_2)^\ell \Gamma(\ell - H)H}{\Gamma(1 - H)\Gamma(N/2 + \ell)}$$

$$\times {}_2F_1\big((N-1)/2 + \ell, \ell - H; N - 1 + 2\ell; 4s_1 s_2/(s_1 + s_2)^2\big).$$

By Euler's reflection formula (6.34) with $z = H$, and (6.31) with $z = H + 1$ we get

$$r_\ell(s_1, s_2) = \frac{\pi^{(N-2)/2}(s_1 + s_2)^{2(H-\ell)}(s_1 s_2)^\ell \Gamma(\ell - H)\Gamma(H + 1)\sin(\pi H)}{\Gamma(N/2 + \ell)}$$

$$\times {}_2F_1\big((N-1)/2 + \ell, \ell - H; N - 1 + 2\ell; 4s_1 s_2/(s_1 + s_2)^2\big),$$

and then, using (6.47) with $z = s^2/t^2$, $a = m - H$, and $b = (N-1)/2 + m$, we finally obtain

$$r_\ell(s_1, s_2) = \frac{c_{NH}^2}{2} s_1^m s_2^{2H-m} G_{2,2}^{1,1}\left(\frac{s_1^2}{s_2^2} \,\middle|\, \begin{matrix} H + 1 - \ell, N/2 + H \\ 0, 1 - N/2 - \ell \end{matrix}\right), \quad \ell \geq 1.$$

This completes the proof. $\qquad\qquad\qquad\qquad\qquad\qquad\qquad\qquad\qquad\qquad\square$

    Thus, for each $\ell \geq 0$ and for each m, $1 \leq m \leq h(S^{N-1}, \ell)$, one may choose a basis $\{e_{\ell mn}(u) : n \geq 1\}$ in the Hilbert space $L^2[0, R]$ and calculate the functions $b_{\ell mn}(s)$ by (2.92), where $a_\ell(s, u)$ are Volterra kernels (2.93). Then, (2.90) is the mean-square convergent spectral expansion of the multiparameter fractional Brownian motion.

    The analytical expression for the functions $b_{\ell mn}(s)$ may be obtained in the closed form using the classical Meijer integral from two G-functions (6.38). Indeed, the Volterra kernels (2.93) as functions of u are expressed as a product of some power of u and the Meijer G-function of argument u^2/s^2. To use (6.38), we choose a basis $\{e_{\ell mn}(u) : n \geq 1\}$ whose elements are expressed in the same manner. Consider the following example.

Example 2.39 Let v be a real number, and let $j_{v,1} < j_{v,2} < \cdots < j_{v,n} < \cdots$ be the positive zeros of the Bessel function $J_v(u)$. For any $v > -1$, the *Fourier–Bessel functions*

$$\varphi_{v,n}(u) = \frac{\sqrt{2u}}{J_{v+1}(j_{v,n})} J_v(j_{v,n}u), \quad n \geq 1$$

form a basis in the space $L^2[0, 1]$ (Watson, 1944, Sect. 18.24). By change of variable we conclude that the functions

$$e_{\ell mn}(u) = \frac{\sqrt{2u}}{R J_{v+1}(j_{v,n})} J_v\big(R^{-1} j_{v,n}u\big), \quad n \geq 1$$

form a basis in the space $L^2[0, R]$. By (6.51) we obtain

$$e_{\ell mn}(u) = \frac{\sqrt{2u}}{R J_{\nu+1}(j_{\nu,n})} G_{0,2}^{1,0}\left(\frac{j_{\nu,n}^2 u^2}{4R^2} \middle| \begin{array}{c} \cdot \\ \nu/2, -\nu/2 \end{array}\right).$$

To calculate $b_{\ell mn}(s)$, use (2.92). First, consider the case of $\ell = 0$:

$$b_{0mn}(s) = \int_0^R c_{NH} u^{H-1/2} G_{2,2}^{2,0}\left(\frac{u^2}{s^2} \middle| \begin{array}{c} N/2, 1 \\ 0, 1-H \end{array}\right)$$

$$\times \frac{\sqrt{2u}}{R J_{\nu+1}(j_{\nu,n})} G_{0,2}^{1,0}\left(\frac{j_{\nu,n}^2 u^2}{4R^2} \middle| \begin{array}{c} \cdot \\ \nu/2, -\nu/2 \end{array}\right) du.$$

By (6.38), we obtain

$$b_{0mn}(s) = \frac{c_{NH} 2^{H+1/2} R^H}{J_{\nu+1}(j_{\nu,n}) j_{\nu,n}^{H+1}} G_{4,2}^{2,1}\left(\frac{4R^2}{s^2 j_{\nu,n}^2} \middle| \begin{array}{c} 1-\frac{H+1+\nu}{2}, \frac{N}{2}, 1, 1-\frac{H+1-\nu}{2} \\ 0, 1-H \end{array}\right).$$

Formula 8.2.2.9 from Prudnikov et al. (1990) states

$$G_{p,q}^{m,n}\left(z \middle| \begin{array}{c} a_1, \ldots, a_{p-1}, b_1 \\ b_1, \ldots, b_q \end{array}\right) = G_{p-1,q-1}^{m-1,n}\left(z \middle| \begin{array}{c} a_1, \ldots, a_{p-1} \\ b_2, \ldots, b_q \end{array}\right). \tag{2.95}$$

If we put $\nu := 1 - H$, then (2.95) decreases the order of the Meijer G-function from 6 to 4. We get

$$b_{0mn}(s) = \frac{c_{NH} 2^{H+1/2} R^H}{J_{2-H}(j_{1-H,n}) j_{1-H,n}^{H+1}} G_{3,1}^{1,1}\left(\frac{4R^2}{s^2 j_{1-H,n}^2} \middle| \begin{array}{c} 0, N/2, 1 \\ 0 \end{array}\right).$$

For further simplification, use (6.39). We get

$$b_{0mn}(s) = \frac{c_{NH} 2^{H+1/2} R^H}{J_{2-H}(j_{1-H,n}) j_{1-H,n}^{H+1}} G_{1,3}^{1,1}\left(\frac{s^2 j_{1-H,n}^2}{4R^2} \middle| \begin{array}{c} 1 \\ 1, 1-N/2, 0 \end{array}\right).$$

Then, use the argument transformation (6.40) with $\alpha = 1$ to obtain

$$b_{0mn}(s) = \frac{c_{NH} 2^{H+1/2} R^H}{J_{2-H}(j_{1-H,n}) j_{1-H,n}^{H+1}} \frac{s^2 j_{1-H,n}^2}{4R^2} G_{1,3}^{1,1}\left(\frac{s^2 j_{1-H,n}^2}{4R^2} \middle| \begin{array}{c} 0 \\ 0, -N/2, -1 \end{array}\right),$$

and use (6.46) with $a_1 = 1$, $b_1 = N/2+1$, $b_2 = 2$, and $z = -s^2 j_{1-H,n}^2/(4R^2)$ to get

$$b_{0mn}(s) = \frac{c_{NH} 2^{H+1/2} R^H}{J_{2-H}(j_{1-H,n}) j_{1-H,n}^{H+1}} \frac{s^2 j_{1-H,n}^2}{4R^2} \frac{1}{\Gamma(N/2+1)}$$

$$\times {}_1F_2\left(1; N/2+1, 2; -s^2 j_{1-H,n}^2/(4R^2)\right).$$

Finally, use (6.55) with $c = N/2 + 1$ and $z = -s^2 j_{1-H,n}^2/(4R^2)$. We obtain

$$
b_{0mn}(s) = \frac{c_{NH} 2^{H+1/2} R^H}{J_{2-H}(j_{1-H,n}) j_{1-H,n}^{H+1} \Gamma(N/2)}
$$

$$
\times \left[2^{(N-2)/2} \Gamma(N/2) \frac{J_{(N-2)/2}(R^{-1} j_{1-H,n} s)}{(R^{-1} j_{1-H,n} s)^{(n-2)/2}} - 1 \right]. \tag{2.96}
$$

Continue with the remaining case of $\ell \geq 1$. By (2.92) and (2.93)

$$
b_{\ell mn}(s) = \int_0^R c_{NH} s^{2H-\ell} u^{\ell-H-1/2} G_{1,1}^{1,0}\left(\frac{u^2}{s^2} \middle| \begin{matrix} N/2 + H \\ 0 \end{matrix} \right)
$$

$$
\times \frac{\sqrt{2u}}{R J_{\nu+1}(j_{\nu,n})} G_{0,2}^{1,0}\left(\frac{j_{\nu,n}^2 u^2}{4R^2} \middle| \begin{matrix} \cdot \\ \nu/2, -\nu/2 \end{matrix} \right) du.
$$

By (6.38), we obtain

$$
b_{\ell mn}(s) = \frac{c_{NH} 2^{\ell-H+1/2} s^{2H-\ell} R^{\ell-H}}{J_{\nu+1}(j_{\nu,n}) j_{\nu,n}^{\ell-H+1}}
$$

$$
\times G_{3,1}^{1,1}\left(\frac{4R^2}{s^2 j_{\nu,n}^2} \middle| \begin{matrix} 1 - \frac{\ell-H+1+\nu}{2}, \frac{N}{2} + H, 1 - \frac{\ell-H+1-\nu}{2} \\ 0 \end{matrix} \right).
$$

To decrease the order of the Meijer G-function from 4 to 2, put $\nu := \ell - 1 - H$ and use (2.95). We get

$$
b_{\ell mn}(s) = \frac{c_{NH} 2^{\ell-H+1/2} s^{2H-\ell} R^{\ell-H}}{J_{\ell-H}(j_{\ell-1-H,n}) j_{\ell-1-H,n}^{\ell-H+1}} G_{2,0}^{0,1}\left(\frac{4R^2}{s^2 j_{\ell-1-H,n}^2} \middle| \begin{matrix} 1 - \ell + H, N/2 + H \\ \cdot \end{matrix} \right).
$$

For further simplification, use the symmetry relation (6.39). We obtain

$$
b_{\ell mn}(s) = \frac{c_{NH} 2^{\ell-H+1/2} s^{2H-\ell} R^{\ell-H}}{J_{\ell-H}(j_{\ell-1-H,n}) j_{\ell-1-H,n}^{\ell-H+1}} G_{0,2}^{1,0}\left(\frac{s^2 j_{\ell-1-H,n}^2}{4R^2} \middle| \begin{matrix} \cdot \\ \ell - H, 1 - N/2 - H \end{matrix} \right).
$$

Then, use the argument transformation (6.40) with $\alpha = \ell/2 + 1/2 - N/4 - H$ to get

$$
b_{\ell mn}(s) = \frac{c_{NH} 2^{H+(N-1)/2} R^{H+(N-2)/2}}{s^{(N-2)/2} J_{\ell-H}(j_{\ell-1-H,n}) j_{\ell-1-H,n}^{H+N/2}}
$$

$$
\times G_{0,2}^{1,0}\left(\frac{s^2 j_{\ell-1-H,n}^2}{4R^2} \middle| \begin{matrix} \cdot \\ \ell/2 + (N-2)/4, -\ell/2 - (N-2)/4 \end{matrix} \right).
$$

Finally, use (6.51) to obtain

$$b_{\ell m n}(s) = \frac{c_{NH} 2^{H+1/2} R^H}{J_{\ell - H}(j_{\ell - 1 - H, n}) j_{\ell - 1 - H, n}^{H+1} \Gamma(N/2)}$$

$$\times 2^{(N-2)/2} \Gamma(N/2) \frac{J_{\ell + (N-2)/2}(R^{-1} j_{\ell - 1 - H, n} s)}{(R^{-1} j_{\ell - 1 - H, n} s)^{(N-2)/2}}.$$

The last formula and (2.96) can be unified as

$$b_{\ell m n}(s) = \frac{c_{NH} 2^{H+1/2} R^H}{J_{|\ell - 1| - H + 1}(j_{|\ell - 1| - H, n}) j_{|\ell - 1| - H, n}^{H+1} \Gamma(N/2)}$$

$$\times \left[g_\ell (R^{-1} j_{|\ell - 1| - H, n} s) - \delta_0^\ell \right]$$

with

$$g_\ell(u) = 2^{(N-2)/2} \Gamma(N/2) \frac{J_{\ell + (N-2)/2}(u)}{u^{(N-2)/2}}.$$

Substituting the value of c_{NH} from (2.94), we obtain

$$b_{\ell m n}(s) = \frac{2^{H+1} \sqrt{\pi^{(N-2)/2} \Gamma(N/2 + H) \Gamma(H + 1) \sin(\pi H)} R^H}{\Gamma(N/2) J_{|\ell - 1| - H + 1}(j_{|\ell - 1| - H, n}) j_{|\ell - 1| - H, n}^{H+1}}$$

$$\times \left[g_\ell (R^{-1} j_{|\ell - 1| - H, n} s) - \delta_0^\ell \right].$$

To obtain the spectral expansion of the multiparameter fractional Brownian motion, substitute this expression into (2.90). The above spectral expansion has the form

$$X(s, \theta_1, \ldots, \theta_{N-2}, \varphi)$$

$$= \frac{2^{H+1} \sqrt{\pi^{(N-2)/2} \Gamma(N/2 + H) \Gamma(H + 1) \sin(\pi H)} R^H}{\Gamma(N/2)}$$

$$\times \sum_{\ell=0}^{\infty} \sum_{m=1}^{h(S^{N-1}, \ell)} \sum_{n=1}^{\infty} X_{\ell m n} \frac{g_\ell (R^{-1} j_{|\ell - 1| - H, n} s) - \delta_0^\ell}{J_{|\ell - 1| - H + 1}(j_{|\ell - 1| - H, n}) j_{|\ell - 1| - H, n}^{H+1}}$$

$$\times S_\ell^m (\theta_1, \ldots, \theta_{N-2}, \varphi). \tag{2.97}$$

This result was proved by Malyarenko (2008) for the case of $R = 1$.

2.7 Bibliographical Remarks

Random fields appeared for the very first time in applied physical papers about turbulence. We would like to mention papers by Friedmann and Keller (1924),

Kampé de Fériet (1939, 1954), Kampé de Fériet and Pai (1954a,b, 1955), T. von Kármán (1937a,b, 1948a,b), T. von Kármán and L. Howarth (1938), T. von Kármán and C.C. Lin (1951), Obukhov (1941a,b). An excellent introduction to probabilistic methods in turbulence may be found in Monin and Yaglom (2007a,b).

Many classical examples of random fields appear as a generalisation of the Wiener process. For example, Lévy (1945) introduced *Lévy's Brownian motion*, i.e., the centred Gaussian random field $X(\mathbf{t})$, $t \in \mathbb{R}^n$ with covariance function

$$R(\mathbf{s}, \mathbf{t}) = \frac{1}{2}(\|\mathbf{s}\| + \|\mathbf{t}\| - \|\mathbf{s} - \mathbf{t}\|).$$

P. Lévy studied properties of Lévy's Brownian motion in his famous book Lévy (1948). See also the second edition Lévy (1965) that contains a note by M. Loève about recent progress on Lévy's Brownian motion.

Yaglom (1957) introduced the multiparameter fractional Brownian motion, i.e. the centred Gaussian random field $X(\mathbf{t})$, $t \in \mathbb{R}^n$ with covariance function

$$R(\mathbf{s}, \mathbf{t}) = \frac{1}{2}(\|\mathbf{s}\|^{2H} + \|\mathbf{t}\|^{2H} - \|\mathbf{s} - \mathbf{t}\|^{2H}),$$

where $0 < H < 1$.

Chentsov (1956) introduced *Čencov's Brownian motion*, i.e., the centred Gaussian random field $X(\mathbf{t})$, $t \in \mathbb{R}^n$ with covariance function

$$R(\mathbf{s}, \mathbf{t}) = \prod_{j=1}^{n} \min\{s_j, t_j\}.$$

Kamont (1996) introduced the *fractional Brownian sheet*, i.e., the centred Gaussian random field $X(\mathbf{t})$, $t \in \mathbb{R}^n$ with covariance function

$$R(\mathbf{s}, \mathbf{t}) = 2^{-n} \prod_{j=1}^{n} (|s_j|^{2H_j} + |t_j|^{2H_j} - |s_j - t_j|^{2H_j}),$$

where $0 < H_j < 1$ for $1 \le j \le n$.

There exist many topics in the theory of random fields that were not discussed in this book. We would like to mention absolute continuity and singularity of measures corresponding to Gaussian homogeneous and isotropic random fields (Yadrenko, 1983), imaging the structure of white matter fibres in the brain (Fletcher and Joshi, 2007), fractal properties of random fields (Adler, 1981; Xiao, 2004, 2008, 2009), geometry of random fields (Adler and Taylor, 2007), limit theorems for random fields (Bulinski and Shashkin, 2007; Leonenko, 1999; Vanmarcke, 2010), Markov random fields (Kindermann and Snell, 1980; Li, 2009; Rozanov, 1982; Rue and Held, 2005), partial differential equations with random initial conditions (Avram et al., 2010; Kelbert et al., 2005; Leonenko, 1999; Leonenko and Ruiz-Medina, 2006, 2010, 2011), probabilistic potential theory (Khoshnevisan, 2002), spin glasses (De Dominicis and Giardina, 2006), statistical mechanics and statistical fluid mechanics (Malyshev and

Minlos, 1991; Monin and Yaglom, 2007a,b; Preston, 1976), statistics of random fields (Guyon, 1995; Ivanov and Leonenko, 1989; Ramm, 1990; Rosenblatt, 2000), stochastic partial differential equations (Rozanov, 1998), stochastic visibility (Zacks, 1994), white noise theory (Hida and Si, 2004).

Definitions of Subsection 2.1.1 are classical. Bachelier (1900) was the first to consider stochastic processes with continuous time. This paper was rediscovered in the 1970s as the first ever paper in financial mathematics and reprinted in Bachelier (1995) and Bachelier (2006).

Definitions of Subsection 2.1.2 appeared in Malyarenko (2011b). In particular, Example 2.30 is an answer to the question of one of the anonymous referees of the above paper.

Let G be a topological group continuously acting on a topological space T. There exist precise links between the following areas:

- the theory of G-invariant positive-definite functions on T;
- the theory of (wide-sense) G-invariant random fields on T;
- the theory of unitary representations of G.

The above links are described in the books by Diaconis (1988) and Hannan (1965).

For example, J. von Neumann (1929) and Wintner (1929) proved that any unitary representation U of the group $\mathbb{Z}$ in a Hilbert space H with an inner product $(\cdot, \cdot)$ has the form

$$U(m) = \int_{\mathbb{T}} e^{im\varphi} \, dE(\varphi), \quad m \in \mathbb{Z},$$

where E is the *projection-valued measure*, i.e., a mapping from the Borel σ-field $\mathfrak{B}(\mathbb{T})$ of the torus $\mathbb{T}$ to the set of self-adjoint projections on H such that $E(\mathbb{T})$ is the identity operator, and for every x, $y \in H$, the set function $A \mapsto (E(A)x, y)$, $A \in \mathfrak{B}(\mathbb{T})$ is a complex-valued measure on $\mathfrak{B}(\mathbb{T})$. The spectral expansion (2.16) follows immediately. However, the above expansion was discovered later by Cramér (1940) and Kolmogorov (1940a). On the left-hand side of (2.16) we have explicit dependence on the discrete time m. In other words, we are in the *time domain*. On the right, for fixed m we have dependence on φ—we are in the *frequency domain*. The celebrated *Kolmogorov Isomorphism Theorem* by Kolmogorov (1941) states that the correspondence between $X(m)$ and $e^{im\cdot}$ extends by linearity and continuity to an isometric isomorphism between the Hilbert space H spanned by $X(m)$ in the L^2-space of the underlying probability space and the space $L^2(\mathbb{T}, d\mu)$, see Bingham (2012a,b) and references cited there.

On the other hand, the spectral expansion of a continuous positive-definite shift-invariant function on $\mathbb{Z}$, Eq. (2.15), follows immediately from (2.16). It was discovered in the PhD thesis by Wold (1938), second edition Wold (1954).

Stone (1930) and J. von Neumann (1932) found the general form of a unitary representation of the group $\mathbb{R}$. Using these results, Kolmogorov (1940a,b) proved the spectral expansion (2.13). Kolmogorov's results have been further developed by Cramér (1942), Loève (1945), Blanc-Lapierre and Fortet (1946a,b) among others.

See also the expository articles by Kolmogorov (1947) and Doob (1949). Cramér's papers were reprinted in his collected works Cramér (1994). The results by Stone were included in his classical book Stone (1932), reprinted in 1990.

We would like to mention a link with *spectral function theory*, see Nikolski (1986, 2002a,b).

The spectral expansion (2.12) immediately follows from (2.13). It was proved by Bochner (1932) and Khintchine (1934), hence the name *Bochner–Khintchine theorem*. The multidimensional generalisation (Eq. (2.30) for the case of $n = 1$) has been proved by Cramér (1940). Cramér also proved the spectral expansion (2.31) (again for the case of $n = 1$).

In a different line of research, the theory of unitary representations of the group $\mathbb{R}$ has been generalised to the case of an arbitrary locally compact commutative topological group by Naĭmark (1943), Ambrose (1944), and Godement (1944). Kampé de Fériet (1948) used the above works to find the spectral expansion of an invariant random field on such a group. We reproduced his result in our Example 2.19. The second part of the above example is due to Malyarenko (1990).

Bochner (1941) found the general form of a continuous left-invariant positive-definite function on a compact topological group G, Eq. (2.18). It follows immediately that the spectral expansion of a left-homogeneous random field $X(g)$ has the form (2.19) and (2.20). This result is due to Yaglom (1961, Theorem 1).

Bochner (1941) also found the general form of a continuous two-way invariant positive-definite function on a compact topological group G:

$$R(g) = \sum_{\lambda \in \hat{G}} F^{(\lambda)} \operatorname{tr}[\lambda(g)],$$

where $F^{(\lambda)}$ are nonnegative real numbers with

$$\sum_{\lambda \in \hat{G}} \dim \lambda F^{(\lambda)} < \infty.$$

It follows immediately that the spectral expansion of a two-side homogeneous random field $X(g)$ has the form (2.22). This result is due to Yaglom (1961, Theorem 2).

Naĭmark (1960, 1961) and Yaglom (1961) proved that the set of all positive-definite functions on a topological group G of type I with countable base is described by (6.10). Using this result, Yaglom (1961) proved Theorem 2.18. In our sketch of proof, we corrected some typos of the original paper.

Theorem 2.22 has been proved by Yaglom (1961, Theorem 8).

Cartan (1926) introduced an important class of homogeneous spaces. Let T be an analytic Riemannian manifold. T is called a *symmetric space* if for any $t \in T$ there exists an isometry g of the space T such that g^2 is the identity mapping and t is an isolated fixed point of g. The stationary subgroups of all points t in the symmetric space T are isomorphic. Fix a point $t_0 \in T$ and let K be a stationary subgroup of the point t_0. Then $T = G/K$. Gel'fand (1950) proved that K is massive in G and all G-invariant positive-definite functions on T can be represented by

Eq. (2.32). A particular case of this result, when $G = \text{ISO}(n)$, $K = \text{SO}(n)$, and $T = \mathbb{R}^n$, has been proved by Schoenberg (1938), who used the language of positive-definite functions. Our Theorem 2.23 shows that (2.32) follows from the fact that K is massive in G and Theorem 2.18. The expansion (2.33) was proved by Yaglom (1961) in his Theorem 7.

Compact and noncompact two-point homogeneous spaces were classified respectively by Wang (1952) and Tits (1955). These spaces are important for us for the reason noted by Malyarenko (2005). A centred random field $X(t)$ on a two-point homogeneous space is G-invariant iff its covariance function $R(s, t)$ depends only on the distance between the points s and t. Invariant random fields on such spaces were considered by Askey and Bingham (1976) and Molchan (1980).

Two-point homogeneous spaces are considered in the books by Besse (1978), Busemann (1955), Chavel (1972), Gromoll et al. (1975), Rosenfeld (1997), Shchepetilov (2006), Volchkov and Volchkov (2009), and Wolf (1974).

The general theory of spherical harmonics on compact homogeneous spaces was developed by Cartan (1929), Weyl (1934), and Weil (1940). In particular, Cartan (1929) determined zonal spherical functions on the spaces S^n and $\mathbb{C}P^n$. Positive-definite $\text{SO}(n + 1)$-invariant continuous functions on the sphere S^n were described by Schoenberg (1942). Zonal spherical functions on the remaining compact two-point homogeneous spaces were determined by Gangolli (1967).

$\text{SO}(3)$-invariant random fields on the sphere S^2 were considered by Obukhov (1947). Yadrenko (1959) found the spectral expansion of the $\text{SO}(n)$-invariant random field on the sphere S^n. G-invariant random fields on compact two-point homogeneous spaces were considered by Askey and Bingham (1976).

Zonal spherical functions on the spaces $\mathbb{R}^n$ and $\mathbb{R}\Lambda^n$ and all spherical functions on $\mathbb{R}^2$ and $\mathbb{R}\Lambda^2$ were obtained by Krein (1949). The case of the space $\mathbb{R}\Lambda^2$ was considered earlier by Gel'fand and Naĭmark (1947). See also Vilenkin (1956, 1963). All spherical functions on the space $\mathbb{R}^n$ were found independently by Vilenkin (1957) and Orihara (1961). The case of the space $\mathbb{R}\Lambda^n$ was considered by Vilenkin (1958). Zonal spherical functions on all noncompact two-point homogeneous spaces with semi-simple group G were determined by Harish-Chandra (1958). All spherical functions for the above spaces were found by Helgason (1974).

Kostant (1975) found the set of all class 1 representations of all semi-simple groups of motions of noncompact two-point homogeneous spaces. In the case of the spaces $\mathbb{R}\Lambda^n$ and $\mathbb{C}\Lambda^n$ more simple proofs were proposed by Takahashi (1963) and Faraut and Harzallah (1972).

The spectral expansion (2.45) for the case of $T = \mathbb{R}^n$ was obtained independently by Yaglom (1961) and by M.Ĭ. Yadrenko in his unpublished PhD thesis. Later on, the particular cases of the above expansion were proved by Ogura (1968) and McNeil (1972). Yaglom (1961) obtained the above expansion for the case of $T = \mathbb{R}\Lambda^n$.

The spectral expansion of G-invariant random fields on noncompact two-point homogeneous spaces, Eq. (2.45) was obtained by Malyarenko (2005).

Let G be the group of all isometries of the real infinite-dimensional separable Hilbert space H. The description of all positive-definite G-invariant continuous functions on H (Eq. (2.46)) is due to Schoenberg (1938). The method of proof was proposed by Malyarenko (1990).

Malyarenko (1994) considered the following problem. Let G be a locally compact topological group of type I, let K be a compact subgroup, let $\mathrm{Aut}(G)$ be the group of continuous automorphisms of the group G equipped with the compact-open topology, and let $\mathrm{Aut}_K(G)$ be the stabiliser of the subgroup K in the group $\mathrm{Aut}(G)$:

$$\mathrm{Aut}_K(G) = \{ u \in \mathrm{Aut}(G) \colon uK \subseteq K \}.$$

The action of any automorphism $u \in \mathrm{Aut}_K(G)$ to G induces an action on the homogeneous space $T = G/K$ by the following formula:

$$ut = \pi u \pi^{-1} t, \quad t \in T,$$

where π is the projection of G to T. The left-hand side of the above formula does not depend on the choice of point in the set $\pi^{-1}t$.

Let M be a compact subgroup of the group $\mathrm{Aut}_K(G)$, and let U be an irreducible unitary representation of the group M in a finite-dimensional complex Hilbert space H. Malyarenko (1994) found the spectral expansion of a centred left G-invariant H-valued random field $\mathbf{X}(t)$ that satisfies the following condition:

$$\mathsf{E}\big[\mathbf{X}(ms) \otimes \mathbf{X}^*(mt)\big] = U(m)\mathsf{E}\big[\mathbf{X}(s) \otimes \mathbf{X}^*(t)\big]U^{-1}(m).$$

All examples of the preceding subsections may be considered as particular cases of the above expansion.

Random fields on vector bundles first appeared in an implicit form in the physical literature. In particular, Kamionkowski et al. (1997) expanded the polarisation field of the Cosmic Microwave Background on the whole sky by tensor spherical harmonics, while Zaldarriaga and Seljak (1997) did the same using spin-weighted spherical harmonics. Geller and Marinucci (2010) stressed that the above polarisation field is an invariant random field in a line bundle. Theorems 2.26, 2.27, and 2.28 were proved by Malyarenko (2011b).

Theorem 2.29 is new. The result of Example 2.30 has been proved by Yadrenko (1963).

Yaglom (1957) proved that the function (2.63) is positive-definite. In other words, he proved the existence of the multiparameter fractional Brownian motion, which we start to study in Example 2.31. Later on, other proofs were proposed by Gangolli (1967), Mandelbrot and van Ness (1968), Wong and Zakai (1974), and Ossiander and Waymire (1989), among others. The covariance functions (2.67) were calculated by Malyarenko (2011a).

Theorem 2.32 and Example 2.33 are new.

Theorem 2.34 was proved by Malyarenko and Olenko (1992). Particular cases of their result were obtained by Itô (1956), and Yaglom (1957). Example 2.35 was calculated by Malyarenko (1985a,b).

Volterra isotropic random fields were introduced by Malyarenko (2011a). He proved that the multiparameter fractional Brownian motion is a Volterra field. For the case of ordinary fractional Brownian motion defined on the interval $[0, \infty)$ this was proved by Decreusefond and Üstünel (1999). The spectral expansion (2.97) of the multiparameter fractional Brownian motion was proved by Malyarenko (2008). For the case of $N = 1$ this expansion was proved by Dzhaparidze and van Zanten (2004). Various spectral expansions of isotropic Gaussian random field with homogeneous increments were obtained by Dzhaparidze et al. (2006).

Random series of functions were studied by Kahane (1985) and Vakhania et al. (1987).

References

R.J. Adler. *The geometry of random fields*. Wiley Ser. Probab. Math. Stat. Wiley, Chichester, 1981.

R.J. Adler and J.E. Taylor. *Random fields and geometry*. Springer Monogr. Math. Springer, New York, 2007.

W. Ambrose. Spectral resolution of groups of unitary operators. *Duke Math. J.*, 11:589–595, 1944.

R. Askey and N.H. Bingham. Gaussian processes on compact symmetric spaces. *Z. Wahrscheinlichkeitstheor. Verw. Geb.*, 37(2):127–143, 1976.

F. Avram, N. Leonenko, and L. Sakhno. On a Szegö type limit theorem, the Hölder–Young–Brascamp–Lieb inequality, and the asymptotic theory of integrals and quadratic forms of stationary fields. *ESAIM Probab. Stat.*, 14:210–255, 2010. doi:10.1051/ps:2008031.

L. Bachelier. Théorie de la spéculation. *Ann. Sci. Éc. Norm. Super.*, 3(17):21–86, 1900.

L. Bachelier. *Théorie de la spéculation*. Grands Class. Gauthier–Villars. Éditions Jacques Gabay, Sceaux, 1995.

L. Bachelier. *Louis Bachelier's theory of speculation: the origins of modern finance*. Princeton University Press, Princeton, 2006. Translated and with an Introduction by Mark Davis & Alison Etheridge. With a foreword by Paul A. Samuelson.

Yu.M. Berezansky. *Expansions in eigenfunctions of selfadjoint operators*, volume 17 of *Transl. Math. Monogr.* Am. Math. Soc., Providence, 1968.

A.L. Besse. *Manifolds all of whose geodesics are closed*, volume 93 of *Ergeb. Math. Grenzgeb.* Springer, Berlin, 1978.

N.H. Bingham. Szegö's theorem and its probabilistic descendants. *Probab. Surv.*, 9:287–324, 2012a. doi:10.1214/11-PS178.

N.H. Bingham. Multivariate prediction and matrix Szegö theory. *Probab. Surv.*, 9:325–339, 2012b. doi:10.1214/12-PS200.

A. Blanc-Lapierre and R. Fortet. Sur la décomposition spectrale des fonctions aléatoires stationaires d'ordre deux. *C. R. Acad. Sci. Paris*, 222(9):467–468, 1946a.

A. Blanc-Lapierre and R. Fortet. Résultats sur la décomposition spectrale des fonctions aléatoires stationnaires d'ordre 2. *C. R. Acad. Sci. Paris*, 222(13):713–714, 1946b.

S. Bochner. *Vorlesungen über Fouriersche Integrale*. Akad. Verlag, Leipzig, 1932.

S. Bochner. Hilbert distances and positively definite functions. *Ann. of Math. (2)*, 42:647–656, 1941.

N. Bourbaki. *Integration. II. Chapters 7–9*. Elem. Math. (Berlin). Springer, Berlin, 2004.

A. Bulinski and A. Shashkin. *Limit theorems for associated random fields and related systems*, volume 10 of *Adv. Ser. Stat. Sci. Appl. Probab.* World Scientific, Hackensack, 2007.

H. Busemann. *The geometry of geodesics*. Academic Press, New York, 1955.

R. Camporesi. The Helgason Fourier transform for homogeneous vector bundles over compact Riemannian symmetric spaces—the local theory. *J. Funct. Anal.*, 220:97–117, 2005.

E. Cartan. Sur une classe remarquable d'espaces de Riemann. *Bull. Soc. Math. Fr.*, 54:214–264, 1926.

E. Cartan. Sur la détermination d'un système orthogonal complet dans un espace de Riemann symétrique clos. *Rend. Circ. Mat. Palermo*, 53:217–252, 1929.

I. Chavel. *Riemannian symmetric spaces of rank one*, volume 5 of *Lect. Notes Pure Appl. Math.* Dekker, New York, 1972.

N.N. Chentsov. Wiener random fields depending on several parameters. *Dokl. Akad. Nauk SSSR (N.S.)*, 106:607–609, 1956. In Russian.

H. Cramér. On the theory of stationary random processes. *Ann. of Math. (2)*, 41:215–230, 1940. URL http://www.jstor.org/stable/1968827.

H. Cramér. On harmonic analysis in certain functional spaces. *Ark. Mat. Astron. Fys.*, 28B(12):1–7, 1942. URL http://www.jstor.org/stable/1968827.

H. Cramér. *Collected works. Vol. I, II*. Springer, Berlin, 1994. Edited and with a preface by Anders Martin-Löf.

C. De Dominicis and I. Giardina. *Random fields and spin glasses. A field theory approach*. Cambridge University Press, New York, 2006.

L. Decreusefond and A.S. Üstünel. Stochastic analysis of the fractional Brownian motion. *Potential Anal.*, 10(2):177–214, 1999.

P. Diaconis. *Group representations in probability and statistics*, volume 11 of *IMS Lect. Notes Monogr. Ser.* Institute of Mathematical Statistics, Hayward, 1988.

J.L. Doob. Time series and harmonic analysis. In J. Neyman, editor, *Proceedings of the Berkeley Symposium on Mathematical Statistics and Probability, 1945, 1946*, pages 303–343. University of California Press, Berkeley, 1949.

J.L. Doob. *Stochastic processes*. Wiley Classics Libr. Wiley, New York 1990. Reprint of the 1953 original.

K. Dzhaparidze and H. van Zanten. A series expansion of fractional Brownian motion. *Probab. Theory Relat. Fields*, 130(1):39–55, 2004.

K. Dzhaparidze, H. van Zanten, and P. Zareba. Representations of isotropic Gaussian random fields with homogeneous increments. *J. Appl. Math. Stoch. Anal.*, 2006: 25 pp., 2006. Art ID 72731.

J. Faraut and K. Harzallah. Fonctions sphériques de type positif sur les espaces hyperboliques. *C. R. Acad. Sci. Paris Sér. A–B*, 274:A1396–A1398, 1972.

M. Flensted-Jensen and T.H. Koornwinder. Positive definite spherical functions on a noncompact, rank one symmetric space. In P. Eymard, J. Faraut, G. Schiffmann and R. Takahashi, editors, *Analyse Harmonique sur les Groupes de Lie, II*, volume 739 of *Lect. Notes Math.*, pages 249–282. Springer, Berlin, 1979.

P.T. Fletcher and S. Joshi. Riemannian geometry for the statistical analysis of diffusion tensor data. *Signal Process.*, 87:250–262, 2007.

A.A. Friedmann and L.P. Keller. Differentialgleichungen fur Turbulente Bewegung einer Kompressiblen Flussigkeit. In *Proceedings of the First International Congress for Applied Mechanics*, Delft, pages 395–405, 1924.

R. Gangolli. Positive definite kernels on homogeneous spaces and certain stochastic processes related to Lévy's Brownian motion of several parameters. *Ann. Inst. H. Poincaré Sect. B (N.S.)*, 3:121–226, 1967.

I.M. Gel'fand. Spherical functions in symmetric Riemann spaces. *Dokl. Akad. Nauk SSSR (N.S.)*, 70:5–8, 1950.

I.M. Gel'fand and M.A. Naǐmark. Unitary representations of the Lorentz group. *Izv. Akad. Nauk SSSR, Ser. Mat.*, 11:411–504, 1947. In Russian.

D. Geller and D. Marinucci. Spin wavelets on the sphere. *J. Fourier Anal. Appl.*, 16:840–884, 2010.

R. Godement. Sur une généralisation d'un théorème de Stone. *C. R. Acad. Sci. Paris*, 218:901–903, 1944.

D. Gromoll, D. Klingenberg, and W. Meyer. *Riemannsche Geometrie im Großen*, volume 55 of *Lect. Notes Math.* Springer, Berlin, second edition, 1975.

K.I. Gross and R.A. Kunze. Bessel functions and representation theory. I. *J. Funct. Anal.*, 22(2):73–105, 1976.

X. Guyon. *Random fields on a network. Modeling, statistics, and applications*. Probab. Appl. (New York). Springer, New York, 1995.

E.J. Hannan. *Group representations and applied probability*, volume 3 of *Methuen's Suppl. Rev. Ser. Appl. Probab.* Methuen, London, 1965.

Harish-Chandra. Spherical functions on a semisimple Lie group. I. *Am. J. Math.*, 80(2):241–310, 1958. URL http://www.jstor.org/stable/2372786.

S. Helgason. Eigenspaces of the Laplacian; integral representations and irreducibility. *J. Funct. Anal.*, 17:328–353, 1974.

T. Hida and S. Si. *An innovation approach to random fields. Application of white noise theory.* World Scientific, River Edge, 2004.

K. Itô. Isotropic random current. In *Proceedings of the Third Berkeley Symposium on Mathematical Statistics and Probability*, volume II, pages 125–132. University of California Press, Berkeley, 1956.

A.V. Ivanov and N.N. Leonenko. *Statistical analysis of random fields*, volume 28 of *Math. Appl. (Sov. Ser.)*. Kluwer Academic, Dordrecht, 1989.

J.-P. Kahane. *Some random series of functions*, volume 5 of *Cambridge Stud. Adv. Math.* Cambridge University Press, Cambridge, second edition, 1985.

M. Kamionkowski, A. Kosowsky, and A. Stebbins. Statistics of cosmic microwave background polarization. *Phys. Rev. D*, 55(12):7368–7388, 1997.

A. Kamont. On the fractional anisotropic Wiener field. *Probab. Math. Stat.*, 16(1):85–98, 1996.

J. Kampé de Fériet. Les fonctions aléatories stationnaries et la théorie statistique de la turbulence homogène. *Ann. Soc. Sci. Bruxelles Sér. I*, 59:145–210, 1939.

J. Kampé de Fériet. Analyse harmonique des fonctions aléatoires stationnaires d'ordre 2 définies sur un groupe abélien localement compact. *C. R. Acad. Sci. Paris*, 226:868–870, 1948.

J. Kampé de Fériet. Introduction to the statistical theory of turbulence. III. *J. Soc. Ind. Appl. Math.*, 2:244–271, 1954.

J. Kampé de Fériet and S.I. Pai. Introduction to the statistical theory of turbulence. I. *J. Soc. Ind. Appl. Math.*, 2:1–9, 1954a.

J. Kampé de Fériet and S.I. Pai. Introduction to the statistical theory of turbulence. II. *J. Soc. Ind. Appl. Math.*, 2:143–174, 1954b.

J. Kampé de Fériet and S.I. Pai. Introduction to the statistical theory of turbulence. IV. *J. Soc. Ind. Appl. Math.*, 3:90–117, 1955.

K. Karhunen. Über lineare Methoden in der Wahrscheinlichkeitsrechnung. *Ann. Acad. Sci. Fennicae. Ser. A. I. Math.-Phys.*, 1(37):3–79, 1947.

T. von Kármán. On the statistical theory of turbulence. *Proc. Natl. Acad. Sci. USA*, 23:98–105, 1937a.

T. von Kármán. The fundamentals of the statistical theory of turbulence. *J. Aeronaut. Sci.*, 4:131–138, 1937b.

T. von Kármán. Sur la théorie statistique de la turbulence. *C. R. Acad. Sci. Paris*, 226:2108–2111, 1948a.

T. von Kármán. Progress in the statistical theory of turbulence. *Proc. Natl. Acad. Sci. USA*, 34:530–539, 1948b.

T. von Kármán and C.C. Lin. On the statistical theory of isotropic turbulence. *Adv. Appl. Mech.*, 2:1–19, 1951.

T. von Kármán and L. Howarth. On the statistical theory of isotropic turbulence. *Proc. R. Soc.*, 164:192–215, 1938.

M.Ya. Kelbert, N. Leonenko, and M.D. Ruiz-Medina. Fractional random fields associated with stochastic fractional heat equations. *Adv. Appl. Probab.*, 37:108–133, 2005.

A. Khintchine. Korrelationstheorie der stationären stochastischen Prozesse. *Math. Ann.*, 109(1):604–615, 1934.

D. Khoshnevisan. *Multiparameter processes. An introduction to random fields.* Springer Monogr. Math. Springer, New York, 2002.

R. Kindermann and J.L. Snell. *Markov random fields and their applications*, volume 1 of *Contemp. Math.* Am. Math. Soc., Providence, 1980.

A.N. Kolmogorov. Kurven im Hilbertschen Raum, die gegenüber einer einparametrigen Gruppe von Bewegungen invariant sind. *C. R. (Dokl.) Acad. Sci. URSS (N.S.)*, 26:6–9, 1940a.

A.N. Kolmogorov. Wienersche Spiralen und einige andere interessante Kurven im Hilbertschen Raum. *C. R. (Dokl.) Acad. Sci. URSS (N.S.)*, 26:115–118, 1940b.

A.N. Kolmogorov. Stationary sequences in Hilbert space. *Bull. Moskov. Gos. Univ. Matematika*, 2:1–41, 1941.

A.N. Kolmogorov. Statistical theory of oscillation with a continuous spectrum. In *Jubilee collection dedicated to the thirties anniversary of the Great October Socialist Revolution*, volume 1, pages 242–252. Akad. Nauk SSSR, Moscow, 1947. In Russian.

B. Kostant. On the existence and irreducibility of certain series of representations. In I. Gel'fand, editor. *Lie groups and their representations*, pages 231–329. Halsted, New York, 1975.

M. Krein. Hermitian positive kernels on homogeneous spaces. *Ukr. Mat. Zh.*, 1(4):64–98, 1949. In Russian.

N. Leonenko. *Limit theorems for random fields with singular spectrum*, volume 465 of *Math. Appl.* Kluwer Academic, Dordrecht, 1999.

N. Leonenko and M.D. Ruiz-Medina. Scaling laws for the multidimensional Burgers equation with quadratic external potential. *J. Stat. Phys.*, 124(1):191–205, 2006. doi:10.1007/s10955-006-9136-5.

N. Leonenko and M.D. Ruiz-Medina. Spatial scalings for randomly initialized heat and Burgers equations with quadratic potentials. *Stoch. Anal. Appl.*, 28(2):303–321, 2010. doi:10.1080/07362990903546561.

N. Leonenko and M.D. Ruiz-Medina. Random fields arising in chaotic systems: Burgers equation and fractal pseudodifferential systems. In E. Porcu, J. Montero, and M. Schlather, editors, *Advances and challenges in space-time modelling of natural events*, Chapter 23, pages 165–220. Springer, Berlin, 2011.

P. Lévy. Sur le mouvement brownien dépendant de plusieurs paramètres. *C. R. Acad. Sci. Paris*, 220:420–422, 1945.

P. Lévy. *Processus stochastiques et mouvement brownien. Suivi d'une note de M. Loève*. Gauthier–Villars, Paris, 1948.

P. Lévy. *Processus stochastiques et mouvement brownien*. Gauthier–Villars, Paris, deuxième édition revue et augmentée edition, 1965. Suivi d'une note de M. Loève.

S.Z. Li. *Markov random field modeling in image analysis*. Springer, Berlin, third edition, 2009.

M. Loève. Analyse harmonique générale d'une fonction aléatoire. *C. R. Acad. Sci. Paris*, 220:380–382, 1945.

A. Malyarenko. Spectral decomposition of multidimensional homogeneous and isotropic random fields. *Dokl. Akad. Nauk Ukrain. SSR Ser. A*, 7:20–22, 1985a. In Russian.

A. Malyarenko. Spectral decomposition of multidimensional homogeneous random fields that are isotropic with respect to some of the variables. *Teor. Veroyatnost. i Mat. Statist.*, 32:66–72, 1985b. In Russian.

A. Malyarenko. Multidimensional covariant random fields on commutative topological groups. *Theory Probab. Math. Stat.*, 41:49–55, 1990.

A. Malyarenko. Spectral decomposition of homogeneous and isotropic random flows on homogeneous space. In *Asymptotic analysis of random evolutions*. Akad. Nauk Ukrainy, Inst. Mat., Kiev, 1994.

A. Malyarenko. Abelian and Tauberian theorems for random fields on two-point homogeneous spaces. *Theory Probab. Math. Stat.*, 69:115–127, 2005.

A. Malyarenko. An optimal series expansion of the multiparameter fractional Brownian motion. *J. Theor. Probab.*, 21(2):459–475, 2008.

A. Malyarenko. A family of series representations of the multiparameter fractional Brownian motion. In R. Dalang, M. Dozzi and F. Russo, editors, *Seminar on stochastic analysis, random fields and applications VI*, Centro Stefano Franscini, Ascona, May 2008, volume 63 of *Prog. Probab.*, pages 209–227. Birkhäuser, Basel, 2011a.

A. Malyarenko. Invariant random fields in vector bundles and application to cosmology. *Ann. Inst. H. Poincaré Probab. Stat.*, 47(4):1068–1095, 2011b. doi:10.1214/10-AIHP409. URL http://projecteuclid.org/euclid.aihp/1317906502.

A. Malyarenko and A.Ya. Olenko. Multidimensional covariant random fields on commutative locally compact groups. *Ukr. Math. J.*, 44(11):1384–1389, 1992.

V.A. Malyshev and R.A. Minlos. *Gibbs random fields. Cluster expansions*, volume 44 of *Math. Appl. (Sov. Ser.)*. Kluwer Academic, Dordrecht, 1991.

B. Mandelbrot and J.W. van Ness. Fractional Brownian motions, fractional noises and applications. *SIAM Rev.*, 10(4):422–437, 1968. URL http://www.jstor.org/stable/2027184.

D.R. McNeil. Representations for homogeneous processes. *Z. Wahrscheinlichkeitstheor. Verw. Geb.*, 22:333–339, 1972.

G.M. Molchan. Homogeneous random fields on symmetric spaces of rank one. *Theory Probab. Math. Stat.*, 21:143–168, 1980.

A.S. Monin and A.M. Yaglom. *Statistical fluid mechanics: mechanics of turbulence*, volume I. Dover, New York, 2007a.

A.S. Monin and A.M. Yaglom. *Statistical fluid mechanics: mechanics of turbulence*, volume II. Dover, New York, 2007b.

M.A. Naĭmark. Positive definite operator functions on a commutative group. *Bull. Acad. Sci. URSS Sér. Math. [Izvestia Akad. Nauk SSSR]*, 7:237–244, 1943.

M.A. Naĭmark. Factor-representations of a locally compact group. *Sov. Math. Dokl.*, 1:1064–1066, 1960.

M.A. Naĭmark. Decomposition into factor representations of unitary representations of locally compact groups. *Sib. Mat. Zh.*, 2:89–99, 1961.

M.A. Naĭmark and A.I. Štern. *Theory of group representations*, volume 246 of *Grundlehren Mat. Wiss.* Springer, New York, 1982. Translated from the Russian by Elizabeth Hewitt. Translation edited by Edwin Hewitt.

J. von Neumann. Allgemeine Eigenverttheorie Hermitischer Funktionaloperatoren. *Math. Ann.*, 102:49–131, 1929.

J. von Neumann. Über einen Satz von Herrn M.H. Stone. *Ann. Math.*, 33:567–573, 1932.

N.K. Nikolski. *Treatise on the shift operator. Spectral function theory. With an appendix by S.V. Hruščev and V.V. Peller*, volume 273 of *Grundlehren Math. Wiss.* Springer, Berlin, 1986.

N.K. Nikolski. *Operators, functions, and systems: an easy reading. Vol. 1. Hardy, Hankel, and Toeplitz*, volume 92 of *Math. Surveys Monogr.* Am. Math. Soc., Providence, 2002a. Translated from the French by Andreas Hartmann.

N.K. Nikolski. *Operators, functions, and systems: an easy reading. Vol. 2. Model operators and systems*, volume 92 of *Math. Surveys Monogr.* Am. Math. Soc., Providence, 2002b. Translated from the French by Andreas Hartmann and revised by the author.

A.M. Obukhov. On the energy distribution in the spectrum of a turbulent flow. *C. R. (Dokl.) Acad. Sci. URSS (N.S.)*, 32(1):19–21, 1941a.

A.M. Obukhov. Über die Energieverteilung im Spektrum des Turbulenzstromes. *Bull. Acad. Sci. URSS. Sér. Géograph. Géophys. [Izvestia Akad. Nauk SSSR]*, 5(4/5):1453–1466, 1941b.

A.M. Obukhov. Statistically homogeneous random fields on a sphere. *Usp. Mat. Nauk*, 2(2):196–198, 1947. In Russian.

H. Ogura. Spectral representation of vector random field. *J. Phys. Soc. Jpn.*, 24:1370–1380, 1968.

A. Orihara. Bessel functions and the Euclidean motion group. *Tôhoku Math. J. (2)*, 13:66–74, 1961.

M. Ossiander and E.C. Waymire. Certain positive-definite kernels. *Proc. Am. Math. Soc.*, 107(2):487–492, 1989. URL http://www.jstor.org/stable/2047839.

C. Preston. *Random fields*, volume 534 of *Lect. Notes Math.* Springer, Berlin, 1976.

A.P. Prudnikov, Yu.A. Brychkov, and O.I. Marichev. *Integrals and series. Vol. 2. Special functions.* Gordon & Breach, New York, second edition, 1988.

A.P. Prudnikov, Yu.A. Brychkov, and O.I. Marichev. *Integrals and series. Vol. 3. More special functions.* Gordon & Breach, New York, 1990.

A.G. Ramm. *Random fields estimation theory*, volume 48 of *Pitman Monogr. Surv. Pure Appl. Math.* Longman Scientific & Technical, Harlow, 1990.

M. Rosenblatt. *Gaussian and non-Gaussian linear time series and random fields.* Springer Ser. Stat. Springer, New York, 2000.

B. Rosenfeld. *Geometry of Lie groups*, volume 393 of *Math. Appl.* Kluwer Academic, Dordrecht, 1997.

Yu.A. Rozanov. *Markov random fields*. Appl. Math. Springer, New York, 1982.

Yu.A. Rozanov. *Random fields and stochastic partial differential equations*, volume 438 of *Math. Appl.* Kluwer Academic, Dordrecht, 1998.

H. Rue and L. Held. *Gaussian Markov random fields. Theory and applications*, volume 104 of *Monogr. Stat. Appl. Probab.* Chapman & Hall/CRC, London, 2005.

I.J. Schoenberg. Metric spaces and completely monotone functions. *Ann. of Math. (2)*, 39(4):811–841, 1938. URL http://www.jstor.org/stable/1968466.

I.J. Schoenberg. Positive definite functions on spheres. *Duke Math. J.*, 9:96–108, 1942.

A.V. Shchepetilov. *Calculus and mechanics on two-point homogeneous Riemannian spaces*, volume 707 of *Lect. Notes Phys.* Springer, Berlin, 2006.

M.H. Stone. Linear transformations in Hilbert space III. Operational methods and group theory. *Proc. Natl. Acad. Sci. USA*, 16:172–175, 1930.

M.H. Stone. *Linear transformations in Hilbert space*, volume 15 of *Am. Math. Soc. Colloq. Publ.* Am. Math. Soc., Providence, 1932. Reprint of the 1932 original, 1990.

R. Takahashi. Sur les représentations unitaires des groupes de Lorentz généralisés. *Bull. Soc. Math. Fr.*, 91:289–433, 1963.

J. Tits. Sur certaines classes d'espaces homogènes de groupes de Lie. *Acad. R. Belg. Cl. Sci. Mém. Coll.*, 29(8):1–268, 1955.

Y. Umemura. Measures on infinite dimensional vector spaces. *Publ. Res. Inst. Math. Sci., Ser. A*, 1:1–47, 1965.

N.N. Vakhania, V.I. Tarieladze, and S.A. Chobanyan. *Probability distributions on Banach spaces*, volume 14 of *Math. Appl. (Sov. Ser.).* Reidel, Dordrecht, 1987.

E. Vanmarcke. *Random fields. Analysis and synthesis*. World Scientific, Hackensack, Revised and expanded new edition, 2010.

N.Ya. Vilenkin. Bessel functions and representations of the group of Euclidean motions. *Usp. Mat. Nauk (N.S.)*, 11(3(69)):69–112, 1956. In Russian.

N.Ya. Vilenkin. The matrix elements of irreducible unitary representations of the group of real orthogonal matrices and group of euclidean $(n − 1)$-dimensional space motions. *Dokl. Akad. Nauk SSSR (N.S.)*, 113:16–19, 1957. In Russian.

N.Ya. Vilenkin. The matrix elements of irreducible unitary representations of a group of Lobachevsky space motions and the generalized Fock–Mehler transformations. *Dokl. Akad. Nauk SSSR (N.S.)*, 118:219–222, 1958. In Russian.

N.Ya. Vilenkin. Special functions associated with class 1 representations of the motion groups of spaces of constant curvature. *Trudy Moskov. Mat. Obšč.*, 12:185–257, 1963. In Russian.

N.Ya. Vilenkin. *Special functions and the theory of group representations*, volume 22 of *Transl. Math. Monogr.* Am. Math. Soc., Providence, 1968. Translated from the Russian by V.N. Singh.

V.V. Volchkov and V.V. Volchkov. *Harmonic analysis of mean periodic functions on symmetric spaces and the Heisenberg group*. Springer Monogr. Math. Springer, London, 2009.

H.-C. Wang. Two-point homogeneous spaces. *Ann. of Math. (2)*, 55(1):177–191, 1952. URL http://www.jstor.org/stable/1969427.

G.N. Watson. *A treatise on the theory of Bessel functions*. Cambridge University Press, Cambridge, 1944. Reprinted in 1995.

A. Weil. *L'intégration dans les groupes topologiques et ses applications*. Number 869 in *Actual. Sci. Ind.* Hermann, Paris, 1940.

H. Weyl. Harmonics on homogeneous manifolds. *Ann. of Math. (2)*, 35(3):486–499, 1934. URL http://www.jstor.org/stable/1968746.

A. Wintner. Zur Theorie der beschränkten Bilinearformen. *Math. Z.*, 30:228–282, 1929.

H. Wold. *A study in the analysis of stationary time series*. PhD thesis, Uppsala University, 1938.

H. Wold. *A study in the analysis of stationary time series*. Almqvist & Wiksell, Stockholm, 2nd edition, 1954. With an appendix by Peter Whittle.

J.A. Wolf. *Spaces of constant curvature*. Publish or Perish, Boston, third edition, 1974.

E. Wong and M. Zakai. Martingales and stochastic integrals for processes with a multidimensional parameter. *Z. Wahrscheinlichkeitstheor. Verw. Geb.*, 29:109–122, 1974.

Y. Xiao. Random fractals and Markov processes. In *Fractal geometry and applications: a jubilee of Benoît Mandelbrot*, volume 72, part 2 of *Proc. Sympos. Pure Math.*, pages 261–338. Am. Math. Soc., Providence, 2004.

Y. Xiao. Strong local nondeterminism and sample path properties of Gaussian random fields. In *Asymptotic theory in probability and statistics with applications*, volume 2 of *Adv. Lect. Math. (ALM)*, pages 136–176. Int. Press, Somerville, 2008.

Y. Xiao. Sample path properties of anisotropic Gaussian random fields. In D. Khoshnevisan and F. Rassoul-Agha, editors. *A minicourse on stochastic partial differential equations*, volume 1962 of *Lect. Notes Math.*, pages 145–212. Springer, Berlin, 2009.

M.Ï. Yadrenko. Isotropic Gauss random fields of the Markov type on a sphere. *Dopovidi Akad. Nauk Ukraïn. RSR*, 3:231–236, 1959. In Ukrainian.

M.Ï. Yadrenko. Isotropic random fields of Markov type in Euclidean space. *Dopovidi Akad. Nauk Ukraïn. RSR*, 1963(3):304–306, 1963. In Ukrainian.

M.Ï. Yadrenko. *Spectral theory of random fields*. Translat. Ser. Math. Eng., Optimization Software, Publications Division, New York, 1983.

A.M. Yaglom. Certain types of random fields in n-dimensional space similar to stationary stochastic processes. *Teor. Veroâtn. Ee Primen.*, 2:292–338, 1957. In Russian.

A.M. Yaglom. Second-order homogeneous random fields. In *Proc. 4th Berkeley Sympos. Math. Statist. and Probab.*, volume II, pages 593–622. University of California Press, Berkeley, 1961.

S. Zacks. *Stochastic visibility in random fields*, volume 95 of *Lect. Notes Stat.* Springer, New York, 1994.

M. Zaldarriaga and U. Seljak. An all-sky analysis of polarisation in the microwave background. *Phys. Rev. D*, 55(4):1830–1840, 1997.

Chapter 3
L² Theory of Invariant Random Fields

Abstract We consider a random field $\mathbf{X}(t)$ as a function taking values in the Hilbert space of all centred finite-dimensional random vectors with finite expectation of the square of the norm. Examples of inversion formulae, which reconstruct the random measures Z when X is known, are given in Sect. 3.1. Linear differential equations whose right-hand side is a random field, are considered in Sect. 3.2. In Sect. 3.3 we introduce rigged Hilbert spaces of ordinary, smooth, and generalised functions, linear functionals, and spectral characteristics connected to the random field $\mathbf{X}(t)$. Using Paley–Wiener theorems, we study how smooth our "smooth functions" are. Bibliographical remarks conclude.

3.1 Inversion Formulae

In Chap. 2 we obtained numerous spectral expansions of random fields X in terms of a set of random measures Z. Assume that X is known. We would like to calculate Z. In other words, we would like to find an *inversion formula*.

Chapter 2 already contains some inversion formulae. For example, under the conditions of Theorems 2.26 and 2.27, the inversion formula has the form (2.52). Under the conditions of Theorem 2.29, the inversion formula has the form (2.57), while under the conditions of Theorem 2.32, it has the form (2.71). Consider more examples.

Example 3.1 Let $X(g)$ be a scalar left-homogeneous random field on a *compact* topological group G. Its spectral expansion has the form (2.19). In coordinate form, the above formula is

$$X(g) = \sum_{\lambda \in \hat{G}} \sum_{j,k=1}^{\dim \lambda} \lambda_{jk}(g) Z_{kj}^{\lambda}.$$

For any $\mu \in \hat{G}$, and for any integers ℓ and m with $1 \le \ell, m \le \dim \mu$, multiply both sides by $\overline{\mu_{\ell m}(g)}$, integrate with respect to the Haar measure dg, and use the Peter–Weyl theorem. We obtain the following inversion formula:

$$Z_{kj}^{\lambda} = \frac{1}{\dim \lambda} \int_G X(g) \overline{\lambda_{jk}(g)} \, dg.$$

A. Malyarenko, *Invariant Random Fields on Spaces with a Group Action*,
Probability and Its Applications, DOI 10.1007/978-3-642-33406-1_3,
© Springer-Verlag Berlin Heidelberg 2013

It is trivial to check that the same inversion formula holds for right-homogeneous and two-way homogeneous fields.

Example 3.2 Let $X(g)$ be a centred homogeneous random field on a commutative locally compact group G with countable base. By (2.11),

$$X(g) = \int_{\hat{G}} \hat{g}(g) \, dZ(\hat{g}).$$

Let dg be a fixed Haar measure on G, and let $d\hat{g}$ be the corresponding Plancherel measure on $\hat{G}$. Multiply both sides of the last display by a function $f(g) \in L^2(G) \cap L^1(G)$ and integrate with respect to dg. We obtain

$$\int_G X(g)f(g) \, dg = \int_G \int_{\hat{G}} \hat{g}(g) \, dZ(\hat{g}) f(g) \, dg.$$

The product $\hat{g}(g)f(g)$ of a bounded function $\hat{g}(g)$ and an integrable function $f(g)$ is integrable over the product space $\hat{G} \times G$ with respect to product measure $dZ(\hat{g}) \, dg$. By Fubini's theorem, we may change the order of integration:

$$\int_G X(g)f(g) \, dg = \int_{\hat{G}} \int_G \hat{g}(g) f(g) \, dg \, dZ(\hat{g}).$$

Let A be a Borel subset of $\hat{G}$ of positive Plancherel measure, and let $f(g)$ be the inverse Fourier transform of the indicator of A:

$$f(g) = \int_A \overline{\hat{g}(g)} \, d\hat{g}.$$

We obtain the following inversion formula:

$$Z(A) = \int_G X(g) \int_A \overline{\hat{g}(g)} \, d\hat{g} \, dg.$$

In particular, let G be the space domain $\mathbb{R}^n$ and let $d\mathbf{x}$ be the Lebesgue measure. The corresponding Plancherel measure on the wavenumber domain is $(2\pi)^{-n} \, d\mathbf{p}$. The inversion formula has the form

$$Z(A) = (2\pi)^{-n} \int_{\mathbb{R}^n} X(\mathbf{x}) \int_A e^{-i(\mathbf{p},\mathbf{x})} \, d\mathbf{p} \, d\mathbf{x}.$$

Example 3.3 Let G be a locally compact topological group of type I. Let dg be a fixed Haar measure on G, and let μ be the corresponding Plancherel measure on $\hat{G}$. Rewrite formula (2.9) in coordinate form:

$$X(g) = \sum_{n=1}^{n=\infty} \int_{\hat{G}_n} \sum_{j,k=1}^n \lambda_{jk}(g) \, dZ_{kj}(\lambda),$$

where $\hat{G}_n$ is the set of all equivalence classes of n-dimensional irreducible unitary representations of G. Multiply both sides of the last display by a function $f(g) \in L^2(G) \cap L^1(G)$ and integrate with respect to dg. We obtain

$$\int_G X(g) f(g) \, dg = \int_G \sum_{n=1}^{n=\infty} \int_{\hat{G}_n} \sum_{j,k=1}^n \lambda_{jk}(g) \, dZ_{kj}(\lambda) f(g) \, dg.$$

Changing the order of integration, we have

$$\int_G X(g) f(g) \, dg = \sum_{n=1}^{n=\infty} \int_{\hat{G}_n} \sum_{j,k=1}^n \int_G f(g) \lambda_{jk}(g) \, dg \, dZ_{kj}(\lambda).$$

Let A be a Borel subset of $\hat{G}_n$, and let j_0 and k_0 be positive integers with $j_0 \leq n$ and $k_0 \leq n$ if $n < \infty$. Let $T(\lambda)$ be the following operator-valued function on $\hat{G}_n$:

$$T_{jk}(\lambda) = \delta_{jj_0} \delta_{kk_0} \chi_A(\lambda),$$

where $\chi_A(\lambda)$ is the indicator of A. Let $f(g)$ be the inverse Fourier transform of $T(\lambda)$:

$$f(g) = \int_{\hat{G}_n} T(\lambda) \lambda^*(g) \, dg$$

$$= \int_A \overline{\lambda_{j_0 k_0}(g)} \, d\mu(\lambda).$$

If the Plancherel measure of A is positive and finite and A is a subset of the support of the above measure, then the right-hand side of the last display is nonzero, and we obtain the following inversion formula:

$$Z_{jk}(A) = \int_G X(g) \int_A \overline{\lambda_{jk}(g)} \, d\mu(\lambda) \, dg$$

(for simplicity, we write j, k in place of j_0, k_0). In coordinate-free form, the last display is

$$Z(A) = \int_G X(g) \int_A \lambda^+(g) \, d\mu(\lambda) \, dg.$$

Under the conditions of Example 2.21, let (φ, τ, ψ) with $-2\pi \leq \varphi \leq 2\pi$, $\tau \geq 0$, and $0 \leq \psi \leq 2\pi$ be the Euler angles on the group $\mathrm{SL}(2, \mathbb{R})$. Fix the following Haar measure:

$$dg := \sinh \tau \, d\varphi \, d\tau \, d\psi. \tag{3.1}$$

The corresponding Plancherel measure has the following form: for the representations $\lambda^{-1/2 + i\varrho, \varepsilon}$ of both principal series,

$$d\mu(\varrho) = \frac{1}{4\pi^2} \varrho \tanh\big(\pi(\varrho + \varepsilon i)\big) \, d\varrho,$$

for the representations $\lambda^{\sigma,\varepsilon,+}$ and $\lambda^{\sigma,\varepsilon,-}$ of the discrete series,

$$\mu\left(\lambda^{\sigma,\varepsilon,+}\right) = \mu\left(\lambda^{\sigma,\varepsilon,-}\right) = \frac{\sigma + 1/2}{4\pi^2},$$

while the measure of the remaining points (the supplementary series, the limiting point of the second principal series, and the trivial representation) is zero. The representation dual to $\lambda^{\sigma,\varepsilon}$ is $\lambda^{-\bar{\sigma}-1,\varepsilon}$.

Let A be a Borel subset of positive finite Plancherel measure in the support of the Plancherel measure. The inversion formula has the form

$$Z(A) = \int_0^{2\pi} \int_0^\infty \int_{-2\pi}^{2\pi} X(\varphi, \tau, \psi) \int_A \lambda^+(g) \, d\mu(\lambda) \sinh\tau \, d\varphi \, d\tau \, d\psi.$$

Fix the following Haar measure on the group $G = SL(2, \mathbb{C})$:

$$dg = \frac{d\alpha \, d\beta \, d\gamma}{|\alpha|^2},$$

where $dz = dx \, dy$ for $z = x + iy \in \mathbb{C}$. The corresponding Plancherel measure lives on the principal series and has the form

$$d\mu(\varrho, n) = (2\pi)^{-4}\left(n^2 + \varrho^2\right) d\varrho.$$

The representation dual to $\lambda^{\sigma,m}$ is $\lambda^{-\sigma-4,-m}$.

Let A be a Borel subset of a positive finite Plancherel measure lying in its support. The inversion formula has the form

$$Z(A) = \int_G X(g) \int_A \lambda^+(g) \, d\mu(\lambda) \, dg.$$

Example 3.4 Let $X(t)$ be a centred left-invariant random field on one of the compact two-point homogeneous spaces T from Table 2.1. This field has the form

$$X(t) = C \sum_{\ell=0}^\infty \sum_{m=1}^{h(T,\ell)} \sqrt{\frac{\mu_\ell}{h(T, \ell)}} Y_{\ell m}(t) X_{\ell m}, \qquad (3.2)$$

where

$$C =: \sqrt{\frac{2\pi^{(n+1)/2}}{\Gamma((n + 1)/2)}}$$

in the case of $T = S^n$ and $C = 1$ otherwise. Recall that $h(T, \ell)$ is a positive integer defined by (2.37), μ_ℓ is a sequence of nonnegative real numbers with

$$\sum_{\ell=0}^\infty \mu_\ell < \infty,$$

$Y_{\ell m}(t)$ form the basis of the space $L^2(T, \mathrm{d}\mu)$, where μ is the Lebesgue measure in the case of $T = S^n$ and the G-invariant probability measure on T otherwise, and $X_{\ell m}$ is a sequence of centred uncorrelated random variables with unit variance.

Multiply both sides of (3.2) by $\overline{Y_{\ell m}(t)}$ and integrate over T with respect to $\mathrm{d}\mu$. We obtain

$$\int_T X(t)\overline{Y_{\ell m}(t)}\,\mathrm{d}\mu(t) = C\sqrt{\frac{\mu_\ell}{h(T,\ell)}}\,X_{\ell m}.$$

Therefore, the inversion formula has the form

$$X_{\ell m} = C^{-1}\sqrt{\frac{h(T,\ell)}{\mu_\ell}}\int_T X(t)\overline{Y_{\ell m}(t)}\,\mathrm{d}\mu(t).$$

Example 3.5 Let $X(t)$ be a centred left-invariant random field on one of the noncompact two-point homogeneous spaces T from Table 2.2. This field has the form (2.45). Multiply both sides of (2.45) by $\overline{Y_{k\ell}^m(\mathbf{n})}$ and integrate over the sphere S^{N-1} with respect to Lebesgue measure $\mathrm{d}\mathbf{n}$. We obtain

$$\int_{S^{N-1}} X(r,\mathbf{n})\overline{Y_{k\ell}^m(\mathbf{n})}\,\mathrm{d}\mathbf{n} = \sqrt{\frac{2\pi^{N/2}}{\pi_{k,\ell}^{(\alpha,\beta)}\Gamma(N/2)}}\int_{\hat{G}_K}\varphi_{k,\ell}(\lambda;r)\,\mathrm{d}Z_{k\ell}^m(\lambda). \qquad (3.3)$$

Let $T = \mathbb{R}^n$. Using (2.42), we can rewrite (3.3) as

$$\int_{S^{n-1}} X(r,\mathbf{n})S_\ell^m(\mathbf{n})\,\mathrm{d}\mathbf{n} = c_n\int_0^\infty \frac{J_{\ell+(n-2)/2}(\lambda r)}{(\lambda r)^{(n-2)/2}}\,\mathrm{d}Z_\ell^m(\lambda),$$

where $c_n^2 = 2^{n-1}\Gamma(n/2)\pi^{n/2}$. Multiply the last equation by a function $f(r)$ and integrate from 0 to ∞ with respect to r:

$$\int_0^\infty\int_{S^{n-1}} X(r,\mathbf{n})S_\ell^m(\mathbf{n})f(r)\,\mathrm{d}\mathbf{n}\,\mathrm{d}r = \int_0^\infty\int_0^\infty c_n f(r)\frac{J_{\ell+(n-2)/2}(\lambda r)}{(\lambda r)^{(n-2)/2}}\,\mathrm{d}r\,\mathrm{d}Z_\ell^m(\lambda).$$

Let γ be a positive real number. Choose a function $f(r)$ in such a way that

$$\int_0^\infty c_n f(r)\frac{J_{\ell+(n-2)/2}(\lambda r)}{(\lambda r)^{(n-2)/2}}\,\mathrm{d}r = \begin{cases}1, & 0\le\lambda\le\gamma, \\ 0, & \lambda\ge\gamma.\end{cases}$$

Rewrite this equation in the following way:

$$\int_0^\infty r^{-n/2}f(a)J_{\ell+(n-2)/2}(\lambda r)r\,\mathrm{d}r = \begin{cases}c_n^{-1}\lambda^{(n-2)/2}, & 0\le\lambda\le\gamma, \\ 0, & \lambda\ge\gamma.\end{cases}$$

It follows that the function $r^{-n/2}f(r)$ is the inverse Hankel transform (6.54) of order $\nu = \ell + (n-2)/2$ of the right-hand side. In other words,

$$f(r) = c_n^{-1}r^{n/2}\int_0^\gamma \lambda^{n/2}J_{\ell+(n-2)/2}(\lambda r)\,\mathrm{d}\lambda.$$

To calculate this integral, use formula 2.12.3.1 from Prudnikov et al. (1988) with $r = \gamma$, $\alpha = n/2 + 1$, $\beta = 1$, $\nu = \ell + (n-2)/2$, $c = r$. We obtain

$$f(r) = c_n^{-1} r^{n/2} \gamma^{n+\ell+1} (r/2)^{\ell+(n-2)/2} \frac{\Gamma(\ell+n)}{\Gamma(\ell+n/2)\Gamma(\ell+n+1)}$$

$$\times {}_2F_3\left(\frac{\ell+n}{2}, \frac{\ell+n+1}{2}; \ell+\frac{n}{2}, \frac{\ell+n+1}{2}, \frac{\ell+n+2}{2}; -\frac{\gamma^2 r^2}{4}\right).$$

This expression can be simplified using (2.65) and (6.31). We have

$$f(r) = \frac{r^{\ell+n-1} \gamma^{\ell+n+1}}{2^{\ell+(n-2)/2} c_n (\ell+n)} {}_1F_2\left(\frac{\ell+n}{2}; \ell+\frac{n}{2}, \frac{\ell+n+2}{2}; -\frac{\gamma^2 r^2}{4}\right).$$

Using formula 7.14.3.3 from Prudnikov et al. (1990) with $a = (\ell+n)/2$, $b = \ell + n/2$, and $z = \gamma r/2$, we obtain

$$f(r) = c_n^{-1} \Gamma(\ell+n/2) \gamma^2 \big[(\ell+n) J_{\ell+(n-2)/2}(\lambda r) s_{n/2-1, \ell+n/2-2}(\gamma r)$$

$$- J_{\ell+(n-4)/2}(\gamma r) s_{n/2, m+(n-2)/2}(\gamma r) \big],$$

where $s_{\mu,\nu}(z)$ is the Lommel function. The inversion formula has the form

$$Z_\ell^m([0,\gamma)) = \int_0^\infty \int_{S^{n-1}} X(r, \mathbf{n}) S_\ell^m(\mathbf{n}) f(r) \, d\mathbf{n} \, dr.$$

In the remaining cases, rewrite (3.3) using (2.42) as follows:

$$\int_{S^{N-1}} X(r, \mathbf{n}) \overline{Y_{k\ell}^m(\mathbf{n})} \, d\mathbf{n} = \sqrt{\frac{2\pi^{N/2}}{\Gamma(N/2)}}$$

$$\times \int_{\hat{G}_K} \frac{\Gamma((\alpha+\beta+1+i\lambda)/2+k)\Gamma((\alpha-\beta+1+i\lambda)/2+\ell)\Gamma(\alpha+1)}{\Gamma((\alpha+\beta+1+i\lambda)/2)\Gamma((\alpha-\beta+1+i\lambda)/2)\Gamma(\alpha+k+\ell+1)}$$

$$\times \sinh^{k+l} r \cosh^{k-l} r \varphi_\lambda^{(\alpha+k+\ell, \beta+k-\ell)}(r) \, dZ_{k\ell}^m(\lambda).$$

Assume that the support of the spectral measure μ of the random field X is a subset of the interval $[0, \infty)$. Multiply the last equation by a function $f(r)$ and integrate from 0 to ∞ with respect to r:

$$\int_0^\infty \int_{S^{N-1}} X(a, \mathbf{n}) \overline{Y_{k\ell}^m(\mathbf{n})} f(r) \, d\mathbf{n} \, dr = \sqrt{\frac{2\pi^{N/2}}{\Gamma(N/2)}}$$

$$\times \int_0^\infty \int_0^\infty f(r) \frac{\Gamma((\alpha+\beta+1+i\lambda)/2+k)\Gamma((\alpha-\beta+1+i\lambda)/2+\ell)\Gamma(\alpha+1)}{\Gamma((\alpha+\beta+1+i\lambda)/2)\Gamma((\alpha-\beta+1+i\lambda)/2)\Gamma(\alpha+k+\ell+1)}$$

$$\times \sinh^{k+l} r \cosh^{k-l} r \varphi_\lambda^{(\alpha+k+\ell, \beta+k-\ell)}(r) \, dr \, dZ_{k\ell}^m(\lambda),$$

or

$$\int_0^\infty \int_{S^{N-1}} X(r,\mathbf{n}) \overline{Y_{k\ell}^m(\mathbf{n})} f(r) \, d\mathbf{n} \, dr = \sqrt{\frac{2\pi^{N/2}}{\Gamma(N/2)} \frac{\Gamma(\alpha+1)}{\Gamma(\alpha+k+\ell+1)}}$$

$$\times \int_0^\infty \tilde{g}(\lambda) \frac{\Gamma(\frac{\alpha+\beta+1+i\lambda}{2}+k)\Gamma(\frac{\alpha-\beta+1+i\lambda}{2}+\ell)}{\Gamma(\frac{\alpha+\beta+1+i\lambda}{2})\Gamma(\frac{\alpha-\beta+1+i\lambda}{2})} \, dZ_{k\ell}^m(\lambda),$$

where $\tilde{g}(\lambda)$ is the Jacobi transform of order $(\alpha+k+\ell, \beta+k-\ell)$ of a function

$$g(r) = (2\sinh r)^{-2\alpha-k-\ell-1}(2\cosh r)^{-2\beta-k+\ell-1} f(a).$$

Let γ be a positive real number. Choose a function $f(r)$ in such a way that

$$\tilde{g}(\lambda) = \begin{cases} c \dfrac{\Gamma((\alpha+\beta+1+i\lambda)/2)\Gamma((\alpha-\beta+1+i\lambda)/2)}{\Gamma((\alpha+\beta+1+i\lambda)/2+k)\Gamma((\alpha-\beta+1+i\lambda)/2+\ell)}, & 0 \le \lambda < \gamma, \\ 0, & \lambda \ge \gamma \end{cases}$$

with

$$c = \sqrt{\frac{\Gamma(N/2)}{2\pi^{N/2}} \frac{\Gamma(\alpha+k+\ell+1)}{\Gamma(\alpha+1)}}.$$

To use the inversion formula for the Jacobi transform of order $(\alpha+k+\ell, \beta+k-\ell)$, we have to make sure that for all spaces in lines 2–5 of Table 2.2,

$$\alpha+k+\ell > -1, \qquad \alpha+\beta+2k+1 \ge 0, \qquad \alpha-\beta+2\ell+1 \ge 0.$$

This is checked by direct calculations, using Table 2.3. By the inversion formula (6.30),

$$g(r) = \frac{1}{2\pi} \int_0^\gamma c \frac{\Gamma((\alpha+\beta+1+i\lambda)/2)\Gamma((\alpha-\beta+1+i\lambda)/2)}{2^{2\alpha+2\beta+4k+2}\Gamma^2(\alpha+k+\ell+1)|\Gamma(i\lambda)|^2}$$

$$\times \overline{\Gamma\left(\frac{\alpha+\beta+1+i\lambda}{2}+k\right)\Gamma\left(\frac{\alpha-\beta+1+i\lambda}{2}+\ell\right)} \varphi_\lambda^{(\alpha+k+\ell,\beta+k-\ell)}(r) \, d\lambda.$$

Simplify this formula using the mirror symmetry (6.35), the absolute value formula (6.36), and (6.31). We obtain

$$g(r) = \frac{1}{2\pi^2} \int_0^\gamma c \frac{|\Gamma((\alpha+\beta+1+i\lambda)/2)\Gamma((\alpha-\beta+1+i\lambda)/2)|^2}{2^{2\alpha+2\beta+4k+2}\Gamma^2(\alpha+k+\ell+1)}$$

$$\times \prod_{j=0}^{k-1}((\alpha+\beta+1+i\lambda)/2+j)\prod_{j=0}^{\ell-1}((\alpha-\beta+1+i\lambda)/2+j)$$

$$\times \lambda \sinh(\pi\lambda)\varphi_\lambda^{(\alpha+k+\ell,\beta+k-\ell)}(r) \, d\lambda,$$

or

$$g(r) = \frac{\sqrt{\Gamma(N/2)}}{2^{2\alpha+2\beta+4k+7/2}\pi^{N/4+2}\Gamma(\alpha+1)\Gamma(\alpha+k+\ell+1)}$$

$$\times \int_0^\gamma \left|\Gamma((\alpha+\beta+1+i\lambda)/2)\Gamma((\alpha-\beta+1+i\lambda)/2)\right|^2$$

$$\times \prod_{j=0}^{k-1}((\alpha+\beta+1+i\lambda)/2+j)\prod_{j=0}^{\ell-1}((\alpha-\beta+1+i\lambda)/2+j)\lambda\sinh(\pi\lambda)$$

$$\times {}_2F_1\left(\frac{\alpha+\beta+1-i\lambda}{2}+k,\frac{\alpha+\beta+1+i\lambda}{2}+k;\alpha+k+\ell+1;-\sinh^2 r\right)d\lambda.$$

It follows that

$$f(r) = \frac{\sqrt{\Gamma(N/2)}(\sinh r)^{2\alpha+k+\ell+1}(\cosh r)^{2\beta+k-\ell+1}}{2^{2k+3/2}\pi^{N/4+2}\Gamma(\alpha+1)\Gamma(\alpha+k+\ell+1)}$$

$$\times \int_0^\gamma \left|\Gamma((\alpha+\beta+1+i\lambda)/2)\Gamma((\alpha-\beta+1+i\lambda)/2)\right|^2$$

$$\times \prod_{j=0}^{k-1}((\alpha+\beta+1+i\lambda)/2+j)\prod_{j=0}^{\ell-1}((\alpha-\beta+1+i\lambda)/2+j)\lambda\sinh(\pi\lambda)$$

$$\times {}_2F_1\left(\frac{\alpha+\beta+1-i\lambda}{2}+k,\frac{\alpha+\beta+1+i\lambda}{2}+k;\alpha+k+\ell+1;-\sinh^2 r\right)d\lambda.$$

The inversion formula has the form

$$Z_{k\ell}^m([0,\gamma)) = \int_0^\infty \int_{S^{N-1}} X(r,\mathbf{n})\overline{Y_{k\ell}^m(\mathbf{n})}f(r)\,d\mathbf{n}\,dr.$$

3.2 Linear Differential Equations with Random Fields

Let G be a Lie group that acts on an analytic manifold T. Let L be a linear differential operator on T, and let $\mathscr{A}$ be the set of all centred wide-sense left G-invariant random fields on T. In this section, we consider the following problem:

- Describe the set $\mathscr{A}_L$ of all random fields $X(t)$, such that the field $LX(t)$ may be correctly defined by mean-square differentiation.
- Find sufficient conditions on L for the inclusion $L\mathscr{A}_L \subset \mathscr{A}$.
- For each $X \in \mathscr{A}_L$, calculate $\mathsf{E}[X(s)\overline{LX(t)}]$.
- Describe the image $L\mathscr{A}_L$ in terms of the spectral measure of X.

Assume that G is a group of type I, let K be a massive compact subgroup, and let $T = G/K$. Then $X(t) \in \mathscr{A}$ iff it has the form (2.33). The associated spherical functions $\varphi_n^{(\lambda)}(t)$ are spherical functions of some type $\delta \in \hat{K}$. Therefore, they are infinitely differentiable. Apply the operator L to both sides of (2.33). We obtain

$$LX(t) = \int_{\hat{G}_K} \sum_{n=1}^{\dim \lambda} \left(L\varphi_n^{(\lambda)}\right)(t) \, dZ_n(\lambda).$$

Then

$$\mathsf{E}\left[\left|LX(t)\right|^2\right] = \int_{\hat{G}_K} \sum_{n=1}^{\dim \lambda} \left|\left(L\varphi_n^{(\lambda)}\right)(t)\right|^2 dF(\lambda).$$

We see that the field $LX(t)$ may be correctly defined by mean-square differentiation iff the integral in the right-hand side converges.

The field $LX(t)$ is in $\mathscr{A}$ if $(L\varphi_n^{(\lambda)})(t) = \lambda_L \varphi_n^{(\lambda)}(t)$ with $\lambda_L \in \mathbb{C}$. This is so if L belongs to $\mathbf{D}(T)$, the algebra of differential operators on T invariant under all the translations $\tau(g) \colon tK \to gtK$ of T. Then we have

$$\mathsf{E}\left[\left|LX(t)\right|^2\right] = \int_{\hat{G}_K} \sum_{n=1}^{\dim \lambda} \left|\lambda_L \varphi_n^{(\lambda)}(t)\right|^2 dF(\lambda)$$

$$= \int_{\hat{G}_K} |\lambda_L|^2 \, dF(\lambda).$$

We have $LX \in \mathscr{A}$ iff

$$\int_{\hat{G}_K} |\lambda_L|^2 \, dF(\lambda) < \infty$$

and the equation $LY(t) = ZX(t)$ has a solution iff

$$\int_{\hat{G}_K} |\lambda_L|^2 \, dF(\lambda) < \infty.$$

Finally,

$$\mathsf{E}\left[X(s)\overline{LX(t)}\right] = \int_{\hat{G}_K} \sum_{n=1}^{\dim \lambda} \varphi_n^{(\lambda)}(s)\overline{\left(L\varphi_n^{(\lambda)}\right)(t)} \, dF(\lambda).$$

The description of the algebra $\mathbf{D}(T)$ is not known in general. However, if T is reductive (Subsection 6.2.2), then $\mathbf{D}(T)$ is described by (6.4). In particular, if T is a two-point homogeneous space, then $\mathbf{D}(T)$ consists of the polynomials in the Laplace–Beltrami operator Δ.

Example 3.6 Let Δ be the Laplace–Beltrami operator on the group SU(2). By Vilenkin (1968, Chap. III, Subsection 4.8, formula (2)) Δ is the second-order differential operator expressed in Euler angles as follows:

$$\Delta = \frac{\partial^2}{\partial\theta^2} + \cot\theta\frac{\partial}{\partial\theta} + \frac{1}{\sin^2\theta}\left(\frac{\partial^2}{\partial\varphi^2} - 2\cos\theta\frac{\partial^2}{\partial\varphi\partial\psi} + \frac{\partial^2}{\partial\psi^2}\right). \tag{3.4}$$

Moreover, the Wigner D-functions $D^\ell_{jk}(\varphi, \theta, \psi)$ are eigenfunctions of Δ, and the corresponding eigenvalues are equal to $-\ell(\ell + 1)$. By Example 2.20, the centred left-homogeneous random field on G has the form

$$X(g) = \sum_{\ell\in\{0,1/2,1,...\}} \sum_{j,k=1}^{2\ell+1} D^\ell_{jk}(g)Z^{(\ell)}_{kj}, \tag{3.5}$$

where the $D^\ell_{jk}(g)$ are Wigner D-functions (6.16). By (2.20) we have

$$\mathsf{E}\big[Z^{(\ell)}_{kj}\overline{Z^{(\ell')}_{k'j'}}\big] = \delta_{\ell\ell'}\delta_{jj'}F^{(\ell)}_{kk'},$$

where $F^{(\ell)}$ are Hermitian nonnegative-definite matrices with

$$\sum_{\ell\in\{0,1/2,1,...\}} \sum_{j=1}^{2\ell+1} F^{(\ell)}_{jj} < \infty. \tag{3.6}$$

Let $p(\Delta)$ be a polynomial in Δ. The differential operator $p(\Delta)$ may be applied to the random field (3.5) iff

$$\sum_{\ell\in\{0,1/2,1,...\}} \sum_{j=1}^{2\ell+1} \big|p(-\ell(\ell+1))\big|^2 F^{(\ell)}_{jj} < \infty$$

and the equation $p(\Delta)X = Y$ has the solution iff the matrices $F^{(\ell)}$ corresponding to the random field Y satisfy the condition

$$\sum_{\ell\in\{0,1/2,1,...\}} \sum_{j=1}^{2\ell+1} \big|p(-\ell(\ell+1))\big|^{-2} F^{(\ell)}_{jj} < \infty.$$

Example 3.7 Consider a centred left-invariant random field $X(t)$ on a two-point compact homogeneous space T. By the results of Subsection 2.3.1 we have

$$X(t) = \sum_{\ell=0}^{\infty} \sum_{m=1}^{h(T,\ell)} C_\ell Y_{\ell m}(t)X_{\ell m}, \tag{3.7}$$

where C_ℓ is a sequence of nonnegative real numbers satisfying

$$\sum_{\ell=0}^{\infty} h(T,\ell)C_\ell^2 < \infty,$$

and where $X_{\ell m}$ is the sequence of centred uncorrelated random variables with unit variance. The spherical harmonics $Y_{\ell m}(t)$ are eigenfunctions of the Laplace–Beltrami operator Δ, and the corresponding eigenvalue is $\lambda_\Delta = -\ell(\ell + \alpha + \beta + 1)$, where α and β are numbers from Table 2.1. Let p be a polynomial with complex coefficients. We obtain: the random field $p(\Delta)X(t)$ is defined correctly as

$$p(\Delta)X(t) := \sum_{\ell=0}^{\infty} \sum_{m=1}^{h(T,\ell)} p\bigl(-\ell(\ell + \alpha + \beta + 1)\bigr) C_\ell Y_{\ell m}(t) X_{\ell m}$$

iff

$$\sum_{\ell=0}^{\infty} h(T, \ell) \bigl| p\bigl(-\ell(\ell + \alpha + \beta + 1)\bigr) C_\ell \bigr|^2 < \infty.$$

Moreover,

$$\mathsf{E}\bigl[X(s)\overline{p(\Delta)X(t)}\bigr] = \sum_{\ell=0}^{\infty} h(T, \ell) C_\ell^2 \overline{p\bigl(-\ell(\ell + \alpha + \beta + 1)\bigr)} \frac{P_\ell^{(\alpha,\beta)}(\cos(d(t_1, t_2)))}{P_\ell^{(\alpha,\beta)}(1)},$$

and equation $p(\Delta)Y = X$ has a solution iff

$$\sum_{\ell:\, C_\ell \neq 0} h(T, \ell) \bigl| p\bigl(-\ell(\ell + \alpha + \beta + 1)\bigr) C_\ell \bigr|^{-2} < \infty.$$

Example 3.8 Let $X(g) = X(\varphi, \tau, \psi)$ be a centred left-invariant random field on the group $SL(2, \mathbb{R})$. Let Δ be the Laplace–Beltrami operator on the group $SL(2, \mathbb{R})$:

$$\Delta = -\frac{1}{\sinh \tau} \frac{\partial}{\partial \tau} \left(\sinh \tau \frac{\partial}{\partial \tau} \right) - \frac{1}{\sinh^2 \tau} \left(\frac{\partial^2}{\partial \varphi^2} - 2 \cosh \tau \frac{\partial^2}{\partial \varphi \partial \psi} + \frac{\partial^2}{\partial \psi^2} \right).$$

It is known that $\Delta \lambda_{mn}^{\sigma,\varepsilon} = \sigma(\sigma + 1)\lambda^{\sigma,\varepsilon}$. Let p be a polynomial with complex coefficients. We obtain: the random field $p(\Delta)X(g)$ is correctly defined as

$$p(\Delta)X(\varphi, \tau, \psi)$$

$$:= \sum_{m,n=-\infty}^{\infty} \int_0^{\infty} p\bigl(-\varrho^2 - 1/4\bigr) \lambda_{mn}^{-1/2+i\varrho,0}(\varphi, \tau, \psi)\, dZ_{nm}^{-1/2+i\varrho,0}(\varrho)$$

$$+ \sum_{m,n=-\infty}^{\infty} \int_{0+}^{\infty} p\bigl(-\varrho^2 - 1/4\bigr) \lambda_{mn}^{-1/2+i\varrho,1/2}(\varphi, \tau, \psi)\, dZ_{nm}^{-1/2+i\varrho,1/2}(\varrho)$$

$$+ \sum_{m,n=-\infty}^{\infty} \int_{-1+}^{-1/2-} p\bigl(\varrho(\varrho + 1)\bigr) \lambda_{mn}^{\varrho,0}(\varphi, \tau, \psi)\, dZ_{nm}^{\varrho,0}(\varrho)$$

$$+ \sum_{\sigma-\varepsilon \leq -1} \sum_{m,n=-\infty}^{\sigma-\varepsilon} p\bigl(\sigma(\sigma + 1)\bigr) \lambda_{mn}^{\sigma,\varepsilon,+}(\varphi, \tau, \psi) Z_{nm}^{\sigma,\varepsilon,+}$$

$$+ \sum_{\sigma-\varepsilon\leq-1} \sum_{m,n=-\sigma-\varepsilon}^{\infty} p\big(\sigma(\sigma+1)\big)\lambda_{mn}^{\sigma,\varepsilon,-}(\varphi,\tau,\psi)Z_{nm}^{\sigma,\varepsilon,-}$$

$$+ p(0)Z^{-1,0}$$

iff

$$\int_0^{\infty}\big|p(\varrho^2+1/4)\big|^2 \,\mathrm{d}\operatorname{tr} F^{-1/2+i\varrho,0} + \int_{0+}^{\infty}\big|p(\varrho^2+1/4)\big|^2 \,\mathrm{d}\operatorname{tr} F^{-1/2+i\varrho,1/2}$$

$$+ \int_{-1+}^{-1/2-}\big|p(\varrho(\varrho+1))\big|^2 \,\mathrm{d}\operatorname{tr} F^{\varrho,0}$$

$$+ \sum_{\sigma-\varepsilon\leq-1}\big|p\big(\sigma(\sigma+1)\big)\big|^2 \operatorname{tr} F^{\sigma,\varepsilon,+}$$

$$+ \sum_{\sigma-\varepsilon\leq-1}\big|p\big(\sigma(\sigma+1)\big)\big|^2 \operatorname{tr} F^{\sigma,\varepsilon,-} < \infty.$$

Equation $p(\Delta)Y = X$ has a solution iff

$$\int_0^{\infty}\big|p(\varrho^2+1/4)\big|^{-2} \,\mathrm{d}\operatorname{tr} F^{-1/2+i\varrho,0} + \int_{0+}^{\infty}\big|p(\varrho^2+1/4)\big|^{-2} \,\mathrm{d}\operatorname{tr} F^{-1/2+i\varrho,1/2}$$

$$+ \int_{-1+}^{-1/2-}\big|p(\varrho(\varrho+1))\big|^{-2} \,\mathrm{d}\operatorname{tr} F^{\varrho,0}$$

$$+ \sum_{\sigma-\varepsilon\leq-1}\big|p\big(\sigma(\sigma+1)\big)\big|^{-2} \operatorname{tr} F^{\sigma,\varepsilon,+}$$

$$+ \sum_{\sigma-\varepsilon\leq-1}\big|p\big(\sigma(\sigma+1)\big)\big|^{-2} \operatorname{tr} F^{\sigma,\varepsilon,-} < \infty.$$

3.3 Linear Functionals of Invariant Random Fields

Let $X(t)$ be a scalar centred G-invariant random field on a space T. Let K be a subgroup of the group G, and let C be an orbit of the group K in T. Assume that there exists a left K-invariant measure $\mathrm{d}\mu$ on C. Let $f(t)$ be a continuous function on C with compact support. The Bochner integral

$$Y := \int_C X(t)f(t)\,\mathrm{d}\mu(t) \tag{3.8}$$

defines a random variable Y, which is a vector of the Hilbert space $H_X^-(C)$—the closed linear span of the vectors $X(t)$, $t \in C$ in the space of all random variables with finite variance.

Is it possible to express *any* element of the space $H_X^-(C)$ in the form of Eq. (3.8)? Consider some examples.

Example 3.9 Let $G = T = K = C = \mathrm{SU}(2)$. It follows from (3.5) that any $Y \in H_X^-(G)$ should have the form

$$Y = \sum_{\ell \in \{0,1/2,1,\dots\}} \sum_{j,k=1}^{2\ell+1} d_{jk}^\ell Z_{kj}^{(\ell)}, \tag{3.9}$$

where d_{jk}^ℓ are complex numbers. The series in the right-hand side converges in mean square iff $\mathsf{E}|Y|^2 < \infty$. We have

$$\mathsf{E}|Y|^2 = \sum_{\ell,\ell' \in \{0,1/2,1,\dots\}} \sum_{j,k=1}^{2\ell+1} \sum_{j',k'=1}^{2\ell'+1} d_{jk}^\ell \overline{d_{j'k'}^{\ell'}} \mathsf{E}\big[Z_{kj}^{(\ell)} \overline{Z_{k'j'}^{(\ell')}}\big]$$

$$= \sum_{\ell \in \{0,1/2,1,\dots\}} \sum_{j,k,k'=1}^{2\ell+1} d_{jk}^\ell \overline{d_{jk'}^\ell} F_{kk'}^{(\ell)}$$

by (3.6).

Let $D_X^-(G)$ be the quotient space of the space of all numerical sequences $d_{jk}^{(\ell)}$ satisfying condition

$$\sum_{\ell \in \{0,1/2,1,\dots\}} \sum_{j,k,k'=1}^{2\ell+1} d_{jk}^\ell \overline{d_{jk'}^\ell} F_{kk'}^{(\ell)} < \infty$$

with respect to the subspace of sequences satisfying condition

$$\sum_{\ell \in \{0,1/2,1,\dots\}} \sum_{j,k,k'=1}^{2\ell+1} d_{jk}^\ell \overline{d_{jk'}^\ell} F_{kk'}^{(\ell)} = 0.$$

Then $D_X^-(G)$ is the Hilbert space with inner product

$$(d_{jk}^\ell, e_{jk}^\ell)_- = \sum_{\ell \in \{0,1/2,1,\dots\}} \sum_{j,k,k'=1}^{2\ell+1} d_{jk}^\ell \overline{e_{jk'}^\ell} F_{kk'}^{(\ell)}$$

and Eq. (3.9) determines an isometric isomorphism between the spaces $H_X^-(G)$ and $D_X^-(G)$.

Let $\mathrm{d}\mu$ be the probability Haar measure on G. By the Peter–Weyl theorem, the functions $\sqrt{2\ell+1} D_{jk}^\ell(g)$ form an orthonormal basis of the Hilbert space $L^2(G, \mathrm{d}\mu)$. For any $f \in L^2(G, \mathrm{d}\mu)$, let $d_{jk}^\ell(f)$ be its Fourier coefficients:

$$d_{jk}^\ell(f) := \sqrt{2\ell+1} \int_G f(g) \overline{D_{jk}^\ell(g)} \, \mathrm{d}\mu(g). \tag{3.10}$$

For any two nonnegative-definite matrices A and B of the same order we have $\operatorname{tr}(AB) \leq \operatorname{tr}(A)\operatorname{tr}(B)$. Put $A := F^{(\ell)}$, $B := F^{(\ell)*}$. Then $\operatorname{tr}(F^{(\ell)}F^{(\ell)*}) \leq [\operatorname{tr}(F^{(\ell)})]^2$. Summing over ℓ, we obtain

$$\sum_{\ell \in \{0,1/2,1,\dots\}} \operatorname{tr}(F^{(\ell)}F^{(\ell)*}) \leq \sum_{\ell \in \{0,1/2,1,\dots\}} [\operatorname{tr}(F^{(\ell)})]^2 < \infty$$

by (3.6). In particular, the set $\{|F_{jk}^{(\ell)}|\}$ is bounded. In Example 6.7, put

$$R := \{ (\ell, j, k) \colon \ell = 0, 1/2, 1, \dots, 1 \leq j, k \leq 2\ell + 1 \}.$$

Let $\mathfrak{R}$ be the σ-field of all subsets of R, and let ν be the counting measure. The kernel

$$K\big((\ell, j, k), (\ell', j', k')\big) = \delta_{\ell\ell'}\delta_{jj'}F_{kk'}^{(\ell)}$$

is positive-definite and bounded. Let $D_X^0(G)$ be the quotient space of the space $L^2(R, d\nu)$ with respect to the subspace of all sequences $d_{jk}^{(\ell)}(f)$ with $\|d_{jk}^{(\ell)}(f)\|_- = 0$. By Example 6.7, the embedding $D_X^-(G) \supset D_X^0(G)$ may be extended to the rigged Hilbert space $D_X^-(G) \supset D_X^0(G) \supset D_X^+(G)$.

On the other hand, Eq. (3.10) determines an isometric isomorphism between $D_X^0(G)$ and $L_0^2(G, d\mu)$, the quotient space of $L^2(G, d\mu)$ with respect to the subspace of all $f \in L^2(G, d\mu)$ with

$$\int_G X(g)f(g)\,d\mu(g) = 0.$$

We have the following diagram.

$$H_X^-(G)$$

$$\uparrow \ (3.9)$$

$$D_X^-(G) \qquad \supset \qquad D_X^0(G) \qquad \supset \qquad D_X^+(G)$$

$$\uparrow \ (3.10)$$

$$L_0^2(G, d\mu)$$

Let $H_X^0(G)$ be the image of the space $D_X^0(G)$ under the isomorphism (3.9). In other words, $H_X^0(G)$ is the Hilbert space of all random variables of the form

$$Y_f := \int_G X(g)f(g)\,d\mu(g), \quad f \in L_0^2(G, d\mu) \tag{3.11}$$

with the inner product

$$(Y_f, Y_h)_0 := (f, h)_0.$$

The remaining space of the first row, $H_X^+(G)$, may be constructed either by extending the embedding $H_X^-(G) \supset H_X^0(G)$ to the rigged Hilbert space or as the image of the space $D_X^+(G)$ under the isomorphism (3.9). Our diagram takes the form

$$
\begin{array}{ccccc}
H_X^-(G) & \supset & H_X^0(G) & \supset & H_X^+(G) \\
\uparrow {\scriptstyle (3.9)} & & \uparrow {\scriptstyle (3.9)} & & \uparrow {\scriptstyle (3.9)} \\
D_X^-(G) & \supset & D_X^0(G) & \supset & D_X^+(G) \\
& & \uparrow {\scriptstyle (3.10)} & & \\
& & L_0^2(G, d\mu) & &
\end{array}
$$

Finally, consider the isomorphism inverse to (3.10):

$$f(g) =: \sum_{\ell \in \{0, 1/2, 1, \dots\}} \sqrt{2\ell + 1} \sum_{j,k=1}^{2\ell+1} d_{jk}^\ell(f) D_{jk}^\ell(g). \tag{3.12}$$

Let $L_+^2(G, d\mu)$ be the image of the space $D_X^+(G)$ under the isomorphism (3.12). The inner product in the Hilbert space $L_+^2(G, d\mu)$ has the form

$$(f, h)_+ = \sum_{\ell \in \{0, 1/2, 1, \dots\}} \sum_{j,k,k'=1}^{2\ell+1} d_{jk}^\ell(f) \overline{d_{jk'}^\ell(h)} \left(F_{kk'}^{(\ell)}\right)^{-1},$$

where $d_{jk}^\ell(f)$ (resp. $d_{jk}^\ell(h)$) is the image of f (resp. h) under the isomorphism (3.10). This sum may contain terms of the form $0 \cdot \infty$. We put them equal to 0 by definition. Extend the embedding $L_0^2(G, d\mu) \supset L_+^2(G, d\mu)$ to the rigged Hilbert space $L_-^2(G, d\mu) \supset L_0^2(G, d\mu) \supset L_+^2(G, d\mu)$. The space $L_-^2(G, d\mu)$ is the space of distributions of the form (3.12) with the following inner product:

$$(f, h)_- := \sum_{\ell \in \{0, 1/2, 1, \dots\}} \sum_{j,k,k'=1}^{2\ell+1} d_{jk}^\ell(f) \overline{d_{jk'}^\ell(h)} F_{kk'}^{(\ell)}, \quad d_{jk}^\ell(f), d_{jk'}^\ell(h) \in D_X^-(G),$$

and the series (3.12) converges in the norm defined by the above inner product. In particular, $d_{jk}^\ell(\delta_{g_0}) = D_{jk}^\ell(g_0)$ are the Fourier coefficients of the Dirac delta-function at the point $g_0 \in G$. The final form of our diagram is

$$
\begin{array}{ccccc}
H_X^-(G) & \supset & H_X^0(G) & \supset & H_X^+(G) \\
\uparrow {\scriptstyle (3.9)} & & \uparrow {\scriptstyle (3.9)} & & \uparrow {\scriptstyle (3.9)} \\
D_X^-(G) & \supset & D_X^0(G) & \supset & D_X^+(G) \\
\downarrow {\scriptstyle (3.12)} & & \uparrow {\scriptstyle (3.10)} & & \downarrow {\scriptstyle (3.12)} \\
L_-^2(G, \mathrm{d}\mu) & \supset & L_0^2(G, \mathrm{d}\mu) & \supset & L_+^2(G, \mathrm{d}\mu)
\end{array}
$$

Now we give names to the elements of all the above spaces. The elements of the space $H_X^0(G)$ are called the *ordinary linear functionals* of the random field $X(g)$. These random variables have the form (3.11). The elements of the space $H_X^+(G)$ are called the *smooth linear functionals* of the random field $X(g)$. They have the form (3.11) with $f \in L_+^2(G, \mathrm{d}\mu)$. The elements of the space $H_X^-(G)$ are called the *generalised linear functionals* of the random field $X(g)$. For generalised linear functionals, the integral in the right-hand side of (3.11) may be defined as

$$
\int_G X(g) f(g) \, \mathrm{d}\mu(g) := \sum_{\ell \in \{0, 1/2, 1, \dots\}} \sum_{j,k=1}^{2\ell+1} d_{jk}^{\ell}(f) Z_{kj}^{(\ell)}, \quad f \in L_-^2(G, \mathrm{d}\mu).
$$

The elements of the space $D_X^0(G)$ are called the *ordinary spectral characteristics* of the random field $X(g)$. The elements of the space $D_X^+(G)$ (resp. $D_X^-(G)$) are called the *smooth spectral characteristics* (resp. the *generalised spectral characteristics*) of the random field $X(g)$. The elements of the space $L_0^2(G, \mathrm{d}\mu)$ are called *ordinary functions* connected to the random field $X(g)$. Finally, the elements of the space $L_0^2(G, \mathrm{d}\mu)$ (resp. $L_-^2(G, \mathrm{d}\mu)$) are called *smooth functions* (resp. *generalised functions* or *distributions*) connected to the random field $X(g)$.

How smooth are our "smooth functions"? Note that their Fourier coefficients are smooth spectral characteristics satisfying the following condition:

$$
\sum_{\ell \in \{0, 1/2, 1, \dots\}} \sum_{j,k,k'=1}^{2\ell+1} d_{jk}^{\ell} \overline{d_{jk'}^{\ell}} \left(F_{kk'}^{(\ell)} \right)^{-1} < \infty.
$$

Let n be a positive integer. Let Δ be the Laplace–Beltrami operator (3.4), and let $f(g)$ be a function with $\Delta^n f \in L^2(G, \mathrm{d}\mu(g))$. It follows from Example 3.6 that the Fourier coefficients of f satisfy

$$
\sum_{\ell \in \{0, 1/2, 1, \dots\}} (\ell(\ell+1))^n \sum_{j,k=1}^{2\ell+1} |d_{jk}^{\ell}|^2 < \infty.
$$

We obtain the following theorem.

Theorem 3.10 *Let $X(g)$ be a centred left-invariant random field on the group $G = SU(2)$, let $F^{(\ell)}$ be the corresponding matrices, and let n be a positive integer. If*

$$F_{kk}^{(\ell)} = O\big([\ell(\ell+1)]^{-n}\big), \quad \ell \to \infty,$$

then all the functions $f(g)$ in the space $L_+^2(G, d\mu)$ are $2n$ times differentiable and $\Delta^n f(g) \in L^2(G, d\mu)$.

Corollary 3.11 *Under conditions of Theorem 3.10, let*

$$F_{kk}^{(\ell)} = O\big([\ell(\ell+1)]^{-n}\big), \quad \ell \to \infty, \; n \geq 1.$$

Then all the functions $f(g)$ in the space $L_+^2(G, d\mu)$ are infinitely differentiable.

Example 3.12 Consider a centred left-invariant random field $X(t)$ on a two-point compact homogeneous space T. It follows from (3.7) that the space $H_X^-(T)$ consists of all random variables of the form

$$Y = \sum_{\ell=0}^{\infty} \sum_{m=1}^{h(T,\ell)} C_\ell d_{\ell m} X_{\ell m} \tag{3.13}$$

with

$$\sum_{\ell=0}^{\infty} C_\ell^2 \sum_{m=1}^{h(T,\ell)} |d_{\ell m}|^2 < \infty. \tag{3.14}$$

The space $D_X^-(T)$ is the quotient space of the space of all complex-valued sequences $d_{\ell m}$ satisfying (3.14) with respect to the subspace of all the above sequences satisfying in addition the following condition:

$$\sum_{\ell=0}^{\infty} C_\ell^2 \sum_{m=1}^{h(T,\ell)} |d_{\ell m}|^2 = 0.$$

In other words, the space $D_X^-(T)$ consists of all complex-valued sequences $d_{\ell m}$ satisfying (3.14) with $d_{\ell m} = 0$ if $C_\ell = 0$. The inner product in $D_X^-(T)$ is

$$(d_{\ell m}, e_{\ell m})_- := \sum_{\ell=0}^{\infty} C_\ell^2 \sum_{m=1}^{h(T,\ell)} d_{\ell m} \overline{e_{\ell m}}.$$

The space $D_X^0(T)$ consists of all complex-valued sequences $d_{\ell m}$ satisfying

$$\sum_{\ell=0}^{\infty} \sum_{m=1}^{h(T,\ell)} |d_{\ell m}|^2 < \infty$$

with $d_{\ell m} = 0$ if $C_\ell = 0$. The inner product in $D_X^0(T)$ is

$$(d_{\ell m}, e_{\ell m})_0 := \sum_{\ell=0}^{\infty} \sum_{m=1}^{h(T,\ell)} d_{\ell m}\overline{e_{\ell m}}.$$

The space $D_X^+(T)$ consists of all complex-valued sequences $d_{\ell m}$ satisfying

$$\sum_{\ell=0}^{\infty} C_\ell^{-2} \sum_{m=1}^{h(T,\ell)} |d_{\ell m}|^2 < \infty$$

with $d_{\ell m} = 0$ if $C_\ell = 0$. The inner product in $D_X^0(T)$ is

$$(d_{\ell m}, e_{\ell m})_+ = \sum_{\ell=0}^{\infty} C_\ell^{-2} \sum_{m=1}^{h(T,\ell)} d_{\ell m}\overline{e_{\ell m}}.$$

Let $Y_{\ell m}$ be the orthonormal basis in the space $L^2(T, d\mu)$, introduced in Example 2.24. Recall that in the case of $T = S^n$, $d\mu$ is Lebesgue measure. In all remaining cases, $d\mu$ is the G-invariant probability measure on T. For any $f \in L^2(T, d\mu)$, let

$$d_{\ell m}(f) := \int_T f(t)\overline{Y_{\ell m}(t)}\, d\mu(t) \tag{3.15}$$

be the Fourier coefficients of f. The space $L_0^2(T, d\mu)$ consists of all functions $f \in L^2(T, d\mu)$ satisfying the following condition: if $C_\ell = 0$, then $d_{\ell m}(f) = 0$, $1 \le m \le h(T, \ell)$. In other words, $D_X^0(T, d\mu)$ is the image of $L_0^2(T, d\mu)$ under the isometric isomorphism (3.15).

The space $H_X^0(T)$ is the Hilbert space of all random variables of the form

$$Y_f = \int_T X(t) f(t)\, d\mu(t), \quad f \in L_0^2(T, d\mu)$$

with the inner product

$$(Y_f, Y_g)_0 := (f, g)_0.$$

The space $H_X^+(T)$ is the image of the space $D_X^+(T)$ under the isomorphism (3.13). The space $L_+^2(T, d\mu)$ is the image of the space $D_X^+(T)$ under the isomorphism

$$f(t) := \sum_{\ell=0}^{\infty} \sum_{m=1}^{h(T,\ell)} d_{\ell m}(f) Y_{\ell m}(t). \tag{3.16}$$

The inner product in the space $L_+^2(T, d\mu)$ has the form

$$(f, g)_+ := \sum_{\ell=0}^{\infty} \sum_{m=1}^{h(T,\ell)} C_\ell^{-1} d_{\ell m}(f)\overline{d_{\ell m}(g)}.$$

The space $L^2_-(T, d\mu)$ is the space of distributions of the form (3.16) with the following inner product:

$$(f, g)_- := \sum_{\ell=0}^{\infty} \sum_{m=1}^{h(T,\ell)} C_\ell d_{\ell m}(f)\overline{d_{\ell m}(g)}, \quad d_{\ell m}(f), d_{\ell m}(g) \in D_X^-(T).$$

Let Δ be the Laplace–Beltrami operator on the space T. The smoothness of the functions $f \in L^2_+(T, d\mu)$ is described as follows.

Theorem 3.13 *Let $X(t)$ be a centred G-invariant random field on a compact two-point homogeneous space T. Let C_ℓ be the corresponding coefficients, and let n be a positive integer. If*

$$C_\ell = O\big([\ell(\ell + \alpha + \beta + 1)]^{-n}\big), \quad \ell \to \infty,$$

where α and β are numbers from Table 2.1, then all the functions $f \in L^2_+(T, d\mu)$ are $2n$ times differentiable, and $\Delta^n f \in L^2(T, d\mu)$.

Corollary 3.14 *Under conditions of Theorem 3.13, let*

$$C_\ell = O\big([\ell(\ell + \alpha + \beta + 1)]^{-n}\big), \quad \ell \to \infty, \, n \geq 1.$$

Then all the functions $f(t)$ in the space $L^2_+(T, d\mu)$ are infinitely differentiable.

Example 3.15 Let $G = K = T = C$ be a noncompact topological group of type I with countable base. By Theorem 2.18, the elements of the space $H_X^-(G)$ should have the form

$$Y = \int_{\hat{G}} \mathrm{tr}\big[D(\lambda) \, dZ(\lambda)\big], \tag{3.17}$$

where $D(\lambda)$ is a bounded linear operator in the space $H^{(\lambda)}$ with

$$\int_{\hat{G}} \mathrm{tr}\big[D(\lambda)D^*(\lambda) \, dF(\lambda)\big] < \infty. \tag{3.18}$$

The space $D_X^-(G)$ is the quotient space of the space of all operator-valued functions $D(\lambda)$ satisfying (3.18) with respect to the subspace of all operator-valued functions for which the left-hand side of (3.18) is equal to 0.

Next, we construct the space $L_0^2(G)$. Let dg be a fixed Haar measure on G, and let $f \in L^2(G, dg)$. The integral $\int_G X(g) f(g) \, dg$ is not well-defined. Let $f \in L^1(G, dg) \cap L^2(G, dg)$. Let $\lambda(f)$ be the Fourier transform of f:

$$\lambda(f) := \int_{\hat{G}} f(g)\lambda(g) \, dg.$$

Consider the following random field:

$$Y(f) := \int_G X(g) f(g) \, dg.$$

The integral in the right-hand side exists as a Bochner integral. By Theorem 2.18, we have

$$Y(f) = \int_G \int_{\hat{G}} \text{tr}\big[\lambda(g) \, dZ(\lambda)\big] f(g) \, dg$$

$$= \int_{\hat{G}} \text{tr}\left[\int_G f(g) \lambda(g) \, dg \, dZ(\lambda)\right]$$

$$= \int_{\hat{G}} \text{tr}\big[\lambda(f) \, dZ(\lambda)\big].$$

The integral in the right-hand side exists for all $f \in L^2(G, dg)$. Indeed,

$$\mathsf{E}\left[\int_G X(g) f(g) \, dg\right]^2 = \int_{\hat{G}} \text{tr}\big[\lambda(f) \lambda^*(f) \, dF(\lambda)\big] < \infty.$$

Therefore, we may define

$$\int_G X(g) f(g) \, dg := \int_{\hat{G}} \text{tr}\big[\lambda(f) \, dZ(\lambda)\big]. \qquad (3.19)$$

The space $L_0^2(G, dg)$ is the quotient space of the space $L^2(G, dg)$ with respect to the subspace of all functions f, for which the right-hand side of (3.19) is equal to 0.

For any $f \in L_0^2(G, dg)$, the right-hand side of (3.19) determines the isometric isomorphism between the spaces $L_0^2(G, dg)$ and $D_X^0(G)$. The embedding $D_X^-(G) \supset D_X^0(G)$ may be extended to the rigged Hilbert space $D_X^-(G) \supset D_X^0(G) \supset D_X^+(G)$. The image of the restriction of the above isomorphism to the space $D_X^+(G)$ is the space $L_+^2(G, dg)$. The embedding $L_0^2(G, dg) \supset L_+^2(G, dg)$ may be extended to the rigged Hilbert space $L_-^2(G, dg) \supset L_0^2(G, dg) \supset L_+^2(G, dg)$. The distribution $f \in L_-^2(G, dg)$ may be considered as the generalised inverse Fourier transform of the corresponding function $D(\lambda) \in D_X^-(G)$.

Finally, Eq. (3.17) determines the isometric isomorphism between the following three pairs of Hilbert spaces: $D_X^-(G)$ and $H_X^-(G)$, $D_X^0(G)$ and $H_X^0(G)$, $D_X^+(G)$ and $H_X^+(G)$.

How smooth are the functions from the space $L_+^2(G, dg)$? We need a more precise description of the space $H_X^+(G)$. Assume that the measure $dF(\lambda)$ is absolutely continuous with respect to the Plancherel measure $d\mu$, and the corresponding density, $f(\lambda)$, is μ-almost everywhere invertible. Condition (3.18) takes the form

$$\int_{\hat{G}} \text{tr}\big[D(\lambda) D^*(\lambda) f(\lambda) \, d\mu(\lambda)\big] < \infty,$$

and the space $H_X^+(G)$ consists of all functions $D(\lambda)$ satisfying

$$\int_{\hat{G}} \mathrm{tr}\big[D(\lambda)D^*(\lambda)f^{-1}(\lambda)\,d\mu(\lambda)\big] < \infty. \tag{3.20}$$

On the other hand, the space $H_X^+(G)$ is the image of the space $L_+^2(G, dg)$ under the Fourier transform. The following result follows directly from Theorem 6.6.

Theorem 3.16 *Let $G = \mathrm{SU}(1, 1)$ and let all the functions satisfying (3.20) belong to the Paley–Wiener space on the group $\mathrm{SU}(1, 1)$. Then all functions from the space $L_+^2(G, dg)$ are infinitely differentiable and have compact support.*

Example 3.17 Consider a centred left-invariant random field $X(t)$ on a two-point noncompact homogeneous space T. Let C be the sphere of radius R in T with centre o. It follows from Eq. (2.45) that the elements of the space $H_X^-(T)$ have the form

$$Y = \sum_{(k,\ell)\in\hat{K}'_M} \sqrt{\frac{2\pi^{N/2}}{\pi_{k,\ell}^{(\alpha,\beta)}\,\Gamma(N/2)}} \sum_{m=1}^{\pi_{k,\ell}^{(\alpha,\beta)}} c_{k\ell}^m(R) \int_{\hat{G}_K} \varphi_{k,\ell}(\lambda; R)\,dZ_{k\ell}^m(\lambda),$$

with

$$\sum_{(k,\ell)\in\hat{K}'_M} \big[\pi_{k,\ell}^{(\alpha,\beta)}\big]^{-1} \sum_{m=1}^{\pi_{k,\ell}^{(\alpha,\beta)}} |c_{k\ell}^m(R)|^2 \int_{\hat{G}_K} |\varphi_{k,\ell}(\lambda; R)|^2\,d\mu(\lambda) < \infty.$$

The further consideration is very similar to that of Example 3.12 and may be left to the reader.

Example 3.18 Consider a centred left-invariant random field $X(t)$ on a two-point noncompact homogeneous space T, and let $C = T$. By Eq. (2.45), the elements of the space $H_X^-(T)$ should have the form

$$Y = \sum_{(k,\ell)\in\hat{K}'_M} \sqrt{\frac{2\pi^{N/2}}{\pi_{k,\ell}^{(\alpha,\beta)}\,\Gamma(N/2)}} \sum_{m=1}^{\pi_{k,\ell}^{(\alpha,\beta)}} \int_{\hat{G}_K} f_{k,\ell}^m(\lambda)\,dZ_{k\ell}^m(\lambda), \tag{3.21}$$

where the functions $f_{k,\ell}^m(\lambda)$ satisfy the following condition:

$$\sum_{(k,\ell)\in\hat{K}'_M} \big[\pi_{k,\ell}^{(\alpha,\beta)}\big]^{-1} \sum_{m=1}^{\pi_{k,\ell}^{(\alpha,\beta)}} \int_{\hat{G}_K} |f_{k,\ell}^m(\lambda)|^2\,d\mu(\lambda) < \infty. \tag{3.22}$$

The space $D_X^-(T)$ is the quotient space of the space of all functions $f_{k,\ell}^m(\lambda)$ satisfying (3.22) with respect to the subspace of all functions for which the left-hand side of (3.22) is equal to 0.

Let $(r, \mathbf{n})$ be the geodesic polar coordinates on T. By Theorem 5.8 from Helgason (2000), $dt = \sinh^{n-1} r\, dr\, d\mathbf{n}$ is a G-invariant measure on T. Let $f \in L^1(T, dt) \cap L^2(T, dt)$. By Eq. (2.45),

$$\int_T X(t) f(t)\, dt = \sum_{(k,\ell) \in \hat{K}'_M} \sqrt{\frac{2\pi^{N/2}}{\pi_{k,\ell}^{(\alpha,\beta)} \Gamma(N/2)}}$$

$$\times \sum_{m=1}^{\pi_{k,\ell}^{(\alpha,\beta)}} \int_{\hat{G}_K} \int_0^\infty f_{k\ell}^m(r) \varphi_{k,\ell}(\lambda; r) \sinh^{n-1} r\, dr\, dZ_{k\ell}^m(\lambda), \quad (3.23)$$

where

$$f_{k\ell}^m(r) = \int_{S^{n-1}} f(r, \mathbf{n}) Y_{k\ell}^m(\mathbf{n})\, d\mathbf{n}.$$

The last formulae make sense for any $f \in L^2(T, dt)$. The space $L_0^2(T, dt)$ is the quotient space of the space $L^2(T, dt)$ with respect to the subspace of all functions f for which the right-hand side of (3.23) is equal to 0.

For any $f \in L_0^2(T, dt)$, the right-hand side of (3.23) determines the isometric isomorphism between the spaces $L_0^2(T, dt)$ and $D_X^0(T)$. The embedding $D_X^-(T) \supset D_X^0(T)$ may be extended to the rigged Hilbert space $D_X^-(T) \supset D_X^0(T) \supset D_X^+(T)$. The image of the restriction of the above isomorphism to the space $D_X^+(T)$ is the space $L_+^2(T, dt)$. The embedding $L_0^2(T, dt) \supset L_+^2(T, dt)$ may be extended to the rigged Hilbert space $L_-^2(T, dt) \supset L_0^2(T, dt) \supset L_+^2(T, dt)$. The distribution $f \in L_-^2(T, dt)$ may be considered as the generalised inverse Fourier transform of the corresponding function $D(\lambda) \in D_X^-(T)$.

Finally, Eq. (3.21) determines the isometric isomorphism between the following three pairs of Hilbert spaces: $D_X^-(T)$ and $H_X^-(T)$, $D_X^0(T)$ and $H_X^0(T)$, $D_X^+(T)$ and $H_X^+(T)$.

3.4 Bibliographical Remarks

Inversion formulae for invariant random fields on compact groups and homogeneous spaces are very easy to prove. As Examples 3.1 and 3.4 show, it is enough to use a suitable version of the Peter–Weyl theorem. Example 3.1 has been considered by Yaglom (1961), while Example 3.4 is new.

Examples 3.2, 3.3, and 3.5 show that inversion formulae for invariant random fields on noncompact groups and homogeneous spaces require a suitable version of the Plancherel theorem. The case of $T = \mathbb{R}^n$ of Example 3.5 is due to Yadrenko

(1983), while the remaining material of the above examples is new. The problem of finding the inversion formula for the case when the set A has non-empty intersection with the complement to the support of the Plancherel measure, remains open.

Yadrenko (1983) considered the action of the Laplace operator to homogeneous and isotropic random fields on the space $\mathbb{R}^n$. The general case has been considered by Malyarenko (1996).

Rigged Hilbert spaces were introduced by Gel'fand and Kostyučenko (1955). Their connection to invariant random fields was established by Malyarenko (2001).

References

I.M. Gel'fand and A.G. Kostyučenko. Expansion in eigenfunctions of differential and other operators. *Dokl. Akad. Nauk SSSR (N.S.)*, 103:349–352, 1955. In Russian.

S. Helgason. *Groups and geometric analysis. Integral geometry, invariant differential operators, and spherical functions*, volume 83 of *Math. Surv. Monogr.* Am. Math. Soc., Providence, 2000. Corrected reprint of the 1984 original.

A. Malyarenko. Linear differential equations with random flows. *Theory Stoch. Process.*, 2(18):110–118, 1996.

A. Malyarenko. Rigged Hilbert spaces connected with random flows. In *Asymptotical and qualitative methods in the theory of nonlinear oscillations*, volume 2, pages 55–58. Akad. Nauk Ukrainy, Inst. Mat., Ukraine, 2001.

A.P. Prudnikov, Yu.A. Brychkov, and O.I. Marichev. *Integrals and series. Vol. 2. Special functions.* Gordon & Breach, New York, second edition, 1988.

A.P. Prudnikov, Yu.A. Brychkov, and O.I. Marichev. *Integrals and series. Vol. 3. More special functions.* Gordon & Breach, New York, 1990.

N.Ya. Vilenkin. *Special functions and the theory of group representations*, volume 22 of *Transl. Math. Monogr.* Am. Math. Soc., Providence, 1968. Translated from the Russian by V.N. Singh.

M.Ĭ. Yadrenko. *Spectral theory of random fields. Translat. Ser. Math. Eng.* Optimization Software, Publications Division, New York, 1983.

A.M. Yaglom. Second-order homogeneous random fields. In *Proc. 4th Berkeley Sympos. Math. Statist. and Probab.*, volume II, pages 593–622. University of California Press, Berkeley, 1961.

Chapter 4
Sample Path Properties of Gaussian Invariant Random Fields

Abstract Let $X(t)$ be a centred Gaussian invariant random field on a two-point homogeneous space T. The corresponding Dudley semi-metric is a function of one real variable. We use Abelian and Tauberian theorems to estimate the Dudley semi-metric and find the uniform moduli of continuity of the random field $X(t)$. The series expansion of the multiparameter fractional Brownian motion previously obtained in Chap. 2 is shown to be rate-optimal. We prove a general functional limit theorem for the multiparameter fractional Brownian motion. The functional law of the iterated logarithm, functional Lévy modulus of continuity and many other results are its particular cases. Bibliographical remarks conclude.

4.1 Introduction

Let $X(t)$ be a centred separable Gaussian random field on the metric space (T, ϱ). Assume that its covariance function, $R(t_1, t_2) = \mathsf{E}[X(t_1)X(t_2)]$, depends only on the distance between t_1 and t_2, $\varrho(t_1, t_2)$. For example, this is true when T is a two-point homogeneous space and $X(t)$ is invariant.

Let $\varrho_X(s, t)$ be the *Dudley semi-metric* generated by X:

$$\varrho_X(s, t) := \sqrt{\mathsf{E}\big[\big(X(s) - X(t)\big)^2\big]}.$$

Let $S \subset T$ be a compact subset of the metric space (T, ϱ_X). Let $U \subset S$ be a ε-net for S. In other words, let the balls of radius ε centred at the points of U, cover all of S. Let $N(S, \varepsilon)$ be the minimal possible number of points in an ε-net for S. Let $H(S, \varepsilon) = \ln N(S, \varepsilon)$ be the *metric entropy* of the set S. Finally, let

$$\mathcal{D}(S, \varepsilon) := \int_0^\varepsilon \sqrt{H(S, u)}\, du$$

be the *Dudley integral*.

Any two balls of the same radius in the space (T, ϱ_X) are isometric. In his Theorem 3 (Sect. 15) Lifshits (1995) proves that the restriction of a random field $X(t)$ to S is bounded and continuous iff the Dudley integral is finite. Moreover, let

$$\theta_{\varrho_X}(\varepsilon) := \sup_{s,t \in S} \big\{ |X(s) - X(t)| : \varrho_X(s, t) \le \varepsilon \big\}$$

A. Malyarenko, *Invariant Random Fields on Spaces with a Group Action*,
Probability and Its Applications, DOI 10.1007/978-3-642-33406-1_4,
© Springer-Verlag Berlin Heidelberg 2013

be the uniform modulus of continuity of the random field $X(t)$ on S. If the Dudley integral is finite, it is a *uniform modulus* for X in the metric ϱ_X, i.e.,

$$\limsup_{\varepsilon \downarrow 0} \frac{\theta_{\varrho_X}(\varepsilon)}{D(S, \varepsilon)} < \infty \quad \text{a.s.}$$

Let ϱ be another metric on the set S with

$$\varrho_X(s, t) \le G\big(\varrho(s, t)\big),$$

where $G(\theta)$ satisfies $G(0) = 0$ and for all $\theta_1, \theta_2 \ge 0$ we have

$$G(\theta_1) \le G(\theta_1 + \theta_2) \le G(\theta_1) + G(\theta_2).$$

Let $H^{(\varrho)}(S, \varepsilon)$ be the metric entropy of the space (S, ϱ). Then the function

$$\varepsilon \mapsto \int_0^\varepsilon \sqrt{H^{(\varrho)}(S, r)} \, dG(r)$$

is a uniform modulus of continuity for X in the metric ϱ.

Let $R(\theta)$ be the covariance between $X(s)$ and $X(t)$ when the distance between s and t is equal to θ. To estimate the Dudley integral, we have to estimate

$$\varrho_X^2(s, t) = 2\big[R(0) - R(\theta)\big]$$

in terms of the spectral measure of the random field $X(t)$. Therefore, we need to prove Abelian and/or Tauberian theorems.

4.2 Elementary Abelian Theorems

In this section, we find the upper estimates of the Dudley distance in the neighbourhood of zero for invariant random fields on two-point homogeneous spaces. These estimates are used to find the uniform moduli of continuity of the above fields.

4.2.1 The Compact Case

Let $X(t)$ be a centred Gaussian isotropic random field on a compact two-point homogeneous space. Its covariance function, $R(t_1, t_2) = \mathsf{E}[X(t_1)X(t_2)]$, depends only on the distance θ between t_1 and t_2. By (2.36), we have

$$R(\theta) = \sum_{\ell=0}^\infty \mu_\ell \frac{P_\ell^{(\alpha, \beta)}(\cos \theta)}{P_\ell^{(\alpha, \beta)}(1)},$$

where

$$\mu_\ell \geq 0, \qquad \sum_{\ell=0}^{\infty} \mu_\ell < \infty.$$

Put $b_\ell := \mu_\ell / h(T, \ell)$. Then

$$R(\theta) = \sum_{\ell=0}^{\infty} h(T, \ell) b_\ell \frac{P_\ell^{(\alpha,\beta)}(\cos\theta)}{P_\ell^{(\alpha,\beta)}(1)},$$

where

$$b_\ell \geq 0, \qquad \sum_{\ell=0}^{\infty} h(T, \ell) b_\ell < \infty.$$

The behaviour of the function $R(\theta)$ in the neighbourhood of 0 is determined by the following theorem.

Theorem 4.1 *Let $X(t)$ be a centred Gaussian isotropic random field on a compact two-point homogeneous space with covariance function $R(\theta)$. Let $\gamma(\lambda) : [0, \infty) \to [0, \infty)$ be a function satisfying the following conditions:*

1. $\lim_{\lambda \to \infty} \gamma(\lambda) = \infty$.
2. *There exists $\lambda_0 \geq 0$ such that the function $\lambda^2 \gamma^{-1}(\lambda)$ is nondecreasing on $[\lambda_0, \infty)$.*
3. $\sum_{\ell=0}^{\infty} h(T, \ell) b_\ell \gamma(\ell) < \infty$.

Then

$$R(0) - R(\theta) = O\big(\gamma^{-1}(\theta^{-1})\big), \quad \theta \downarrow 0.$$

Proof Note that

$$R(0) - R(\theta) = \sum_{\ell=0}^{\infty} h(T, \ell) b_\ell \left[1 - \frac{P_\ell^{(\alpha,\beta)}(\cos\theta)}{P_\ell^{(\alpha,\beta)}(1)} \right].$$

Yadrenko (1983) proved that

$$1 - \frac{P_\ell^{(\alpha,\alpha)}(\cos\theta)}{P_\ell^{(\alpha,\alpha)}(1)} \leq \frac{1}{2} \ell^2 \theta^2. \tag{4.1}$$

Consider the following expansion:

$$P_\ell^{(\alpha,\beta)}(u) = \sum_{k=0}^{\ell} b_{k\ell} P_\ell^{(\beta,\beta)}(u).$$

Askey (1975, formula (7.33)) proved that if $\alpha > \beta > -1$ (this is so for all projective spaces of Table 2.1), then the coefficients $b_{0\ell}, b_{1\ell}, \ldots, b_{\ell\ell}$ are nonnegative. It follows from (6.42) that the coefficients

$$a_{k\ell} = \frac{b_{k\ell} P_\ell^{(\beta,\beta)}(1)}{P_\ell^{(\alpha,\beta)}(1)}$$

of the expansion

$$\frac{P_\ell^{(\alpha,\beta)}(u)}{P_\ell^{(\alpha,\beta)}(1)} = \sum_{k=0}^\ell a_{k\ell} \frac{P_\ell^{(\beta,\beta)}(u)}{P_\ell^{(\beta,\beta)}(1)} \tag{4.2}$$

are again nonnegative. When $u = 1$, we obtain

$$\sum_{k=0}^\ell a_{k\ell} = 1.$$

Therefore,

$$1 - \frac{P_\ell^{(\alpha,\beta)}(\cos\theta)}{P_\ell^{(\alpha,\beta)}(1)} = \sum_{k=0}^\ell a_{k\ell} - \sum_{k=0}^\ell a_{k\ell} \frac{P_\ell^{(\beta,\beta)}(\cos\theta)}{P_\ell^{(\beta,\beta)}(1)}$$

$$= \sum_{k=0}^\ell a_{k\ell} \left[1 - \frac{P_\ell^{(\beta,\beta)}(\cos\theta)}{P_\ell^{(\beta,\beta)}(1)} \right]$$

$$\leq \frac{1}{2}\theta^2 \sum_{k=0}^\ell k^2 a_{k\ell}$$

by (4.1). Differentiate both sides of (4.2) at the point $u = 1$. By (6.42) and (6.43), we obtain

$$\frac{\ell(\ell + \alpha + \beta + 1)}{\alpha + 1} = \frac{1}{\beta + 1} \sum_{k=0}^\ell k(k + 2\beta + 1)a_{k\ell}.$$

For all projective spaces of Table 2.1 we have $2\beta + 1 > 0$. Then

$$\sum_{k=0}^\ell k^2 a_{k\ell} \leq \frac{\ell(\ell + \alpha + \beta + 1)(\beta + 1)}{\alpha + 1}$$

$$\leq \frac{(\alpha + \beta + 2)(\beta + 1)}{\alpha + 1}\ell^2,$$

and finally

$$1 - \frac{P_\ell^{(\alpha,\beta)}(\cos\theta)}{P_\ell^{(\alpha,\beta)}(1)} \leq \frac{(\alpha + \beta + 2)(\beta + 1)}{2\alpha + 2}\ell^2\theta^2.$$

This inequality holds also for spheres by (4.1).

Fix a positive integer k such that the function $\lambda^2 \gamma^{-1}(\lambda)$ is nondecreasing on $[k, \infty)$. Choose θ with $\theta^{-1} > k$ and let m be the integer part of θ^{-1}. Then

$$R(0) - R(\theta) = \sum_{\ell=0}^{k} h(T, \ell) b_\ell \left[1 - \frac{P_\ell^{(\alpha,\beta)}(\cos\theta)}{P_\ell^{(\alpha,\beta)}(1)} \right]$$

$$+ \sum_{\ell=k+1}^{m} h(T, \ell) b_\ell \left[1 - \frac{P_\ell^{(\alpha,\beta)}(\cos\theta)}{P_\ell^{(\alpha,\beta)}(1)} \right]$$

$$+ \sum_{\ell=m+1}^{\infty} h(T, \ell) b_\ell \left[1 - \frac{P_\ell^{(\alpha,\beta)}(\cos\theta)}{P_\ell^{(\alpha,\beta)}(1)} \right].$$

We prove that all three terms in the right-hand side are $O(\gamma^{-1}(\theta^{-1}))$. For the first sum,

$$\left| \sum_{\ell=0}^{k} h(T, \ell) b_\ell \left[1 - \frac{P_\ell^{(\alpha,\beta)}(\cos\theta)}{P_\ell^{(\alpha,\beta)}(1)} \right] \right| \leq \frac{(\alpha + \beta + 2)(\beta + 1)}{\alpha + 1} \sum_{\ell=0}^{k} h(T, \ell) b_\ell \ell^2 \theta^2$$

$$= O(\gamma^{-1}(\theta^{-1})),$$

by condition 2.

For the second sum, we have

$$\sum_{\ell=k+1}^{m} h(T, \ell) b_\ell \left[1 - \frac{P_\ell^{(\alpha,\beta)}(\cos\theta)}{P_\ell^{(\alpha,\beta)}(1)} \right] \leq \frac{(\alpha + \beta + 2)(\beta + 1)}{\alpha + 1} \theta^2 \sum_{\ell=k+1}^{m} h(T, \ell) b_\ell \ell^2.$$

$$(4.3)$$

Put

$$c_\ell := \frac{\ell^2}{\gamma(\ell)}, \quad d_\ell = \gamma(l) h(T, \ell) b_\ell, \quad D_j = \sum_{\ell=k+1}^{j} d_\ell.$$

The sum in the right-hand side of (4.3) may be estimated using the following summation by parts:

$$\sum_{\ell=k+1}^{m} c_\ell d_\ell = c_m d_m - c_{k+1} D_{k+1} - \sum_{\ell=k+1}^{m-1} (c_{\ell+1} - c_\ell) D_\ell$$

$$\leq c_m D_m$$

again by condition 2. Therefore,

$$\frac{(\alpha + \beta + 2)(\beta + 1)}{\alpha + 1} \theta^2 \sum_{\ell=k+1}^{m} h(T, \ell) b_\ell \ell^2$$

$$\leq \frac{(\alpha+\beta+2)(\beta+1)}{\alpha+1}\theta^2\frac{m^2}{\gamma(m)}\sum_{\ell=k+1}^{m}h(T,\ell)b_\ell\gamma(\ell)$$

$$\leq \frac{(\alpha+\beta+2)(\beta+1)}{\alpha+1}\theta^2\frac{\theta^{-2}}{\gamma(\theta)}\sum_{\ell=k+1}^{m}h(T,\ell)b_\ell\gamma(\ell)$$

$$= O\left(\gamma^{-1}\left(\theta^{-1}\right)\right)$$

by conditions 2 and 3.

For the third sum, use the estimate

$$\left|1-\frac{P_\ell^{(\alpha,\beta)}(\cos\theta)}{P_\ell^{(\alpha,\beta)}(1)}\right|\leq 2$$

to see that

$$\sum_{\ell=m+1}^{\infty}h(T,\ell)b_\ell\left[1-\frac{P_\ell^{(\alpha,\beta)}(\cos\theta)}{P_\ell^{(\alpha,\beta)}(1)}\right]\leq 2\gamma^{-1}\left(\theta^{-1}\right)\sum_{\ell=m+1}^{\infty}h(T,\ell)b_\ell\gamma\left(\theta^{-1}\right)$$

$$\leq 2\gamma^{-1}\left(\theta^{-1}\right)\sum_{\ell=m+1}^{\infty}h(T,\ell)b_\ell\gamma(m)$$

$$= O\left(\gamma^{-1}\left(\theta^{-1}\right)\right). \qquad \square$$

Corollary 4.2 *Let $X(t)$ and $\gamma(\lambda)$ satisfy conditions of Theorem 4.1. Put $G(\theta):=\sqrt{\gamma^{-1}(\theta^{-1})}$ and assume that*

$$\lim_{\varepsilon\downarrow 0}\ln^{1/2}\left(\varepsilon^{-1}\right)G(\varepsilon)=0, \qquad \int_0^\varepsilon\frac{G(\theta)\,d\theta}{\theta\sqrt{\ln(\theta^{-1})}}=O\left(\ln^{1/2}\left(\varepsilon^{-1}\right)G(\varepsilon)\right), \quad \varepsilon\downarrow 0.$$

Then the function $\ln^{1/2}(\varepsilon^{-1})G(\varepsilon)$ is a uniform modulus of the random field $X(t)$.

Proof The metric entropy of the compact n-dimensional manifold X from Table 2.1 is $O(n\ln(\varepsilon^{-1}))$. It follows that the function

$$\Theta(\varepsilon)=\int_0^\varepsilon\ln^{1/2}\left(r^{-1}\right)dG(r)$$

is a uniform modulus of the random field $X(t)$. Integrating by parts, we obtain

$$\Theta(\varepsilon)=\ln^{1/2}\left(\varepsilon^{-1}\right)G(\varepsilon)-\int_0^\varepsilon G(r)\,d\left(\ln\left(r^{-1}\right)^{1/2}\right)$$

$$=\ln^{1/2}\left(\varepsilon^{-1}\right)G(\varepsilon)+O\left(\ln^{1/2}\left(\varepsilon^{-1}\right)G(\varepsilon)\right).$$

Therefore, $\ln^{1/2}(\varepsilon^{-1})G(\varepsilon)$ is a uniform modulus for X. $\qquad\square$

Example 4.3 The following functions satisfy the conditions of Corollary 4.2:

$$\gamma_1(\lambda) = \ln^{1+2\delta}(1+\lambda), \quad \delta > 0,$$

$$\gamma_2(\lambda) = \lambda^{2\delta}, \quad 0 < \delta < 1.$$

Let $X_j(t)$ be a centred Gaussian isotropic random field on a compact two-point homogeneous space T, satisfying condition 3 of Theorem 4.1 with $\gamma(\lambda) = \gamma_j(\lambda)$, $j = 1, 2$. Then the functions

$$\Theta_1(\varepsilon) = \ln^{-\delta}(\varepsilon^{-1}),$$

$$\Theta_2(\varepsilon) = \varepsilon^{\delta} \ln^{1/2}(\varepsilon^{-1})$$

are uniform moduli of the corresponding random fields.

4.2.2 The Noncompact Case

Let $X(t)$ be a centred Gaussian isotropic random field on a noncompact two-point homogeneous space T. Its covariance function, $R(t_1, t_2) = \mathsf{E}[X(t_1)X(t_2)]$, depends only on the distance r between t_1 and t_2 and has the form

$$R(t_1, t_2) = \int_{\hat{G}_K} \varphi_\lambda(r) \, d\nu(\lambda),$$

where ν is a finite measure on $\hat{G}_K$.

Lemma 4.4 *There exists a positive constant $C = C(T)$ such that for $0 \le r \le 1$ and for $\lambda \in \hat{G}_K$*

$$1 - \varphi_\lambda(r) \le \min\{C(|\lambda|r)^2, 2\}.$$

Proof The function $\varphi_\lambda(r)$ is an element of a unitary matrix. Therefore, $|\varphi_\lambda(r)| \le 1$ and $1 - \varphi_\lambda(r) \le 2$. To prove the inequality $1 - \varphi_\lambda(r) \le C|\lambda|^2 r^2$, consider two cases.
   In the case of $T = \mathbb{R}^n$ the zonal spherical function has the following representation (Vilenkin, 1968, Chap. XI, Sect. 3, Subsection 2, formula (2)):

$$\varphi_\lambda(r) = \frac{\Gamma(n/2)}{\sqrt{\pi}\,\Gamma((n-1)/2)} \int_0^\pi \cos(\lambda r \cos\varphi) \sin^{n-2}\varphi \, d\varphi.$$

We have

$$1 - \varphi_\lambda(r) = \varphi_\lambda(0) - \varphi_\lambda(r)$$

$$= \frac{\Gamma(n/2)}{\sqrt{\pi}\,\Gamma((n-1)/2)} \int_0^\pi \left(1 - \cos(\lambda r \cos\varphi)\right) \sin^{n-2}\varphi \, d\varphi$$

$$\leq \frac{2\Gamma(n/2)}{\sqrt{\pi}\,\Gamma((n-1)/2)} \int_0^\pi \sin^2(\lambda r \cos\varphi/2) \sin^{n-2}\varphi \, d\varphi$$

$$\leq C\lambda^2 r^2.$$

In the remaining cases, the zonal spherical function has the following Mehler-type representation (Koornwinder, 1984): for all $\alpha > -1/2$ and $-1/2 \leq \beta \leq \alpha$

$$\varphi_\lambda^{(\alpha,\beta)}(r) = \frac{2}{(2\sinh r)^{2\alpha+1}(2\cosh r)^{2\beta+1}} \int_0^r \cos\lambda s \, A_{\alpha,\beta}(s,r) \, ds,$$

where

$$A_{\alpha,\beta}(s,r)$$

$$= \frac{2^{3\alpha+3/2}\Gamma(\alpha+1)}{\sqrt{\pi}\,\Gamma(\alpha-\beta)\Gamma(\beta+1/2)} \sinh(2r)(\cosh r)^{\beta-1/2}(\cosh r - \cosh s)^{\alpha-1/2}$$

$$\times {}_2F_1\left(1/2+\beta, 1/2-\beta; \alpha+1/2; \frac{\cosh r - \cosh s}{2\cosh r}\right),$$

where $0 < s < r$ and $-1/2 < \beta < \alpha$, with degenerate cases $(\alpha > -1/2)$

$$A_{\alpha,-1/2}(s,r) = \frac{2^{3\alpha+1/2}\Gamma(\alpha+1)}{\Gamma(\alpha+1/2)\sqrt{\pi}} \sinh r(\cosh r - \cosh s)^{\alpha-1/2},$$

$$A_{\alpha,\alpha}(s,r) = \frac{2^{3\alpha+3/2}\Gamma(\alpha+1)}{\Gamma(\alpha+1/2)\sqrt{\pi}} \sinh(2r)\left[\cosh(2r)r - \cosh(2s)\right]^{\alpha-1/2}.$$

Use the above representation and the fact that $\varphi_\lambda^{(\alpha,\beta)}(0) = 1$,

$$1 - \varphi_\lambda^{(\alpha,\beta)}(r) = \frac{2}{(2\sinh r)^{2\alpha+1}(2\cosh r)^{2\beta+1}} \int_0^r (1 - \cos\lambda s) A_{\alpha,\beta}(s,r) \, ds.$$

Substitute the above value of $A_{\alpha,\beta}(s,r)$. We obtain

$$1 - \varphi_\lambda^{(\alpha,\beta)}(r) = \frac{2^{\alpha-2\beta+3/2}\Gamma(\alpha+1)}{\sqrt{\pi}\,\Gamma(\alpha-\beta)\Gamma(\beta+1/2)(\sinh r)^{2\alpha}(\cosh r)^{\beta+1/2}}$$

$$\times \int_0^r (1 - \cos\lambda s)(\cosh r - \cosh s)^{\alpha-1/2}$$

$$\times {}_2F_1\left(1/2+\beta, 1/2-\beta; \alpha+1/2; \frac{\cosh r - \cosh s}{2\cosh r}\right) ds.$$

The hypergeometric function in the kernel $A_{\alpha,\beta}(s,r)$ is bounded above. Therefore,

$$1 - \varphi_\lambda^{(\alpha,\beta)}(r) \le \frac{C}{(\sinh r)^{2\alpha}} \int_0^r (1 - \cos \lambda s)(\cosh r - \cosh s)^{\alpha-1/2}\,ds.$$

For $0 \le s \le r \le 1$, power series show that $\cosh r - \cosh s \le c(r^2 - s^2)$. This implies that

$$1 - \varphi_\lambda^{(\alpha,\beta)}(r) \le \frac{C}{(\sinh r)^{2\alpha}} \int_0^r (1 - \cos \lambda s)(r^2 - s^2)^{\alpha-1/2}\,ds.$$

Compare the last formula with the last display at p. 2271 of Bray and Pinsky (2008):

$$1 - 2^\alpha \Gamma(\alpha + 1)\frac{J_\alpha(\lambda r)}{(\lambda r)^\alpha} = \frac{4\Gamma(\alpha + 1)r^{-2\alpha}}{\sqrt{\pi}\,\Gamma(\alpha + 1/2)} \int_0^r (1 - \cos \lambda s)(r^2 - s^2)^{\alpha-1/2}\,ds.$$

For $\lambda \in [0, \infty)$ we obtain

$$1 - \varphi_\lambda^{(\alpha,\beta)}(r) \le \frac{Cr^{2\alpha}}{(\sinh r)^{2\alpha}}\left(1 - 2^\alpha \Gamma(\alpha + 1)\frac{J_\alpha(\lambda r)}{(\lambda r)^\alpha}\right)$$

$$\le C\lambda^2 r^2,$$

by the first part of the proof. For $\lambda = i\omega \in (0, is_0) \cup \{i\varrho\}$ and $r \in [0, 1]$ we have

$$\left|1 - 2^\alpha \Gamma(\alpha + 1)\frac{J_\alpha(i\omega r)}{(i\omega r)^\alpha}\right| \le \frac{2\Gamma(\alpha + 1)}{\sqrt{\pi}\,\Gamma(\alpha)} \int_0^\pi \sinh^2(\omega r \cos \varphi/2) \sin^{n-2} \varphi\,d\varphi$$

$$\le C \sinh^2 \varrho(\omega r)^2,$$

because the graph of the convex function $y = \sinh x$ with $0 \le x \le \varrho$ lies under the tangent line $y = \sinh \varrho x$. $\qquad\square$

Theorem 4.5 *Let $X(t)$ be a centred Gaussian isotropic random field on a noncompact two-point homogeneous space T with covariance function $R(r)$. Let $\gamma(\lambda)$ be a nonnegative differentiable function on $[0, \infty)$ satisfying the following conditions.*

1. $\lim_{\lambda\to\infty} \gamma(\lambda) = \infty$.
2. *There exists $\lambda_0 \ge 0$ such that the function $\lambda^2\gamma^{-1}(\lambda)$ is nondecreasing on $[\lambda_0, \infty)$.*
3. $\int_0^\infty \gamma(\lambda)\,d\nu(\lambda) < \infty$.

Then

$$R(0) - R(r) = O\bigl(\gamma^{-1}(r^{-1})\bigr), \quad r \downarrow 0.$$

Proof For the case of $T = \mathbb{R}^n$ this assertion has been proved by Yadrenko (1983, Chap. 2, Sect. 2, Theorem 6). For the remaining cases, assume that $r < \lambda_0^{-1}$ and write

$$R(0) - R(r) = \int_0^{\lambda_0} \left[1 - \varphi_\lambda^{(\alpha,\beta)}(r)\right] d\upsilon(\lambda) + \int_{\lambda_0}^{r^{-1}} \left[1 - \varphi_\lambda^{(\alpha,\beta)}(r)\right] d\upsilon(\lambda)$$

$$+ \int_{r^{-1}}^\infty \left[1 - \varphi_\lambda^{(\alpha,\beta)}(r)\right] d\upsilon(\lambda) + \int_{+0}^\varrho \left[1 - \varphi_{i\omega}^{(\alpha,\beta)}(r)\right] d\upsilon(\omega).$$

By Lemma 4.4 and condition 2, the first integral for small r is estimated as

$$\int_0^{\lambda_0} \left[1 - \varphi_\lambda^{(\alpha,\beta)}(r)\right] d\upsilon(\lambda) \le Cr^2 \int_0^{\lambda_0} \lambda^2 \, d\upsilon(\lambda)$$

$$= O\left(\gamma^{-1}\left(r^{-1}\right)\right).$$

The fourth integral is estimated similarly.

For the second integral, we have

$$\int_{\lambda_0}^{r^{-1}} \left[1 - \varphi_\lambda^{(\alpha,\beta)}(r)\right] d\upsilon(\lambda) \le Cr^2 \int_{\lambda_0}^{r^{-1}} \lambda^2 \, d\upsilon(\lambda).$$

Integrating by parts, we obtain

$$\int_{\lambda_0}^{r^{-1}} \lambda^2 \, d\upsilon(\lambda) = \int_{\lambda_0}^{r^{-1}} \lambda^2 \gamma^{-1}(\lambda) \, d\left[\int_0^\lambda \gamma(u) \, d\upsilon(u)\right]$$

$$= \lambda^2 \gamma^{-1}(\lambda) \int_0^\lambda \gamma(u) \, d\upsilon(u)\big|_{\lambda_0}^{r^{-1}} - \int_{\lambda_0}^{r^{-1}} \gamma(\lambda) \frac{d}{d\lambda}\left(\lambda^2 \gamma^{-1}\right) d\upsilon(\lambda).$$

It follows from condition 2 that

$$\frac{d}{d\lambda}\left(\lambda^2 \gamma^{-1}\right) \ge 0, \quad \lambda > \lambda_0.$$

Therefore,

$$r^2 \int_{\lambda_0}^{r^{-1}} \lambda^2 \, d\upsilon(\lambda) \le \gamma^{-1}\left(r^{-1}\right) \int_{\lambda_0}^{r^{-1}} \gamma(\lambda) \, d\upsilon(\lambda).$$

It follows from the above inequality and condition 3 that

$$\int_{\lambda_0}^{r^{-1}} \left[1 - \varphi_\lambda^{(\alpha,\beta)}(r)\right] d\upsilon(\lambda) = O\left(\gamma^{-1}\left(r^{-1}\right)\right).$$

By Lemma 4.4,

$$\int_{r^{-1}}^{\infty} \left[1 - \varphi_\lambda^{(\alpha,\beta)}(r)\right] d\nu(\lambda) \leq 2 \int_{r^{-1}}^{\infty} d\nu(\lambda)$$

$$\leq 2\gamma^{-1}(r^{-1}) \int_{r^{-1}}^{\infty} \gamma(\lambda) d\nu(\lambda)$$

$$= O(\gamma^{-1}(r^{-1})). \qquad \square$$

Corollary 4.2 and Example 4.3 are transferred to the case of noncompact space with the only change: instead of referring to Theorem 4.1 we refer to Theorem 4.5.

4.3 Advanced Abelian and Tauberian Theorems

In this section, we consider more advanced Abelian and Tauberian theorems in terms of regularly varying functions. The above theorems help us to refine the moduli of continuity found in the previous section.

4.3.1 The Compact Case

Let $X(t)$ be a centred Gaussian isotropic random field on a compact two-point homogeneous space T with covariance function

$$R(\theta) = \sum_{\ell=0}^{\infty} h(T, \ell) b_\ell R_\ell^{(\alpha,\beta)}(\cos\theta),$$

where

$$R_\ell^{(\alpha,\beta)}(\cos\theta) = \frac{P_\ell^{(\alpha,\beta)}(\cos\theta)}{P_\ell^{(\alpha,\beta)}(1)}.$$

Note that $h(T, \ell) = w_\ell^{(\alpha,\beta)}$, where the right-hand side is calculated by (6.60). The incremental variance of the random field $X(t)$ is

$$\sigma^2(\theta) = 2 \sum_{\ell=1}^{\infty} w_\ell^{(\alpha,\beta)} b_\ell \left[1 - R_\ell^{(\alpha,\beta)}(\cos\theta)\right]. \qquad (4.4)$$

Introduce the following notation:

$$B_\ell := 2 \sum_{j=\ell}^{\infty} w_j^{(\alpha,\beta)} b_j.$$

Suppose that the coefficients B_ℓ satisfy the following condition: there exist a real number $\gamma \in [0, 1]$ and a slowly varying function $L(x)$ such that

$$B_\ell \sim \ell^{-2\gamma} L(\ell) \tag{4.5}$$

as $\ell \to \infty$. The symbol $\sim$ means that the ratio of the two expressions connected with $\sim$ tends to 1 as $\ell \to \infty$. Moreover, if $\gamma = 0$, then $L(x) \to 0$ as $x \to \infty$ (otherwise the remainder of the convergent series, B_ℓ, cannot converge to 0 as $\ell \to \infty$).

Lemma 4.6 *The incremental variance of the random field $X(t)$ satisfies the asymptotic relation*

$$\sigma^2(\theta) \sim \theta^2 \left(\frac{B_1(\alpha + \beta + 2)}{4(\alpha + 1)} + \sum_{\ell=1}^{\infty} \frac{c_\ell w_\ell^{(\alpha+1,\beta)}}{\ell^{2\alpha+4}} R_\ell^{(\alpha+1,\beta)}(\cos\theta) \right) \tag{4.6}$$

as $\theta \downarrow 0$, where

$$c_\ell := \frac{(2\ell + \alpha + \beta + 2)B_{\ell+1} \cdot \ell^{2\alpha+4}}{4(\alpha + 1)w_\ell^{(\alpha+1,\beta)}}. \tag{4.7}$$

Proof Rewrite Eq. (4.4) as

$$\sigma^2(\theta) = \sum_{\ell=1}^{\infty} (B_\ell - B_{\ell+1})\left[1 - R_\ell^{(\alpha,\beta)}(\cos\theta)\right].$$

Consider the partial sum by ℓ from 1 to k of the terms in the right-hand side and use the following summation-by-parts formula:

$$\sum_{\ell=1}^{k} (B_\ell - B_{\ell+1})\left[1 - R_\ell^{(\alpha,\beta)}(\cos\theta)\right] = B_1\left[R_0^{(\alpha,\beta)}(\cos\theta) - R_1^{(\alpha,\beta)}(\cos\theta)\right]$$

$$+ \sum_{\ell=1}^{k-1} B_{\ell+1}\left[R_\ell^{(\alpha,\beta)}(\cos\theta) - R_{\ell+1}^{(\alpha,\beta)}(\cos\theta)\right]$$

$$- B_{k+1}\left[1 - R_k^{(\alpha,\beta)}(\cos\theta)\right]. \tag{4.8}$$

The term B_{k+1} tends to 0 as $k \to \infty$ as a remainder of the convergent series. For $\alpha \geq \beta \geq -1/2$ by Szegö (1975), (7.32.2):

$$|R_k^{(\alpha,\beta)}(\cos\theta)| \leq 1.$$

Table 4.1 The regions corresponding to different types of convergence of (4.6) for compact spaces T

T	Absolute convergence	Conditional convergence	Abel summability
S^1	$(1/2, 1)$	$(0, 1/2]$	$\{0\}$
$S^2, \mathbb{R}P^2$	$(1/4, 1)$	$[0, 1/4]$	$\varnothing$
$S^3, \mathbb{R}P^3$	$(0, 1)$	$\{0\}$	$\varnothing$
Others	$[0, 1)$	$\varnothing$	$\varnothing$

It follows that the last term in the right-hand side of (4.8) tends to 0 as $k \to \infty$. Passing to the limit, we obtain

$$\sigma^2(\theta) = B_1\left[R_0^{(\alpha,\beta)}(\cos\theta) - R_1^{(\alpha,\beta)}(\cos\theta)\right]$$

$$+ \sum_{\ell=1}^{\infty} B_{\ell+1}\left[R_\ell^{(\alpha,\beta)}(\cos\theta) - R_{\ell+1}^{(\alpha,\beta)}(\cos\theta)\right]. \tag{4.9}$$

According to Eq. 10.8(32) from Erdélyi et al. (1953) we have

$$R_\ell^{(\alpha,\beta)}(\cos\theta) - R_{\ell+1}^{(\alpha,\beta)}(\cos\theta) = \frac{(2\ell + \alpha + \beta + 2)\sin^2(\theta/2)}{\alpha + 1} R_\ell^{(\alpha+1,\beta)}(\cos\theta).$$

Substitute this Eq. into (4.9) and use the asymptotic relation $\sin^2(\theta/2) \sim \theta^2/4$ as $\theta \downarrow 0$. We obtain (4.6) and (4.7). $\qquad\square$

Next, we determine the asymptotic behaviour of the sequence $\{c_\ell : \ell \geq 1\}$. Using (4.7) and Stirling's formula, we obtain

$$c_\ell \sim \frac{\Gamma(\alpha+1)\Gamma(\alpha+\beta+3)}{4\Gamma(\beta+1)}\ell^{2-2\gamma}L(\ell)$$

as $\ell \to \infty$. Using the results of Sect. 6.6, we obtain the following lemma.

Lemma 4.7 *If $0 < 2 - 2\gamma < \alpha + 3/2$, then the series in the right-hand side of (4.6) converges absolutely. If $\alpha + 3/2 \leq 2 - 2\gamma < \alpha + 5/2$, then the above series converges conditionally. If $\alpha + 5/2 \leq 2 - 2\gamma < 2\alpha + 4$, then the above series is Abel summable.*

Using values of α for various compact two-point homogeneous spaces (Table 2.1), we obtain the result given in Table 4.1.

Apply the Abelian theorem for Jacobi series in the case of absolute convergence (Theorem 6.8). The following theorem follows.

Theorem 4.8 *Let $X(t)$ be a centred Gaussian isotropic random field on a compact two-point homogeneous space T satisfying (4.5). Assume that γ lies in the region of* *absolute convergence (Table 4.1, column 2). Then the incremental variance of the random fields $X(t)$ has the following asymptotic behaviour:*

$$\sigma^2(\theta) \sim \frac{\Gamma(1-\gamma)\Gamma(\alpha+1)}{2^{2\gamma}\Gamma(\alpha+\gamma+1)}\theta^{2\gamma}L(\theta^{-1}) \tag{4.10}$$

as $\theta \downarrow 0$.

The following corollary may be deduced from Theorem 4.8 in the same way as Corollary 4.2 has been deduced from Theorem 4.1.

Corollary 4.9 *Let $X(t)$ be a random field satisfying conditions of Theorem 4.8. Assume that the function $L(x)$ satisfies the following conditions:*

$$\lim_{\varepsilon \downarrow 0} \varepsilon^\gamma \sqrt{L(\varepsilon^{-1})\ln(\varepsilon^{-1})} = 0,$$

$$\int_0^\varepsilon \frac{\sqrt{L(\theta^{-1})}\,d\theta}{\theta^{1-\gamma}\sqrt{\ln(\theta^{-1})}} = O\left(\varepsilon^\gamma \sqrt{L(\varepsilon^{-1})\ln(\varepsilon^{-1})}\right). \tag{4.11}$$

Then the function $\theta^\gamma \sqrt{L(\theta^{-1})\ln(\theta^{-1})}$ is a uniform modulus of the random field $X(t)$.

Example 4.10 Under conditions of Corollary 4.9, let $\gamma > 0$ and $L(x) = 1$. Both conditions in (4.11) are satisfied. Therefore, the function $\theta^\gamma \sqrt{\ln(\theta^{-1})}$ is a uniform modulus of the random field $X(t)$.

Let $\gamma = 0$ and $L(x) = \ln^{2\delta} x$, $\delta \in \mathbb{R}$. The first condition in (4.11) is satisfied when $\delta < -1/2$. The integral in the left-hand side of the second condition is calculated as follows:

$$\int_0^\varepsilon \ln^{\delta-1/2}(\theta^{-1})\theta^{-1}\,d\theta = -\frac{\ln^{\delta+1/2}\theta}{\delta+1/2}\Big|_0^\varepsilon$$

$$= \frac{\ln^{\delta+1/2}(\varepsilon^{-1})}{\delta+1/2}.$$

It follows that the second condition in (4.11) is also satisfied and the function $\ln^{\delta+1/2}(\theta^{-1})$ is a uniform modulus of the random field $X(t)$.

On the other hand, if $\gamma = 0$ and $\delta \geq -1/2$, then the Dudley integral diverges and the corresponding random field is discontinuous.

The Tauberian theorem for Jacobi series in the case of absolute convergence (Theorem 6.9) implies the following result.

Theorem 4.11 *Let $X(t)$ be a centred Gaussian isotropic random field on a compact two-point homogeneous space. Assume that γ lies in the region of absolute convergence (Table 4.1, column 2) and that the incremental variance of the random fields*

$X(t)$ satisfies (4.10). If the coefficients c_ℓ satisfy any of the Tauberian conditions 1–3 of Theorem 6.9, then the numbers B_ℓ satisfy (4.5).

Apply the Abelian theorem for Jacobi series in the case of conditional convergence (Theorem 6.12). The following theorem follows.

Theorem 4.12 *Let $X(t)$ be a centred Gaussian isotropic random field on a compact two-point homogeneous space T satisfying the following condition: there exist a real number γ lying in the region of conditional convergence (Table 4.1, column 3) and a slowly varying and quasi-monotone function $L(x)$ such that*

$$c_\ell = \ell^{2-2\gamma} L(\ell).$$

Then the incremental variance of the random fields $X(t)$ satisfies (4.10).

Corollary 4.13 *Let $X(t)$ be a random field satisfying conditions of Theorem 4.12. Assume that the function $L(x)$ satisfies (4.11). Then the function $\theta^\gamma \sqrt{L(\theta^{-1}) \ln(\theta^{-1})}$ is a uniform modulus of the random field $X(t)$.*

Example 4.14 Under conditions of Corollary 4.13, put $L(x) := \ln^{2\delta}(1 + x)$, $\delta \in \mathbb{R}$. Hence $L(x)$ is monotone and slowly varying, so it is quasi-monotone (Bingham et al., 1987, Subsection 2.7.1). It follows from (4.7) that

$$B_{\ell+1} = \frac{4(\alpha + 1)w_\ell^{(\alpha+1,\beta)} \ln^{2\delta}(\ell + 1)}{(2\ell + \alpha + \beta + 2)\ell^{2(1+\gamma+\alpha)}}$$

and

$$b_\ell = 2(\alpha + 1)\left(\frac{\ln^{2\delta}(\ell)}{(2\ell + \alpha + \beta)w_\ell^{(\alpha+1,\beta)}(\ell - 1)^{2(1+\gamma+\alpha)}}\right.$$
$$\left. - \frac{\ln^{2\delta}(\ell + 1)}{(2\ell + \alpha + \beta + 2)\ell^{2(1+\gamma+\alpha)}}\right).$$

If $\gamma > 0$, then the function $\theta^\gamma \ln^{\delta+1/2}(\theta^{-1})$ is a uniform modulus of the random field $X(t)$. If $\gamma = 0$ and $\delta < -1/2$, then the function $\ln^{\delta+1/2}(\theta^{-1})$ is a uniform modulus of the random field $X(t)$.

On the other hand, if $\gamma = 0$ and $-1/2 \le \delta < 0$, then the Dudley integral diverges and the corresponding random field is discontinuous. This statement can be proved in exactly the same way as in Example 4.10.

The Tauberian theorem for Jacobi series in the case of conditional convergence (Theorem 6.13) implies the following result.

Theorem 4.15 *Let $X(t)$ be a centred Gaussian isotropic random field on a compact two-point homogeneous space. Assume that γ lies in the region of conditional* convergence (*Table* 4.1, *column* 3) *and that the incremental variance of the random fields* $X(t)$ *satisfies* (4.10). *If the coefficients* c_ℓ *satisfy the Tauberian condition of Theorem* 6.13, *then the numbers* B_ℓ *satisfy* (4.5).

Apply the Abelian theorem for Jacobi series in the case of Abel summability (Theorem 6.14). The following theorem follows.

Theorem 4.16 *Let* $X(t)$ *be a centred Gaussian isotropic random field on* S^1 *satisfying the following condition*:

$$c_\ell = \ell^2 L(\ell),$$

where $L(\ell)$ *and* $\ell[L(\ell + 1) - L(\ell)]$ *are slowly varying and quasi-monotone sequences. Then the incremental variance of the random field* $X(t)$ *satisfies*

$$\sigma^2(\theta) \sim 4L(\theta^{-1}).$$

Corollary 4.17 *Let* $X(t)$ *be a random field satisfying conditions of Theorem* 4.16. *Assume that the function* $L(x)$ *satisfies the following condition*:

$$\int_0^\varepsilon \frac{\sqrt{L(\theta^{-1})}\,d\theta}{\theta\sqrt{\ln(\theta^{-1})}} = o(\varepsilon).$$

Then the function $\sqrt{L(\theta^{-1})\ln(\theta^{-1})}$ *is a uniform modulus of the random field* $X(t)$.

Example 4.18 In his Complement 4, Bingham (1978) proves that if $M(u)$ is slowly varying, then

$$L(x) := \int_1^x \frac{M(u)\,du}{u}$$

satisfies the following condition: $n[L(n + 1) - L(n)]$ is slowly varying. Put $M(u) := \ln(u)/2$. It follows that $L(\ell) = \ln^2(\ell)$ satisfies the above condition. To check that both $\ln^2(x)$ and $x[\ln^2(x + 1) - \ln^2(x)]$ are quasi-monotone, use the Quasi-Monotonicity Theorem 6.11. For the first function, put $f_1(x) := \ln^2(x)$ and $f_2(x) := 1$. For the second function, put $f_1(x) := \sqrt{x}\ln(x^2 + x)$ and $f_2(x) := [\sqrt{x}\ln(1 + x^{-1})]^{-1}$.

Let $X(t)$ be a centred Gaussian isotropic random field on S^1 with $c_\ell = \ell^2 \ln^2(\ell)$. Its incremental variance has the following asymptotic behaviour:

$$\sigma^2(\theta) \sim 4\ln^2(\theta^{-1}).$$

The Dudley integral

$$\int_0^\varepsilon \ln^{1/2}(\theta^{-1})\,d\ln(\theta^{-1})$$

diverges, therefore $X(t)$ is discontinuous.

The Tauberian theorem for Jacobi series in the case of Abel summability (Theorem 6.15) implies the following result.

Theorem 4.19 *Let $X(t)$ be a centred Gaussian isotropic random field on S^1. Let its incremental variance satisfy*

$$\sigma^2(\theta) \sim L(\theta^{-1})$$

as $\theta \downarrow 0$. Then the numbers B_ℓ satisfy

$$B_\ell \sim L(\ell)$$

as $\ell \to \infty$.

Consider the limiting case of $\gamma = 1$. It is easy to see that if the series $\sum_{\ell=1}^{\infty} L(\ell)/\ell$ converges, then the series in the right-hand side of (4.6) absolutely converges. In this case we have

$$\sigma^2(\theta) \sim \frac{B_1(\alpha + \beta + 2)}{4(\alpha + 1)}\theta^2$$

as $\theta \downarrow 0$.

Example 4.20 Let $X(t)$ be a centred Gaussian isotropic random field on a compact two-point homogeneous space with $B_0 = B_1 = B_2 = (2\ln(2))^{-2}$ and $B_\ell = (\ell \ln(\ell))^{-2}$, $\ell \geq 3$. In this case $L(\ell) = \ln^{-2}(\ell)$, the series $\sum_{\ell=1}^{\infty} L(\ell)/\ell$ converges. We have

$$\sigma^2(\theta) \sim \frac{\alpha + \beta + 2}{16 \ln^2(2)(\alpha + 1)}\theta^2.$$

Because

$$\int_0^\varepsilon \frac{d\theta}{\sqrt{\ln(\theta^{-1})}} = o(\varepsilon), \quad \varepsilon \downarrow 0,$$

we obtain that the function $\theta \ln^{1/2}(\theta^{-1})$ is a uniform modulus of the random field $X(t)$.

Assume that the series $\sum_{\ell=1}^{\infty} L(\ell)/\ell$ diverges. The Abelian theorem in the limiting case (Theorem 6.16) implies the following result.

Theorem 4.21 *Let $X(t)$ be a centred Gaussian isotropic random field on a compact two-point homogeneous space T satisfying*

$$B_\ell \sim \ell^{-2}L(\ell)$$

as $\ell \to \infty$. Assume that the series $\sum_{\ell=1}^{\infty} L(\ell)/\ell$ diverges. Then the incremental variance of the random fields $X(t)$ has the following asymptotic behaviour:

$$\sigma^2(\theta) \sim \frac{\theta^2}{2(\alpha+1)} \int_0^{\theta^{-1}} \frac{L(u)\, du}{u}$$

as $\theta \downarrow 0$.

Example 4.22 Let $X(t)$ be a centred Gaussian isotropic random field on a compact two-point homogeneous space T satisfying $B(0) = B(1) = B(2) = \ln(2)/4$ and $B(\ell) = \ell^{-2}\ln(\ell)$, $\ell \geq 3$. Its incremental variance has the following asymptotic behaviour:

$$\sigma^2(\theta) \sim \frac{\theta^2 \ln^2(\theta^{-1})}{4(\alpha+1)}.$$

Because

$$\int_0^{\varepsilon} \sqrt{\ln(\theta^{-1})}\, d\theta = o(\varepsilon), \quad \varepsilon \downarrow 0,$$

we obtain that the function $\theta \ln^{3/2}(\theta^{-1})$ is a uniform modulus of the random field $X(t)$.

4.3.2 The Noncompact Case

Let $X(t)$ be a centred Gaussian isotropic random field on a noncompact two-point homogeneous space T. Its autocorrelation function has the form

$$R(t_1, t_2) = \int_{\hat{G}_K} \varphi_\lambda\left(a_2^{-1} k_2^{-1} k_1 a_1\right) d\nu(\lambda),$$

where $t_j = g_j K \in T$, $g_j = k_j a_j$, $j = 1, 2$. Using formula (2.40), we obtain

$$R(t_1, t_2) = \sum_{\mu \in \hat{K}'_M} \int_{\hat{G}_K} \varphi_{\lambda,\mu}(a_1)\overline{\varphi_{\lambda,\mu}(a_2)}\, \mathrm{Re}\, \varphi_\mu\left(k_2^{-1} k_1\right) d\nu(\lambda),$$

where $\varphi_{\lambda,\mu}(a)$ are associated spherical functions (multiplied by $\sqrt{c_T}$, where c_T is given by (2.41)), and $\varphi_\mu(k)$ are the zonal spherical functions of the representation μ.

Let $(r, \mathbf{n})$ be the geodesic polar coordinates on T, and let $\theta = \theta(t_1, t_2)$ be the length of the geodesic line drawn between the points t_1 and t_2. For θ small enough we can find such an element $g \in G$ that the points gt_1 and gt_2 lie on the surface of the sphere $S^{N-1} = \{r = 1\}$. By definition of the isotropic random field we have $R(gt_1, gt_2) = R(t_1, t_2)$. Therefore, we can suppose that $t_1, t_2 \in S^{N-1}$. In other words, we consider the *restriction* $X(\mathbf{n})$ of the random field $X(t)$ to the sphere S^{N-1}. The above restriction is invariant under the group K (Table 2.2, column 3). Its incremental variance is

$$\sigma^2(\theta) = 2 \sum_{\mu \in \hat{K}'_M} \int_{\hat{G}_K} \varphi_{\lambda,\mu}(1)\overline{\varphi_{\lambda,\mu}(1)}\left[1 - \operatorname{Re}\varphi_\mu\left(k_2^{-1}k_1\right)\right] d\nu(\lambda).$$

The set $\hat{K}'_M$ may be identified with some set of pairs (k, ℓ) of nonnegative integers. In the case of $T = \mathbb{R}^n$ (Table 2.3, line 1) we have $\hat{K}'_M = \{(k, 0) : k \geq 0\}$. The zonal spherical function is

$$\varphi_k(\psi) = R_k^{(n-3)/2,(n-3)/2}(\cos \psi),$$

and the incremental variance may be written as

$$\sigma^2(\theta) = \sum_{k=0}^{\infty} b_k\left(1 - R_k^{(n-3)/2,(n-3)/2}(\cos \psi)\right),$$

where $\psi = \psi(\theta)$ is the distance between the points t_1 and t_2 along the geodesic line on the sphere. To calculate b_k, use (2.39):

$$b_k = 2^{n-1}\Gamma^2(n/2) \int_0^\infty \frac{J_{(n-2)/2}^2(\lambda)}{\lambda^{n-2}} d\nu(\lambda). \tag{4.12}$$

It is clear that $\psi(\theta) \sim \theta$ as $\theta \downarrow 0$. By Lemma 4.6, we obtain

$$\sigma^2(\theta) \sim \theta^2\left(\frac{B_1(\alpha' + \beta' + 2)}{4(\alpha' + 1)} + \sum_{k=1}^{\infty} \frac{c_k w_k^{(\alpha'+1,\beta')}}{k^{2\alpha'+4}} R_k^{(\alpha'+1,\beta')}(\cos \theta)\right) \tag{4.13}$$

as $\theta \downarrow 0$, where

$$c_k = \frac{(2k + \alpha' + \beta' + 2)B_{k+1} \cdot k^{2\alpha'+4}}{4(\alpha' + 1)w_k^{(\alpha'+1,\beta')}}$$

with $\alpha' = \beta' = (n - 3)/2$.

In the remaining cases, consider the following region:

$$\Omega = \left\{(x, y) \in \mathbb{R}^2 : x^2 \leq y \leq 1\right\}$$

with measure

$$dm_{\alpha,\beta}(x, y) = \frac{\Gamma(\alpha + \beta + 5/2)}{\Gamma(\alpha + 1)\Gamma(\beta + 1)\sqrt{\pi}}(1 - y)^\alpha\left(y - x^2\right)^\beta dx\, dy,$$

where $\alpha > -1$, $\beta > -1$. The polynomials

$$R_{k,\ell}^{(\alpha,\beta)}(x, y) = R_\ell^{(\alpha,\beta+k-\ell+1/2)}(2y - 1)y^{(k-\ell)/2} R_{k-\ell}^{(\beta,\beta)}(x/\sqrt{y}), \quad k \in \mathbb{Z}_+, \ 0 \leq \ell \leq k$$

form a complete orthogonal system in the space $L^2(\Omega, dm_{\alpha,\beta})$.

As $\alpha \downarrow -1$, the measure $dm_{\alpha,\beta}$ weakly converges to a measure supported on the edge $y = 1$ of Ω, and we get back Jacobi polynomials $R_k^{(\beta,\beta)}(x)$. As $\beta \downarrow -1$, the measure $dm_{\alpha,\beta}$ weakly converges to a measure supported on the edge $y = x^2$ of Ω, and we get back Jacobi polynomials $R_{k+\ell}^{(\alpha,\alpha)}(x)$ with $\ell = k$ or $\ell = k - 1$.

By Theorem 5 from Flensted-Jensen and Koornwinder (1979), for any space T lying in lines 2–5 of Table 2.2 there exists a set Ω' such that the orthogonal system $\{\operatorname{Re} \varphi_\mu : \mu \in \hat{K}_M'\}$ coincides with the orthogonal system $R_{k,\ell}^{(\alpha-\beta-1,\beta-\frac{1}{2})}(u \cos \psi, u^2)$, where (k, ℓ) are given in the second column of Table 2.3.

In particular, when $T = \mathbb{C}\Lambda^1$, the set Ω' is described by $u = 1$, $\psi \in [0, \pi]$. We have $\alpha - \beta - 1 = -1$, so

$$\varphi_k(\psi) = R_k^{0,0}(\cos \psi),$$

and the asymptotic behaviour of the incremental variance is described by (4.13) with $\alpha' = \beta' = 0$. The coefficients b_k are given by

$$b_k := 2 \int_{\hat{G}_K} \left| \varphi_{k,0}^{(0,0)}(\lambda; 1) \right|^2 d\nu(\lambda). \tag{4.14}$$

When $T = \mathbb{R}\Lambda^n$, the set Ω' is described by $u \in [0, 1]$, $\psi \in \{0, \pi\}$. We have $\beta - 1/2 = -1$, so

$$\varphi_k(\psi) = R_k^{((n-2)/2,(n-2)/2)}(\cos \psi),$$

and the asymptotic behaviour of the incremental variance is described by (4.13) with $\alpha' = \beta' = (n-2)/2$. The coefficients b_k are given by

$$b_{2k} := 2 \int_{\hat{G}_K} \left| \varphi_{k,k}^{((n-2)/2,-1/2)}(\lambda; 1) \right|^2 d\nu(\lambda), \quad k \geq 0,$$

$$b_{2k-1} := 2 \int_{\hat{G}_K} \left| \varphi_{k,k-1}^{((n-2)/2,-1/2)}(\lambda; 1) \right|^2 d\nu(\lambda), \quad k \geq 1. \tag{4.15}$$

In the remaining cases, we have $\Omega' = [0, 1] \times [0, \pi]$. Choose the points t_1 and t_2 at a distance θ in such a way that $u = 1$. Then

$$\varphi_{k,\ell}(\psi) = R_{k-\ell}^{(\beta-1/2,\beta-1/2)}(\cos \psi),$$

and the asymptotic behaviour of the incremental variance is described by (4.13) with $\alpha' = \beta' = \beta - 1/2$. To calculate the coefficients b_k, we have to group the terms with the same value of $k - \ell$:

$$b_k := 2 \sum_{\ell=0}^{\infty} \int_{\hat{G}_K} \left| \varphi_{k+\ell,\ell}^{(\alpha,\beta)}(\lambda; 1) \right|^2 d\nu(\lambda). \tag{4.16}$$

The results of our investigation are shown in Table 4.2. Using values of α' from Table 4.2, we obtain the following result (Table 4.3).

Table 4.2 The coefficients α', β', and b_k for noncompact spaces T

T	α', β'	b_k
$\mathbb{R}^n$	$(n-3)/2$	(4.12)
$\mathbb{C}\Lambda^1$	0	(4.14)
$\mathbb{R}\Lambda^n$	$(n-2)/2$	(4.15)
Others	$\beta - 1/2$	(4.16)

Table 4.3 The regions corresponding to different types of convergence of (4.13) for noncompact spaces T

T	Absolute convergence	Conditional convergence	Abel summability
$\mathbb{R}^2, \mathbb{R}\Lambda^2$	$(1/2, 1)$	$(0, 1/2]$	$\{0\}$
$\mathbb{R}^3, \mathbb{R}\Lambda^3, \mathbb{C}\Lambda^n$	$(1/4, 1)$	$[0, 1/4]$	$\varnothing$
$\mathbb{R}^4, \mathbb{R}\Lambda^4$	$(0, 1)$	$\{0\}$	$\varnothing$
Others	$[0, 1)$	$\varnothing$	$\varnothing$

The results of Subsection 4.3.1 may be transferred to the case of noncompact space T, using Table 4.3. Put

$$B_\ell := 2 \sum_{k=\ell}^{\infty} w_j^{((N-2)/2,(N-2)/2)} b_k,$$

where b_k is calculated by the corresponding formula shown at the third column of Table 4.2. For example, the analogue of Theorem 4.8 has the following form.

Theorem 4.23 *Let $X(t)$ be a centred Gaussian isotropic random field on a noncompact two-point homogeneous space T satisfying (4.5). Assume that γ lies in the region of absolute convergence (Table 4.3, column 2). Then the asymptotic behaviour of the incremental variance of the random field $X(t)$ is described by (4.10).*

Note that our methods do not work for the case of $T = \mathbb{R}^1$. Abelian and Tauberian theorems for this case can be found in Pitman (1968).

It would be interesting to obtain conditions equivalent to (4.5) directly in terms of the asymptotic behaviour of the measure ν near infinity.

4.4 An Optimal Series Expansion of the Multiparameter Fractional Brownian Motion

In Example 2.39 we proved the following series expansion:

$$
X(s, \theta_1, \ldots, \theta_{N-2}, \varphi)
$$

$$
= \frac{2^{H+1} \sqrt{\pi^{(N-2)/2} \Gamma(N/2 + H) \Gamma(H + 1) \sin(\pi H)} R^H}{\Gamma(N/2)}
$$

$$
\times \sum_{\ell=0}^{\infty} \sum_{m=1}^{h(S^{N-1}, \ell)} \sum_{n=1}^{\infty} X_{\ell m n} \frac{g_\ell(R^{-1} j_{|\ell-1|-H,n} s) - \delta_0^\ell}{J_{|\ell-1|-H+1}(j_{|\ell-1|-H,n}) j_{|\ell-1|-H,n}^{H+1}}
$$

$$
\times S_\ell^m(\theta_1, \ldots, \theta_{N-2}, \varphi), \tag{4.17}
$$

where $0 \leq s \leq R$,

$$
g_\ell(u) := 2^{(N-2)/2} \Gamma(N/2) \frac{J_{\ell+(N-2)/2}(u)}{u^{(N-2)/2}}, \tag{4.18}
$$

$X_{\ell m n}$ are independent standard normal random variables, and $j_{|\ell-1|-H,n}$ is the nth zero of the Bessel function $J_{|\ell-1|-H}(x)$. In what follows, we will call this expansion *local*, because it is valid only inside the centred ball of radius R.

The above expansion is not unique. In this section, we describe all possible expansions, and discuss the optimality of the above expansion.

Let $(B, \| \cdot \|)$ be a real separable Banach space, and let μ be a centred Gaussian measure in B. Kwapień and Szymański (1980) and Tarieladze (1980) independently proved that there exist an a.s. convergent series $\sum \mathbf{x}_k X_k$ with $\mathbf{x}_k \in B$ and X_k a sequence of independent standard normal random variables such that $\sum \mathbf{x}_k X_k$ is distributed as μ and $\sum \| \mathbf{x}_k \|^2 < \infty$. In particular, let T be a compact metric space, let $B = C(T)$ be the space of continuous functions on T equipped by the sup-norm, let μ be a Gaussian measure in B, and let $X(t)$ be the Gaussian random field on T distributed as μ. The above result guarantees the existence of an a.s. convergent expansion

$$
X(t) = \sum_k f_k(t) X_k, \quad t \in T. \tag{4.19}
$$

Luschgy and Pagès (2009) call a sequence of elements $\{f_k \in B : k \geq 1\}$ *admissible* for X if the expansion (4.19) holds true. By part a) of Corollary 1 of Luschgy and Pagès (2009), if for every $s, t \in T$,

$$
\mathsf{E}[X(s) X(t)] = \sum_{j \geq 1} f_j(s) f_j(t), \tag{4.20}
$$

then the sequence $\{f_j \in E : j \geq 1\}$ is admissible for X.

The right-hand side of the series (4.17) is the sum of products of independent standard normal random variables by continuous functions. Equation (4.20) follows from Karhunen's theorem. Therefore, the expansion (4.17) converges uniformly in the centred closed ball of radius R in the space $\mathbb{R}^N$.

By the Riesz representation theorem, the dual space, $C^*(T)$, coincides with the space of finite signed Borel measures on T. The *covariance operator* of X is defined as a linear operator acting from B^* to B by

$$C\mu := \mathsf{E}\big[\mu(X)X\big], \quad \mu \in B^*.$$

When $B = C(T)$, we have

$$(C\mu)(t) = \int_T \mathsf{E}\big[X(s)X(t)\big]\,d\mu(s).$$

The operator C is symmetric and nonnegative-definite. By Proposition III.1.6 of Vakhania et al. (1987), there exists a unique Hilbert subspace $(H, (\cdot,\cdot))$ of the space B such that $C(B^*)$ is dense in H and for all $\mu, \nu \in B^*$

$$(C\mu, C\nu) = \nu(C\mu). \tag{4.21}$$

The space H is called the *reproducing kernel Hilbert space* of the field X. Equation (4.21) is called the *reproducing property*.

A sequence of elements $\{f_j \in H : j \geq 1\}$ is called a *Parseval frame* for H if for every $h \in H$ the series $\sum_{j\geq1}(f_j,h)h$ converges to h in H. An excellent introduction to frame theory may be found in Christensen (2003). All possible expansions are described by Theorem 1 of Luschgy and Pagès (2009): a sequence $\{f_j \in E : j \geq 1\}$ is admissible for X iff the sequence $\{f_j \in E : j \geq 1\}$ is a Parseval frame for the reproducing kernel Hilbert space of X.

Introduce the following notation. If a_n, $n \geq 1$, and b_n, $n \geq 1$ are sequences of positive real numbers, we write $a_n \preceq b_n$ provided that $a_n \leq cb_n$ for a certain $c > 0$ and for any positive integer n. Then $a_n \approx b_n$ means that $a_n \preceq b_n$ as well as $b_n \preceq a_n$. In the same way, we write $f(\mathbf{u}) \preceq g(\mathbf{u})$ provided that $f(\mathbf{u}) \leq cg(\mathbf{u})$ for a certain $c > 0$ and uniformly for all $\mathbf{u}$, and $f(\mathbf{u}) \approx g(\mathbf{u})$ if $f(\mathbf{u}) \preceq g(\mathbf{u})$ as well as $g(\mathbf{u}) \preceq f(\mathbf{u})$.

Put

$$\ell_p(X) := \inf \mathsf{E}\left[\left\|\sum_{j\geq p} X_j f_j\right\|\right],$$

where the infimum is taken over all admissible sequences for X. The numbers $\ell_p(X)$ are called the *approximation numbers* of the random field X. The series expansion

$$\sum_{j\geq1} X_j f_j$$

is called *rate-optimal* if

$$\mathsf{E}\left[\left\|\sum_{j\geq p} X_j f_j\right\|\right] \approx \ell_p(X).$$

Let X be the multiparameter fractional Brownian motion on the space $\mathbb{R}^N$. We need a lemma.

Lemma 4.24 *The Bessel functions, the functions g_ℓ defined by (4.18), the numbers $h(S^{N-1}, \ell)$, and the spherical harmonics $S_\ell^m(\theta_1, \ldots, \theta_{N-2}, \varphi)$ have the following properties.*

1. $j_{v,n} \approx n + v/2 - 1/4$.
2. $J_{v+1}^2(j_{v,n}) \approx \frac{1}{j_{v,n}}$.
3. $|g_0(u)| \le 1$.
4. $|g_m(u)| \preceq \frac{1}{[m(m+N-2)]^{(N-1)/4}}$, $m \ge 1$, $N \ge 2$.
5. $|g_\ell'(u)| \preceq \frac{1}{[\ell(\ell+N-2)]^{(N-3)/4}}$, $\ell \ge 1$, $N \ge 2$.
6. $h(S^{N-1}, \ell) \preceq \ell^{N-2}$, $N \ge 2$.
7. $|S_\ell^m(\theta_1, \ldots, \theta_{N-2}, \varphi)| \preceq \ell^{(N-2)/2}$, $N \ge 2$.

Proof Property 1 is proved in Watson (1944, Sect. 15.53). It is shown in Watson (1944, Sect. 7.21) that

$$J_v^2(u) + J_{v+1}^2(u) \approx \frac{1}{u}, \qquad (4.22)$$

since Property 2. Property 3 follows from the fact that $g_0(u)$ is the element of the unitary matrix (Vilenkin, 1968).

Let $\mu_1 < \mu_2 < \cdots$ be the sequence of positive stationary values of the Bessel function $J_{\ell+(N-2)/2}(u)$. It is shown in Watson (1944, Sect. 15.31) that

$$\left| J_{\ell+(N-2)/2}(\mu_1) \right| > \left| J_{\ell+(N-2)/2}(\mu_2) \right| > \cdots .$$

Let u_1 be the first positive maximum of the function $g_\ell(u)$. Let x_1 denote the maximal value of the function $|g_\ell(u)|$ in the interval $(0, j_{\ell+(N-2)/2,1})$, let x_2 denote the maximal value of $|g_\ell(u)|$ in the interval $(j_{\ell+(N-2)/2,1}, j_{\ell+(N-2)/2,2})$, and so on. Then we have

$$x_1 \preceq \frac{|J_{\ell+(N-2)/2}(\mu_1)|}{u_1^{(N-2)/2}},$$

$$x_2 \preceq \frac{|J_{\ell+(N-2)/2}(\mu_2)|}{j_{\ell+(N-2)/2,1}^{(N-2)/2}},$$

$$x_3 \preceq \frac{|J_{\ell+(N-2)/2}(\mu_3)|}{j_{\ell+(N-2)/2,2}^{(N-2)/2}},$$

and so on. The right-hand sides form a decreasing sequence. Then

$$|g_\ell(u)| \preceq \frac{|J_{\ell+(N-2)/2}(\mu_1)|}{u_1^{(N-2)/2}}.$$

It follows from (4.22) that

$$|J_\nu(u)| \le \frac{1}{\sqrt{u}},$$

and we obtain

$$|g_\ell(u)| \le \frac{1}{\mu_1 u_1^{(N-2)/2}}.$$

To estimate u_1, consider the differential equation

$$u^2 \frac{d^2 f}{du^2} + (N-1)u \frac{df}{du} + [u^2 - \ell(\ell+N-2)]f = 0. \tag{4.23}$$

It follows from formulae (3) and (4) in Watson (1944, Sect. 4.31) that the function g_ℓ satisfies this equation. It follows from the series representation (6.53) of the Bessel function and the definition (4.18) of the function $g_\ell(u)$ that for $\ell \ge 1$ we have $g_\ell(0) = 0$ and g_ℓ increases in some right neighbourhood of zero. Therefore we have $g_\ell(u_1) > 0$, $g_\ell'(u_1) = 0$ and $g_\ell''(u_1) \le 0$. It follows from Eq. (4.23) that $u_1 \ge \sqrt{\ell(\ell+N-2)}$.

For μ_1, we have the estimate $\mu_1 > \ell + (N-2)/2$ (Watson, 1944, Sect. 15.3). Using the inequality $\ell + (N-2)/2 \ge \sqrt{\ell(\ell+N-2)}$, we obtain Property 4.

In any extremum of $g_\ell'(u)$ we have $g_\ell''(u) = 0$. It follows from (4.23) that

$$|g_\ell'(u)| = \left| \frac{[u^2 - \ell(\ell+N-2)]g_\ell(u)}{(N-1)u} \right| \le |ug_\ell(u)|.$$

Property 5 follows from this formula and Property 4.

Property 6 follows from (6.56) and Stirling's formula.

It follows from (6.58) that

$$\left(S_\ell^m(\theta_1, \ldots, \theta_{N-2}, \varphi) \right)^2 \le \frac{\Gamma(N/2)h(S^{N-1}, \ell)}{2\pi^{N/2}} \frac{C_\ell^{(N-2)/2}(\cos\varphi)}{C_\ell^{(N-2)/2}(1)}.$$

By Erdélyi et al. (1953, formula 10.18(7)),

$$\max_{0 \le \varphi \le \pi} |C_\ell^{(N-2)/2}(\cos\varphi)| = C_\ell^{(N-2)/2}(1).$$

It follows that $|S_\ell^m(\theta_1, \ldots, \theta_{N-2}, \varphi)| \le \sqrt{h(S^{N-1}, \ell)}$. Now Property 7 follows from Property 6. $\qquad\square$

We prove the following theorem.

Theorem 4.25 *The series expansion* (2.97) *is rate-optimal.*

Proof For simplicity, put $R = 1$. In their Theorem 1, Ayache and Linde (2008) proved that $\ell_p(X) \approx p^{-H/N}(\ln p)^{1/2}$. Therefore, it is enough to prove that the rate of convergence in (2.97) is not more than $p^{-H/N}(\ln p)^{1/2}$, where p denote the number of terms in a suitable truncation of the series. Since for $N > 1$ we have a triple sum in our expansion (2.97), it is not clear a priori how we should truncate the series.

Denote

$$u_{mn}^{\ell}(s, \theta_1, \ldots, \theta_{N-2}, \varphi) := \frac{2^{H+1}\sqrt{\pi}^{(N-2)/2}\Gamma(N/2 + H)\Gamma(H + 1)\sin(\pi H)}{\Gamma(N/2)}$$

$$\times \frac{g_\ell(j_{|\ell-1|-H,n}s) - \delta_0^\ell}{J_{|\ell-1|-H+1}(j_{|\ell-1|-H,n})j_{|\ell-1|-H,n}^{H+1}}$$

$$\times S_\ell^m(\theta_1, \ldots, \theta_{N-2}, \varphi).$$

$$(4.24)$$

Using Lemma 4.24, we can write the following estimate:

$$\mathsf{E}\left[\sum_{m=1}^{h(S^{N-1},\ell)} u_{mn}^{\ell}(s, \theta_1, \ldots, \theta_{N-2}, \varphi)X_{\ell mn}\right]^2 \leq \frac{1}{(\ell+1)(\ell/2+n)^{2H+1}}. \quad (4.25)$$

For any positive integer q, consider the truncation

$$\sum_{(\ell+1)(\ell/2+n)^{2H+1}\leq q}\;\sum_{m=1}^{h(S^{N-1},\ell)} u_{mn}^{\ell}(s, \theta_1, \ldots, \theta_{N-2}, \varphi)X_{\ell mn}$$

of the expansion (2.97). The number of terms in this sum is asymptotically equal to the integral

$$\iint_{(u+1)(u/2+v)\leq q,u\geq 0,v\geq 1} u^{N-2}\,du\,dv,$$

and therefore is bounded by a constant times $q^{N/(2H+2)}$. So we need to prove that

$$\left(\mathsf{E}\left\|\sum_{(\ell+1)(\ell/2+n)^{2H+1}>q}\;\sum_{m=1}^{h(S^{N-1},\ell)} u_{mn}^{\ell}(s, \theta_1, \ldots, \theta_{N-2}, \varphi)X_{\ell mn}\right\|^2\right)^{1/2}$$

$$\leq q^{(N/(2H+2))(-H/N)}\left(\ln q^{N/(2H+2)}\right)^{1/2} \leq q^{-H/(2H+2)}(\ln q)^{1/2}.$$

By equivalence of moments (Kühn and Linde, 2002, Proposition 2.1), the last formula is equivalent to the following asymptotic relation:

$$\mathsf{E}\sup_{\|x\|\leq 1}\left|\sum_{(\ell+1)(\ell/2+n)^{2H+1}>q}\;\sum_{m=1}^{h(S^{N-1},\ell)} u_{mn}^{\ell}(x)X_{\ell mn}\right| \leq q^{-H/(2H+2)}(\ln q)^{1/2}.$$

To prove this relation, consider the partial sum $\eta_k(\mathbf{x})$ defined by

$$\eta_k(\mathbf{x}) := \sum_{2^{k-1} < (\ell+1)(\ell/2+n)^{2H+1} \le 2^k} \sum_{m=1}^{h(S^{N-1},\ell)} u_{mn}^{\ell}(\mathbf{x}) X_{\ell mn}.$$

For a given $\varepsilon_k > 0$, let $\mathbf{x}_1, \ldots, \mathbf{x}_{P_k} \in S^{N-1}$ be a maximal ε_k-net in S^{N-1}, i.e., the angle $\varphi(\mathbf{x}_j, \mathbf{x}_k)$ between any two different vectors $\mathbf{x}_j$ and $\mathbf{x}_k$ is greater than ε_k and the addition of any new point breaks this property. Then $P_k \approx \varepsilon_k^{-N+1}$. A proof of this fact for the case of $N = 3$ by Baldi et al. (2009, Lemma 5), is easy generalised to higher dimensions. The *Voronoi cell* $S(\mathbf{x}_j)$ is defined as

$$S(\mathbf{x}_j) := \left\{ \mathbf{x} \in S^{N-1} : \varphi(\mathbf{x}, \mathbf{x}_j) \le \varphi(\mathbf{x}, \mathbf{x}_k), k \ne j \right\}.$$

Divide the ball $\mathcal{B} = \{\mathbf{x} \in \mathbb{R}^N : \|\mathbf{x}\| \le 1\}$ into $[\varepsilon_k^{-1}]$ concentric spherical layers of thickness $\preceq \varepsilon_k$. Voronoi cells determine the division of each layer into P_k segments. The angle $\varphi(\mathbf{x}, \mathbf{y})$ between any two vectors $\mathbf{x}$ and $\mathbf{y}$ in the same segment is $\preceq \varepsilon_k$ (we never choose the point $\mathbf{0}$). Call all segments in all layers $\mathcal{B}_j$, $1 \le j \le M_k = P_k[\varepsilon_k^{-1}] \preceq \varepsilon_k^{-N}$. Choose a point $\mathbf{x}_j$ inside each segment. Then we have

$$\mathsf{E} \sup_{\mathbf{x} \in \mathcal{B}} |\eta_k(\mathbf{x})| \le \mathsf{E} \sup_{1 \le j \le M_k} |\eta_k(\mathbf{x}_j)| + \mathsf{E} \sup_{1 \le j \le M_k} \sup_{\mathbf{x}, \mathbf{y} \in \mathcal{B}_j} |\eta_k(\mathbf{x}) - \eta_k(\mathbf{y})|.$$

By a maximal inequality for Gaussian sequences (see Van der Vaart and Wellner, 1996, Lemma 2.2.2), the first term is bounded by a positive multiple of

$$\sqrt{1 + \ln M_k} \sup_{1 \le j \le M_k} \sqrt{\mathsf{E}(\eta_k(\mathbf{x}_j))^2}.$$

Using the estimate (4.25), we obtain

$$\mathsf{E}(\eta_k(\mathbf{x}))^2 \preceq \iint_{2^{k-1} < (u+1)(u/2+v)^{2H+1} \le 2^k, u \ge 0, v \ge 1} \frac{du\, dv}{(u+1)(u/2+v)^{2H+1}}.$$

The integral in the right-hand side is easily seen to be $\preceq 2^{-kH/(H+1)}$. It follows that

$$\mathsf{E} \sup_{1 \le j \le M_k} |\eta_k(\mathbf{x}_j)| \le 2^{-kH/(2H+2)} \sqrt{1 + \ln M_k}. \tag{4.26}$$

To estimate the second term we write

$$\mathsf{E} \sup_{1 \le j \le M_k} \sup_{\mathbf{x}, \mathbf{y} \in \mathcal{B}_j} |\eta_k(\mathbf{x}) - \eta_k(\mathbf{y})|$$

$$\le \mathsf{E} \sup_{1 \le j \le M_k} \sup_{\mathbf{x}, \mathbf{y} \in \mathcal{B}_j} \sum_{2^{k-1} < (\ell+1)(\ell/2+n)^{2H+1} \le 2^k} \sum_{m=1}^{h(S^{N-1},\ell)} |u_{mn}^{\ell}(\mathbf{x}) - u_{mn}^{\ell}(\mathbf{y})| \cdot |X_{\ell mn}|.$$

The difference $|u_{mn}^\ell(\mathbf{x}) - u_{mn}^\ell(\mathbf{y})|$ can be estimated as

$$|u_{mn}^\ell(\mathbf{x}) - u_{mn}^\ell(\mathbf{y})| \leq |g_\ell(j_{|\ell-1|-H,n}\|\mathbf{x}\|) - \delta_\ell^0| \cdot |S_\ell^m(\mathbf{x}/\|\mathbf{x}\|) - S_\ell^m(\mathbf{y}/\|\mathbf{y}\|)|$$
$$+ |S_\ell^m(\mathbf{y}/\|\mathbf{y}\|)| \cdot |g_\ell(j_{|\ell-1|-H,n}\|\mathbf{x}\|) - g_\ell(j_{|\ell-1|-H,n}\|\mathbf{y}\|)|.$$

Let Δ_0 denote the angular part of the Laplace operator in the space $\mathbb{R}^N$. Using the Mean Value Theorem for g_ℓ, Lemma 4.24 and formulae

$$\Delta_0 S_\ell^m(\mathbf{x}/\|\mathbf{x}\|) = -\ell(\ell + N - 2) S_\ell^m(\mathbf{x}/\|\mathbf{x}\|),$$
$$|S_\ell^m(\mathbf{x}/\|\mathbf{x}\|) - S_\ell^m(\mathbf{y}/\|\mathbf{y}\|)| \preceq \sup_{\mathbf{z}} |\sqrt{-\Delta_0} S_\ell^m(\mathbf{z}/\|\mathbf{z}\|)| \varphi(\mathbf{x}, \mathbf{y}),$$

where the sup is taken over all points $\mathbf{z}$ belonging to the segment of the geodesic circle connecting the points $\mathbf{x}/\|\mathbf{x}\|$ and $\mathbf{y}/\|\mathbf{y}\|$, we obtain

$$|u_{mn}^\ell(\mathbf{x}) - u_{mn}^\ell(\mathbf{y})|$$
$$\preceq \varepsilon_k \frac{2^{H+1}\sqrt{\pi^{(N-2)/2}}\Gamma(H + N/2)\Gamma(H + 1)\sin(\pi H)}{\Gamma(N/2)J_{|\ell-1|-H+1}(j_{|\ell-1|-H,n})j_{|\ell-1|-H,n}^{H+1}} \sqrt{\ell(\ell/2 + n)}.$$

Hence, we have

$$\mathsf{E} \sup_{1\leq j\leq M_k} \sup_{\mathbf{x},\mathbf{y}\in\mathcal{B}_j} |\eta_k(\mathbf{x}) - \eta_k(\mathbf{y})| \preceq \varepsilon_k \sum_{2^{k-1}<(\ell+1)(\ell/2+n)^{2H+1}\leq 2^k} h(S^{N-1}, \ell)$$
$$\times \frac{2^{H+1}\sqrt{\pi^{(N-2)/2}}\Gamma(H + N/2)\Gamma(H + 1)\sin(\pi H)}{\Gamma(N/2)J_{|\ell-1|-H+1}(j_{|\ell-1|-H,n})j_{|\ell-1|-H,n}^{H+1}} \sqrt{m(m/2 + n)}.$$

Using Lemma 4.24 once more, the last formula may be rewritten as

$$\mathsf{E} \sup_{1\leq j\leq M_k} \sup_{\mathbf{x},\mathbf{y}\in\mathcal{B}_j} |\eta_k(\mathbf{x}) - \eta_k(\mathbf{y})|$$
$$\preceq \varepsilon_k \sum_{2^{k-1}<(\ell+1)(\ell/2+n)^{2H+1}\leq 2^k} \ell^{N-3/2}(\ell/2 + n)^{1/2-H}.$$

The integral which corresponds to the sum in the right-hand side is easily seen to be asymptotically equal to $2^{kN/(2H+2)}$. Therefore,

$$\mathsf{E} \sup_{1\leq j\leq M_k} \sup_{\mathbf{x},\mathbf{y}\in\mathcal{B}_j} |\eta_k(\mathbf{x}) - \eta_k(\mathbf{y})| \preceq \varepsilon_k \cdot 2^{kN/(2H+2)}.$$

Combining this with the estimate (4.26) for the first term, we obtain the asymptotic relation

$$\mathsf{E} \sup_{\mathbf{x}\in\mathcal{B}} |\eta_k(\mathbf{x})| \preceq 2^{-kH/(2H+2)}\sqrt{1 + \ln M_k} + \varepsilon_k \cdot 2^{kN/(2H+2)}.$$

Now set $\varepsilon_k := 2^{-k(H+N)/(2H+2)}$ and recall that $M_k \preceq \varepsilon_k^{-N} = 2^{kN(H+N)/(2H+2)}$. Then we see that the first term is bounded by a multiple of $2^{-kH/(2H+2)}\sqrt{k}$ and the second one is of lower order. This proves that

$$\mathsf{E} \sup_{\mathbf{x} \in B} |\eta_k(\mathbf{x})| \preceq 2^{-kH/(2H+2)}\sqrt{k}. \tag{4.27}$$

To complete the proof, it suffices to show that

$$\mathsf{E} \sup_{\mathbf{x} \in B} \left| \sum_{(\ell+1)(\ell/2+n)^{2H+1}>q} \sum_{m=1}^{h(S^{N-1},\ell)} u_{mn}^\ell(\mathbf{x}) X_{\ell m n} \right| \preceq q^{-H/(2H+2)}(\ln q)^{1/2}. \tag{4.28}$$

Let r be the positive integer such that $2^{r-1} < q \leq 2^r$. Then by the triangle inequality

$$\left| \sum_{(\ell+1)(\ell/2+n)^{2H+1}>q} \sum_{m=1}^{h(S^{N-1},\ell)} u_{mn}^\ell(\mathbf{x}) X_{\ell m n} \right|$$

$$\leq \left| \sum_{(\ell+1)(\ell/2+n)^{2H+1}>2^r} \sum_{m=1}^{h(S^{N-1},\ell)} u_{mn}^\ell(\mathbf{x}) X_{\ell m n} \right|$$

$$+ \left| \sum_{q<(\ell+1)(\ell/2+n)^{2H+1}\leq 2^r} \sum_{m=1}^{h(S^{N-1},\ell)} u_{mn}^\ell(\mathbf{x}) X_{\ell m n} \right|$$

$$\leq \sum_{k>r} |\eta_k(\mathbf{x})| + \left| \sum_{q<(\ell+1)(\ell/2+n)^{2H+1}\leq 2^r} \sum_{m=1}^{h(S^{N-1},\ell)} u_{mn}^\ell(\mathbf{x}) X_{\ell m n} \right|.$$

Using (4.27) and the fact that

$$\sum_{k\geq r} \sqrt{k} 2^{-kN/(2H+2)} \preceq \sqrt{r} 2^{-rN/(2H+2)},$$

we obtain

$$\sum_{k>r} \mathsf{E} \sup_{\mathbf{x} \in B} |\eta_k(\mathbf{x})| \preceq \sum_{k>r} \sqrt{k} 2^{-kN/(2H+2)}$$

$$\preceq \sqrt{r-1} \cdot 2^{-(r-1)H/(2H+2)}$$

$$\preceq q^{-H/(2H+2)}(\ln q)^{1/2}.$$

The arguments in the proof of (4.27) show that since $2^{r-1} < q \leq 2^r$, we also have

$$
\mathrm{E} \sup_{\mathbf{x} \in \mathcal{B}} \left| \sum_{q < (\ell+1)(\ell/2+n)^{2H+1} \leq 2^r} \sum_{m=1}^{h(S^{N-1}, \ell)} u_{mn}^\ell(\mathbf{x}) X_{\ell mn} \right| \leq \sqrt{r} 2^{-rH/(2H+2)}
$$

$$
\leq q^{-H/(2H+2)} (\ln q)^{1/2}.
$$

Relation (4.28) is proved. $\square$

4.5 Functional Limit Theorems for the Multiparameter Fractional Brownian Motion

Let $T = [0, \infty)$, and let $W(t) = W(t, \omega)$, $t \in T$, $\omega \in \Omega$ be a measurable version of the Brownian motion on a probability space $(\Omega, \mathfrak{F}, \mathrm{P})$. This process has homogeneous increments, i.e., for any $s, t \in T$,

$$
W(s + t) - W(s) \overset{d}{=} W(s),
$$

where the sign $\overset{d}{=}$ means that the two random functions connected with this sign have the same finite-dimensional distributions. Therefore, $W(t)$ has the same local modulus of continuity at each point. In particular, the celebrated *local law of the iterated logarithm* states that the function $t \mapsto \sqrt{t \ln \ln t^{-1}}$ is the local modulus of continuity of the Wiener process at point 0 almost surely (a.s.). In other words,

$$
\mathrm{P}\left\{ \omega : \limsup_{t \downarrow 0} \frac{W(t, \omega)}{\sqrt{2t \ln \ln t^{-1}}} = 1 \right\} = 1.
$$

Consider the following measurable subset of the set $[0, 1] \times \Omega$:

$$
A := \left\{ (t, \omega) \in [0, 1] \times \Omega : \limsup_{s \to t} \frac{W(s, \omega)}{\sqrt{2|s - t| \ln \ln |s - t|^{-1}}} \neq 1 \right\}.
$$

The double integral of the indicator $\chi_A(t, \omega)$ converges:

$$
\int_0^1 \int_\Omega \chi_A(t, \omega) \, \mathrm{dP}(\omega) \, \mathrm{d}t < \infty.
$$

By Fubini's theorem, the above integral is equal to the iterated integral:

$$
\int_0^1 \int_\Omega \chi_A(t, \omega) \, \mathrm{dP}(\omega) \, \mathrm{d}t = \int_0^1 \left(\int_\Omega \chi_A(t, \omega) \, \mathrm{dP}(\omega) \right) \mathrm{d}t.
$$

For any fixed $t \in [0, 1]$, the inner integral in the right-hand side is equal to 0. It follows that both sides are equal to 0. Again by Fubini's theorem, we have

$$\int_{\Omega} \left(\int_0^1 \chi_A(t, \omega) \, dt \right) dP(\omega) = 0.$$

It follows that for any fixed $\omega_0 \in \Omega$, the set of *fast* points $t \in [0, 1]$, where the local law of the iterated logarithm fails, has Lebesgue measure 0.

What is the local behaviour of the Wiener process at the fast points? $W(t)$ escapes from these points faster than the local modulus of continuity but slower than the global modulus of continuity, the function $t \mapsto \sqrt{2t \ln(t^{-1})}$. The *Lévy modulus of continuity* states that

$$\limsup_{t \downarrow 0} \sup_{s \in [0,1]} \frac{W(s+t) - W(s)}{\sqrt{2t \ln t^{-1}}} = 1 \quad \text{a.s.}$$

The *global law of iterated logarithm* states that

$$\limsup_{t \to \infty} \frac{W(t, \omega)}{\sqrt{2t \ln \ln t}} = 1 \quad \text{a.s.} \tag{4.29}$$

All the three limit theorems described above have functional counterparts. Consider the restriction of the Wiener process $W(t)$ to the interval $[0, 1]$. The reproducing-kernel Hilbert space of this restriction is the space $\mathcal{H}$ of all absolutely continuous functions $f: [0, 1] \to \mathbb{R}$ with $f(0) = 0$ and finite *Strassen norm*

$$\|f\|_{\mathcal{H}} := \left(\int_0^1 \left(f'(t) \right)^2 dt \right)^{1/2}.$$

The centred unit ball of the space $\mathcal{H}$

$$\mathcal{K} := \left\{ f \in \mathcal{H} : \|f\|_H \leq 1 \right\}$$

is called the *Strassen ball*. Define the *Strassen cloud* as the following subset of the space $C[0, 1]$:

$$\mathcal{S} := \left\{ Y_u(t) = \frac{W(ut)}{\sqrt{2u \ln \ln u}} : u > e \right\}.$$

The *global functional law of the iterated logarithm*, or the Strassen law, states that, in the uniform topology, the set of P-a.s. limit points of the Strassen cloud as $u \to \infty$ is the Strassen ball. In what follows, we describe the above situation briefly as follows: *the Strassen cloud is attracted to the Strassen ball as $u \to \infty$.* The notation $\mathcal{S} \to\to \mathcal{K}$ is often used.

To deduce the global law of the iterated logarithm from its functional counterpart, consider a continuous functional $F: C[0, 1] \to \mathbb{R}$. Then

$$\limsup_{u \to \infty} F\left(Y_u(t) \right) = \sup_{f \in \mathcal{K}} F(f) \quad \text{a.s.} \tag{4.30}$$

In particular, put $F(f) := f(1)$. The supremum $\sup_{f \in \mathcal{K}} f(1)$ is equal to 1 and is attained on the function $f(t) = t$. Therefore, (4.30) transforms to (4.29).

To formulate the functional counterpart to the Lévy modulus of continuity, define the family of *Mueller clouds* indexed by a parameter $u \in (0, 1]$ as

$$S(u) := \left\{ Y_u(t) = \frac{W(s + ut) - W(s)}{\sqrt{2u \ln u^{-1}}} : 0 \le s \le 1 - u \right\}.$$

Mueller (1981) proved that the family of Mueller clouds $S(u)$ is attracted to the Strassen ball as $u \downarrow 0$. The Lévy modulus of continuity follows from its functional counterpart in the same way as the global law of the iterated logarithm follows from the Strassen law.

In fact, Mueller (1981) proved a general functional limit theorem that includes the functional law of the iterated logarithm, the functional Lévy modulus of continuity and many other results as particular cases. In this section, we will prove a similar theorem for the case of the multiparameter fractional Brownian motion $X(\mathbf{t}), \mathbf{t} \in \mathbb{R}^N$.

The field $X(\mathbf{t})$ is isotropic, i.e., SO(N)-invariant. This property prompts us to use the SO(N)-invariant closed unit ball

$$\mathcal{B} := \left\{ \mathbf{x} \in \mathbb{R}^N : \|\mathbf{x}\| \le 1 \right\}$$

(but *not* the cube $[0, 1]^N$, which is not SO(N)-invariant!) in the case of the multiparameter fractional Brownian motion instead of the interval $[0, 1]$ in the case of the Wiener process. The description of the corresponding reproducing-kernel Hilbert space $\mathcal{H}$ follows easily from the local spectral expansion (2.97) with $R = 1$ and Theorem 1 by Luschgy and Pagès (2009). Denote

$$f_{\ell mn}(r, \theta_1, \ldots, \theta_{N-2}, \varphi) := \frac{2^{H+1} \sqrt{\pi^{(N-2)/2} \Gamma(N/2 + H) \Gamma(H + 1) \sin(\pi H)} R^H}{\Gamma(N/2)}$$

$$\times \frac{g_\ell(j_{|\ell-1|-H,n} r) - \delta_0^\ell}{J_{|\ell-1|-H+1}(j_{|\ell-1|-H,n}) j_{|\ell-1|-H,n}^{H+1}}$$

$$\times S_\ell^m(\theta_1, \ldots, \theta_{N-2}, \varphi).$$

The space $\mathcal{H}$ consists of all functions $f \in C(\mathcal{B})$ with $f(\mathbf{0}) = 0$ of the form

$$f(r, \theta_1, \ldots, \theta_{N-2}, \varphi) := \sum_{\ell=0}^{\infty} \sum_{m=1}^{h(S^{N-1}, \ell)} \sum_{n=1}^{\infty} c_{\ell mn}(f) f_{\ell mn}(r, \theta_1, \ldots, \theta_{N-2}, \varphi) \quad (4.31)$$

with

$$\|f\|_{\mathcal{H}}^2 := \sum_{\ell=0}^{\infty} \sum_{m=1}^{h(S^{N-1}, \ell)} \sum_{n=1}^{\infty} c_{\ell mn}^2(f) < \infty. \quad (4.32)$$

The Strassen ball $\mathcal{K}$ is the centred closed unit ball of the space $\mathcal{H}$.

Let t_0 be a real number. Let there exist for every $t \geq t_0$ a non-empty set of indices $\mathcal{J}(t)$. Let every element $j \in \mathcal{J}(t)$ define the vector $\mathbf{y}_j \in \mathbb{R}^N$ and the positive real number u_j. Let $R_r(\mathbf{y}_j, u_j)$ be the cylinder

$$R_r(\mathbf{y}_j, u_j) := \left\{ (\mathbf{y}, u) : \| \mathbf{y} - \mathbf{y}_j \| \leq r u_j, \, e^{-r} u_j \leq u \leq e^r u_j \right\}, \quad r > 0.$$

Now we define the function $F_r(t)$. In words: this is the volume of the union of all cylinders $R_r(\mathbf{y}_j, u_j)$ that are defined before the moment t, with respect to the measure $u^{-N-1} \, d\mathbf{y} \, du$. Formally,

$$F_r(t) := \int_{\bigcup_{t_0 \leq v \leq t} \bigcup_{j \in \mathcal{J}(v)} R_r(\mathbf{y}_j, u_j)} u^{-N-1} \, d\mathbf{y} \, du.$$

Finally, we define

$$\mathcal{P}(t) := \left\{ (\mathbf{y}_j, u_j) : j \in \mathcal{J}(t) \right\}$$

and the family of Mueller clouds

$$\mathcal{S}(t) := \left\{ Y(\mathbf{x}) = \frac{X(\mathbf{y} + u\mathbf{x}) - X(\mathbf{y})}{\sqrt{2h(t)} u^H} : (\mathbf{y}, u) \in \mathcal{P}(t) \right\} \subset C(\mathcal{B}).$$

Theorem 4.26 *Let the function $h(t) \colon [t_0, \infty) \mapsto \mathbb{R}$ satisfy the following conditions:*

1. *$h(t)$ is increasing and $\lim_{t \to \infty} h(t) = \infty$.*

2. *The integral $\displaystyle\int_{t_0}^{\infty} e^{-ah(t)} \, dF_1(t)$ converges for $a > 1$ and diverges for $a < 1$.*

Then the family of Mueller clouds $\mathcal{S}(t)$ is attracted to the Strassen ball $\mathcal{K}$ as $t \to \infty$.

Proof We proceed in steps.

Step 1. We formulate two lemmas and show how Theorem 4.26 follows.

Lemma 4.27 *The family of Mueller clouds $\mathcal{S}(t)$ is a.s. almost inside $\mathcal{K}$, i.e.,*

$$\lim_{t \to \infty} \sup_{\eta \in \mathcal{S}(t)} \inf_{f \in \mathcal{K}} \| \eta - f \|_{C(\mathcal{B})} = 0 \quad a.s. \tag{4.33}$$

Lemma 4.28 *Any neighbourhood of any element $f \in \mathcal{K}$ is caught by the family of Mueller clouds $\mathcal{S}(t)$ infinitely often, i.e.,*

$$\sup_{f \in \mathcal{K}} \liminf_{t \to \infty} \inf_{\eta \in \mathcal{S}(t)} \| \eta - f \|_{C(\mathcal{B})} = 0. \tag{4.34}$$

On the one hand, it follows from (4.33) that the set of a.s. limit points of $\mathcal{S}(t)$ is contained in the closure of $\mathcal{K}$. On the other hand, it follows from (4.34) that $\mathcal{K}$ is contained in the set of a.s. limit points of $\mathcal{S}(t)$. According to general theory (Lifshits, 1995), $\mathcal{K}$ is compact. Therefore it is closed, and we are done.

Step 2. We construct the auxiliary sequences. Divide the set $\mathbb{R}^N \times (0, \infty)$ into parallelepipeds

$$R_{\mathbf{k}p} := \left\{ (\mathbf{y}, u) \colon k_j re^{pr} \le y_j \le (k_j + 1)re^{pr} \text{ for } 1 \le j \le N, e^{pr} \le u \le e^{(p+1)r} \right\},$$

where $\mathbf{k} \in \mathbb{Z}^N$ and $p \in \mathbb{Z}$. The following Lemma describes the most important property of the parallelepipeds $R_{\mathbf{k}p}$.

Lemma 4.29 *For any $t \in [t_0, \infty)$ the union of all cylinders $R_r(\mathbf{y}_j, u_j)$ that are defined before the moment t is contained in the union of finitely many parallelepipeds $R_{\mathbf{k}p}$.*

Proof It follows from condition 2 of Theorem 4.26 that for any $t \in [t_0, \infty)$ the volume of all cylinders $R_r(\mathbf{y}_j, u_j)$ that are defined before the moment t with respect to the measure $u^{-N-1} \, dy \, du$ is finite. So it is enough to prove that the volume of any parallelepiped $R_{\mathbf{k}p}$ with respect to the above mentioned measure is also finite. We have

$$\int_{R_{\mathbf{k}p}} u^{-N-1} \, dy \, du \sim \frac{r^N e^{Npr} [e^{(p+1)r} - e^{pr}]}{e^{(N+1)pr}}$$

$$\sim r^N (e^r - 1)$$

$$\sim r^{N+1} \quad (r \downarrow 0). \qquad \square$$

Lemma 4.30 *There exist a sequence of real numbers t_q and a sequence of parallelepipeds $R_{\mathbf{k}_q p_q}$, $q \ge 0$, that satisfy the following conditions.*

1. *For any $q \ge 0$ and for any $\varepsilon > 0$ there exists a real number $t \in (t_q, t_q + \varepsilon)$ such that*

$$\mathcal{P}(t) \cap R_{\mathbf{k}_q p_q} \ne \varnothing.$$

2. *If $r < 2/\sqrt{N}$ and $a > 1$, then*

$$\sum_{q=0}^{\infty} \exp(-ah(t_q)) < \infty.$$

Proof We use mathematical induction.

The real number t_0 is already constructed (it is involved in the formulation of Theorem 4.26). According to Lemma 4.29, the union of all cylinders $R_r(\mathbf{y}_j, u_j)$ that are defined before the moment $t_0 + 1$, is contained in the union of finitely many parallelepipeds $R_{\mathbf{k}p}$. Therefore there exists a parallelepiped $R_{\mathbf{k}_0 p_0}$ which intersects infinitely many sets from the sequence $\mathcal{P}(t_0 + 1), \mathcal{P}(t_0 + 1/2), \dots, \mathcal{P}(t_0 + 1/n), \dots$.

Assume that the real numbers $t_0, t_1, \dots, t_q$, and the parallelepipeds $R_{\mathbf{k}_0 p_0}, R_{\mathbf{k}_1 p_1}, \dots, R_{\mathbf{k}_q p_q}$ are already constructed. Define

$$t_{q+1} := \inf\{t > t_q : \mathcal{P}(t) \not\subseteq R_{\mathbf{k}_0 p_0} \cup R_{\mathbf{k}_1 p_1} \cup \dots \cup R_{\mathbf{k}_q p_q}\}.$$

Parallelepiped $R_{\mathbf{k}_{q+1}P_{q+1}}$ is defined as a parallelepiped that intersects infinitely many sets from the sequence $\mathcal{P}(t_q + 1), \mathcal{P}(t_q + 1/2), \ldots, \mathcal{P}(t_q + 1/n), \ldots$. This means that condition 1 is satisfied.

In order to prove condition 2, define the function $F_r'(t)$ as the volume of the union of all the parallelepipeds $R_{\mathbf{k}p}$ that intersect at least one set $\mathcal{P}(v)$ for $v \in [t_0, t]$, with respect to the measure $u^{-N-1}\, d\mathbf{y}\, du$. Formally,

$$F_r'(t) := \int_{\bigcup_{(\mathbf{k},p)\in\mathbb{Z}^{N+1}:\, R_{\mathbf{k}p}\cap(\bigcup_{t_0\leq v\leq t} \mathcal{P}(v))\neq\varnothing} R_{\mathbf{k}p}} u^{-N-1}\, d\mathbf{y}\, du.$$

The length of a side of a cube, which is inscribed in the ball of radius 1 in the space $\mathbb{R}^N$, is equal to $2/\sqrt{N}$. It follows that if $r < 2/\sqrt{N}$ and $(\mathbf{y}, u) \in R_{\mathbf{k}p}$, then $R_{\mathbf{k}p} \subset R_1(\mathbf{y}, u)$. Therefore we have $F_r'(t) \leq F_1(t)$ and

$$\sum_{q=0}^{\infty} \exp(-ah(t_q)) \sim \frac{1}{r^{N+1}} \int_{t_0}^{\infty} \exp(-ah(t))\, dF_r'(t)$$

$$\leq \frac{1}{r^{N+1}} \int_{t_0}^{\infty} \exp(-ah(t))\, dF_1(t)$$

$$< \infty. \qquad\qquad \square$$

Step 3. We prove Lemma 4.27.
Denote

$$Y_{\mathbf{s},u}(\mathbf{t}) := \frac{X(\mathbf{s} + u\mathbf{t}) - X(\mathbf{s})}{\sqrt{2}u^H}, \qquad \mathbf{t} \in \mathcal{B}.$$

It is well known that the multiparameter fractional Brownian motion $X(\mathbf{t})$ has homogeneous increments, i.e., for any $\mathbf{s} \in \mathbb{R}^N$

$$X(\mathbf{t} + \mathbf{s}) - X(\mathbf{s}) \stackrel{d}{=} X(\mathbf{t}),$$

and is self-similar, i.e., for any $u \in \mathbb{R}$

$$X(u\mathbf{t}) \stackrel{d}{=} u^H X(\mathbf{t}).$$

Using these properties, we obtain

$$Y_{\mathbf{s},u}(\mathbf{t}) \stackrel{d}{=} \frac{1}{\sqrt{2}} X(\mathbf{t}).$$

Let $(\mathbf{y}_q, u_q)$ be the centre of the parallelepiped $R_{\mathbf{k}_q P_q}$. Let $(r, \vartheta_1, \ldots, \vartheta_{N-2}, \varphi)$ be the spherical coordinates of a point $\mathbf{t} \in \mathcal{B}$. Let $(r_2, \vartheta_{1,2}, \ldots, \vartheta_{N-2,2}, \varphi_2)$ be the spherical coordinates of a point $\mathbf{v} \in \mathcal{B}$. Let $b_{m'n'm''n''qs}^{\ell'\ell''}$ be the Fourier coefficients of the function

$$b_{qs}(\mathbf{t}, \mathbf{v}) = \mathsf{E}\left[Y_{\mathbf{s}_q, u_q}(\mathbf{t}) Y_{\mathbf{s}_s, u_s}(\mathbf{v})\right]$$

with respect to the orthonormal basis

$$f_{\ell'm'n'}(r, \theta_1, \ldots, \theta_{N-2}, \varphi) f_{\ell''m''n''}(r_2, \theta_{1,2}, \ldots, \theta_{N-2,2}, \varphi_2).$$

Let $\{X_{mn}^{\ell q}\}$, $q \geq 0$, be a sequence of series of standard normal random variables independent in each series, with the following covariance between series:

$$\mathsf{E}\big[X_{m'n'}^{\ell' q} X_{m''n''}^{\ell'' s}\big] := b_{m'n'm''n''}^{\ell'\ell''}, \quad q \neq s.$$

Then we have

$$Y_{s_q, u_q}(\mathbf{t}) = \frac{1}{\sqrt{2}} \sum_{\ell=0}^{\infty} \sum_{m=1}^{h(S^{N-1}, \ell)} \sum_{n=1}^{\infty} X_{mn}^{\ell q} f_{\ell mn}(r, \theta_1, \ldots, \theta_{N-2}, \varphi).$$

Let ℓ_0 and n_0 be two positive integers. Denote

$$Y_{s_q, u_q}^{(\ell_0, n_0)}(\mathbf{t}) := \frac{1}{\sqrt{2}} \sum_{\ell=0}^{\ell_0} \sum_{m=1}^{h(S^{N-1}, \ell)} \sum_{n=1}^{n_0} X_{mn}^{\ell q} f_{\ell mn}(r, \theta_1, \ldots, \theta_{N-2}, \varphi).$$

For any $\varepsilon > 0$ consider the following three events:

$$A_{1q}(\varepsilon) := \left\{ \frac{Y_{s_q, u_q}^{(\ell_0, n_0)}}{\sqrt{h(t_q)}} \notin \mathcal{K}_{\varepsilon/3} \right\},$$

$$A_{2q}(\varepsilon) := \left\{ \frac{\|Y_{s_q, u_q} - Y_{s_q, u_q}^{(\ell_0, n_0)}\|_\infty}{\sqrt{h(t_q)}} > \frac{\varepsilon}{3} \right\},$$

$$A_{3q}(\varepsilon) := \left\{ \sup_{(s,u) \in R_{k_q P q}} \frac{\|Y_{s_q, u_q} - Y_{s,u}\|_\infty}{\sqrt{h(t_q)}} > \frac{\varepsilon}{3} \right\},$$

where $\mathcal{K}_{\varepsilon/3}$ denotes the $\varepsilon/3$-neighbourhood of Strassen ball $\mathcal{K}$ in the space $C(\mathcal{B})$. To prove Lemma 4.27, it is enough to prove that for any $\varepsilon > 0$ there exist natural numbers $\ell_0 = \ell_0(\varepsilon)$ and $n_0 = n_0(\varepsilon)$ such that the events $A_{1q}(\varepsilon)$, $A_{2q}(\varepsilon)$, and $A_{3q}(\varepsilon)$ occur only finitely many times a.s. In other words,

$$\mathsf{P}\Big\{\limsup_{q \to \infty} A_{1q}(\varepsilon)\Big\} = \mathsf{P}\Big\{\limsup_{q \to \infty} A_{2q}(\varepsilon)\Big\} = \mathsf{P}\Big\{\limsup_{q \to \infty} A_{3q}(\varepsilon)\Big\} = 0.$$

By the Borel–Cantelli lemma, it is enough to prove that

$$\sum_{q=1}^{\infty} \mathsf{P}\{A_{1q}(\varepsilon)\} < \infty, \tag{4.35a}$$

$$\sum_{q=1}^{\infty} \mathsf{P}\{A_{2q}(\varepsilon)\} < \infty, \tag{4.35b}$$

$$\sum_{q=1}^{\infty} \mathsf{P}\{A_{3q}(\varepsilon)\} < \infty. \tag{4.35c}$$

We prove (4.35b) first. Denote

$$\sigma_{\ell_0 n_0}^2(\mathbf{t}) := \mathsf{E}\big[Y_{s_q, u_q}(\mathbf{t}) - Y_{s_q, u_q}^{(\ell_0, n_0)}(\mathbf{t})\big]^2, \qquad \sigma_{\ell_0 n_0}^2 = \max_{\mathbf{t} \in B} \sigma_{\ell_0 n_0}^2(\mathbf{t}).$$

Using the large-deviation estimate (see Lifshits, 1995, Sect. 12, (11)), we obtain

$$\mathsf{P}\bigg\{\big\|Y_{s_q, u_q} - Y_{s_q, u_q}^{(\ell_0, n_0)}\big\|_{C(B)} > \frac{\varepsilon\sqrt{h(t_q)}}{3}\bigg\} \le \exp\bigg[-\frac{\varepsilon^2 h(t_q)}{18\sigma_{\ell_0 n_0}^2} + o\bigg(\frac{\varepsilon\sqrt{h(t_q)}}{3}\bigg)\bigg].$$

By Lemma 4.30, condition 2, it is sufficient to prove that for any $\varepsilon > 0$ there exist positive integers $\ell_0 = \ell_0(\varepsilon)$ and $n_0 = n_0(\varepsilon)$ such that, say,

$$\frac{\varepsilon^2}{9\sigma_{\ell_0 n_0}^2} < 1.$$

Denote

$$Y_{s_q, u_q}^{(\ell_0)}(\mathbf{t}) := \frac{1}{\sqrt{2}} \sum_{\ell=0}^{\ell_0} \sum_{m=1}^{h(S^{N-1}, \ell)} \sum_{n=1}^{\infty} X_{mn}^{\ell q} f_{\ell m n}(r, \theta_1, \ldots, \theta_{N-2}, \varphi)$$

and

$$\sigma_{\ell_0}^2(\mathbf{t}) := \mathsf{E}\big[Y_{s_q, u_q}(\mathbf{t}) - Y_{s_q, u_q}^{(\ell_0)}(\mathbf{t})\big]^2, \qquad \sigma_{\ell_0}^2 = \max_{\mathbf{t} \in B} \sigma_{\ell_0}^2(\mathbf{t}).$$

The sequence $\sigma_{\ell_0}^2(\mathbf{t})$ converges to zero as $\ell_0 \to \infty$ for all $\mathbf{t} \in B$. Moreover, $\sigma_{\ell_0}^2(\mathbf{t}) \ge \sigma_{\ell_0+1}^2(\mathbf{t})$ for all $\mathbf{t} \in B$ and all positive integers ℓ_0. Functions $\sigma_{\ell_0}^2(\mathbf{t})$ are nonnegative and continuous. By Dini's theorem, the sequence $\sigma_{\ell_0}^2(\mathbf{t})$ converges to zero uniformly on B, i.e.,

$$\lim_{\ell_0 \to \infty} \sigma_{\ell_0}^2 = 0,$$

and we choose such an ℓ_0 that for any $\ell > \ell_0$, $\sigma_\ell^2 < \varepsilon^2/18$.

In the same way, we can apply Dini's theorem to the sequence of functions

$$\mathsf{E}\big[Y_{s_q, u_q}^{(\ell_0)}(\mathbf{t}) - Y_{s_q, u_q}^{(\ell_0, n_0)}(\mathbf{t})\big]^2, \quad n_0 \ge 1$$

and find such an n_0 that

$$\sup_{\mathbf{t} \in B} \mathsf{E}\big[Y_{s_q, u_q}^{(\ell_0)}(\mathbf{t}) - Y_{s_q, u_q}^{(\ell_0, n)}(\mathbf{t})\big]^2 \le \varepsilon^2/18$$

for all $n > n_0$. (4.35b) is proved.

Now we prove (4.35a). In what follows we denote by C a constant depending only on N, H and that may vary at each occurrence. Specific constants will be denote by C_1, C_2, ...

Consider the finite-dimensional subspace E of the space $C(\mathcal{B})$ spanned by the functions

$$f_{\ell mn}(r, \theta_1, \ldots, \theta_{N-2}, \varphi),$$

for $1 \le n \le n_0$, $0 \le \ell \le \ell_0$, and $1 \le m \le h(S^{N-1}, \ell)$. All norms on E are equivalent. In particular, there exists a constant $C_1 = C_1(\ell_0, n_0)$ such that the $\varepsilon/3$-neighbourhood of Strassen ball $\mathcal{K}$ in the space E equipped with the uniform norm is contained in the ball of radius $1 + C_1 \varepsilon$ with respect to Strassen norm. Then we have

$$P\{A_{1q}(\varepsilon)\} \le P\left\{ \left\| \frac{Y^{(\ell_0, n_0)}_{s_q, u_q}}{\sqrt{h(t_q)}} \right\|^2_{\mathcal{H}} > (1 + C_1\varepsilon)^2 \right\}$$

$$= P\left\{ \left\| Y^{(\ell_0, n_0)}_{s_q, u_q} \right\|^2_{\mathcal{H}} > (1 + C_1\varepsilon)^2 h(t_q) \right\}$$

$$= P\left\{ \sum_{\ell=0}^{\ell_0} \sum_{m=1}^{h(S^{N-1}, \ell)} \sum_{n=1}^{n_0} (X^{\ell q}_{mn})^2 > 2(1 + C_1\varepsilon)^2 h(t_q) \right\}.$$

Denote

$$M := n_0 \sum_{\ell=0}^{\ell_0} h(S^{N-1}, \ell)$$

and let χ_M denote a random variable that has χ^2 distribution with M degrees of freedom. Using standard probability estimates for χ_M, we can write

$$P\{A_{1q}(\varepsilon)\} \le P\{\chi_M > 2(1 + C_1\varepsilon)^2 h(t_q)\}$$
$$\le \exp\{-(1 + C_1\varepsilon)^2 h(t_q)\}$$

for large enough $h(t_q)$. Applying Lemma 4.30, condition 2 concludes the proof.

Now we prove (4.35c). Denote

$$Z_1 := \sup_{\substack{(s_1, u_1) \in R_r(s, u) \\ (s_2, u_2) \in R_r(s, u)}} \frac{|X(s_1) - X(s_2)|}{\sqrt{2}u_1^H},$$

$$Z_2 := \sup_{\substack{(s_1, u_1) \in R_r(s, u) \\ (s_2, u_2) \in R_r(s, u)}} \frac{\sup_{t \in \mathcal{B}} |X(s_1 + u_1 t) - X(s_2 + u_2 t)|}{\sqrt{2}u_1^H},$$

$$Z_3 := \sup_{\substack{(s_1, u_1) \in R_r(s, u) \\ (s_2, u_2) \in R_r(s, u)}} \left| \frac{1}{\sqrt{2}u_1^H} - \frac{1}{\sqrt{2}u_2^H} \right| \sup_{t \in \mathcal{B}} |X(s_2 + u_2 t) - X(s_2)|.$$

It is easy to see that $Z_1 \le Z_2$ and $\|Y_{s_q, u_q} - Y_{s,u}\|_{C(\mathcal{B})} \le Z_1 + Z_2 + Z_3$. It follows that

$$P\{A_{3q}(\varepsilon)\} \le P\{Z_2 > \varepsilon \sqrt{h(t_q)}/12\} + P\{Z_3 > \varepsilon \sqrt{h(t_q)}/6\}. \tag{4.36}$$

The second term in the right-hand side may be estimated as

$$P\{Z_3 > \varepsilon \sqrt{h(t_q)}/6\} = P\left\{ \sup_{\substack{t \in B \\ (s_1,u_1) \in R_r(s,u) \\ (s_2,u_2) \in R_r(s,u)}} |X(t)|[(u_2/u_1)^H - 1] > \frac{\varepsilon \sqrt{2h(t_q)}}{6} \right\}$$

$$\le P\left\{ \sup_{t \in B} X(t) > \frac{\varepsilon \sqrt{2h(t_q)}}{6\delta} \right\},$$

$$\tag{4.37}$$

where $\delta = \delta(r) := \max\{r, e^{2Hr} - 1, e^r - 1\}$.

Another large-deviation estimate (see Lifshits, 1995, Sect. 14, (12)) states that there exists a constant $C = C(H)$ such that for all $K > 0$

$$P\left\{ \sup_{t \in B} X(t) > K \right\} \le C K^{N/H-1} \exp(-K^2/2). \tag{4.38}$$

Using this fact, we can continue estimate (4.37) as follows:

$$P\{Z_3 > \varepsilon \sqrt{h(t_q)}/6\} \le C \frac{\varepsilon^{N/H-1}[h(t_q)]^{(N-H)/(2H)}}{\delta^{N/H-1}} \exp\left(-\frac{2\varepsilon^2 h(t_q)}{72\delta^2} \right).$$

If we choose such a small r that $\delta < \varepsilon/6$, then by Lemma 4.30, condition 2 the series

$$\sum_{q=1}^{\infty} P\{Z_3 > \varepsilon \sqrt{h(t_q)}/6\}$$

converges.

Using the self-similarity of the multiparameter fractional Brownian motion, we write random variable Z_2 as follows:

$$Z_2 = \sup_{\substack{t \in B \\ (s_1,u_1) \in R_r(s,u) \\ (s_2,u_2) \in R_r(s,u)}} \left| X\left(\frac{s_1 + u_1 t}{2^{1/(2H)} u_1} \right) - X\left(\frac{s_2 + u_2 t}{2^{1/(2H)} u_1} \right) \right|.$$

The right-hand side can be estimated as follows:

$$\frac{\|(s_1 + u_1 t) - (s_2 + u_2 t)\|}{2^{1/(2H)} u_1} \leq \frac{\|s_1 - s_2\|}{2^{1/(2H)} u_1} + \frac{\|t\| \cdot |u_1 - u_2|}{2^{1/(2H)} u_1}$$

$$\leq \frac{2ru}{2^{1/(2H)} u_1} + 2^{-1/(2H)} \left(1 - \frac{u_2}{u_1}\right)$$

$$\leq 2^{1-1/(2H)} r e^r + 2^{-1/(2H)} \left(e^{2r} - 1\right)$$

$$\leq 2^{1-1/(2H)} \delta e^r + 2^{-1/(2H)} \left(\delta^2 + 2\delta\right)$$

$$\leq 2^{1-1/(2H)} \left(\delta^2 + \delta\right) + 2^{-1/H} \left(\delta^2 + \delta\right)$$

$$\leq C_2 \delta.$$

Let v_j, $1 \leq j \leq C(C_2 \delta)^{-N}$ be the $C_1 \delta$-net in $\mathcal{B}$. A standard entropy estimate for the first term in the right-hand side of (4.36) gives

$$P\{Z_2 > \varepsilon \sqrt{h(t_q)}/12\} \leq P\left\{ \sup_{\substack{t \in \mathcal{B} \\ \|s\| \leq C_2 \delta}} |X(t + s) - X(t)| > \frac{\varepsilon \sqrt{h(t_q)}}{12} \right\}$$

$$\leq P\left\{ \sup_{\substack{1 \leq j \leq C(C_2 \delta)^{-N} \\ \|s\| \leq C_2 \delta}} |X(v_j + s) - X(v_j)| > \frac{\varepsilon \sqrt{h(t_q)}}{24} \right\}$$

$$\leq C\delta^{-N} P\left\{ \sup_{\|s\| \leq 1} |X(s)| > \frac{\varepsilon \sqrt{h(t_q)}}{24 (C_2 \delta)^H} \right\}$$

$$\leq C\delta^{-N} \cdot \frac{\varepsilon^{\frac{N}{H}-1} [h(t_q)]^{\frac{N-H}{2H}}}{\delta^{N-H}} \exp\left(-\frac{\varepsilon^2 h(t_q)}{1152 (C_2 \delta)^{2H}}\right).$$

Here we used the properties of $X(t)$ and (4.38). If we choose such a small r that

$$\frac{\varepsilon^2}{1152 (C_2 \delta)^{2H}} > 1,$$

then by Lemma 4.30, condition 2 the series

$$\sum_{q=1}^{\infty} P\{Z_2 > \varepsilon \sqrt{h(t_q)}/12\}$$

converges. This concludes the proof of Lemma 4.27.

Step 4. We prove Lemma 4.28.

By Lemma 4.30, condition 1, for any $q \geq 0$ there exists a number $t'_q \in [t_q, t_{q+1})$ such that $\mathcal{P}(t) \cap R_{k_q p_q} \neq \emptyset$. Choose arbitrary points $(s'_q, u'_q) \in \mathcal{P}(t) \cap R_{k_q p_q}$. It is easy to see that the sequence t'_q satisfies Lemma 4.30, condition 2 as well.

The set of all $f \in C(\mathcal{B})$ with $0 < \|f\|_{\mathcal{H}} < 1$ is dense in $\mathcal{K}$. It is enough to prove that for any such f we have

$$\liminf_{q \to \infty} \left\| \frac{X(\mathbf{s}'_q + u'_q \mathbf{t}) - X(\mathbf{s}'_q)}{(u'_q)^H \sqrt{2h(t'_q)}} - f \right\|_{C(\mathcal{B})} = 0 \quad \text{a.s.}$$

According to Li and Shao (2001, Theorem 5.1), there exists a constant $C_3 = C_3(N, H)$ such that for all $\varepsilon \in (0, 1]$

$$\mathsf{P}\left\{ \sup_{\mathbf{t} \in \mathcal{B}} |X(\mathbf{t})| \leq \varepsilon \right\} \geq \exp\left(-C_3 \varepsilon^{-N/H} \right).$$

Denote $\beta := (C_3)^{H/N} (1 - \|f\|_{\mathcal{S}}^2)^{-H/N}$. We will prove that

$$\liminf_{q \to \infty} \left[h(t'_q) \right]^{(N+2H)/(2N)} \left\| \frac{X(\mathbf{s}'_q + u'_q \mathbf{t}) - X(\mathbf{s}'_q)}{(u'_q)^H \sqrt{2h(t'_q)}} - f \right\|_{C(\mathcal{B})} \leq \frac{1}{\sqrt{2}} \beta \quad \text{a.s.}$$

Consider the event

$$\tilde{A}_{1q}(\varepsilon) := \left\{ \left\| \frac{X(\mathbf{s}'_q + u'_q \mathbf{t}) - X(\mathbf{s}'_q)}{(u'_q)^H} - \sqrt{2h(t'_q)} f \right\|_{C(\mathcal{B})} \leq \beta(1 + \varepsilon) \left[h(t'_q) \right]^{-H/N} \right\}.$$

Using Monrad and Rootzén (1995, Proposition 4.2), we obtain

$$\ln \mathsf{P}\{\tilde{A}_{1q}(\varepsilon)\} \geq 2h(t'_q)(-1/2) \|f\|_{\mathcal{H}}^2 - C_3 \beta^{-N/H} (1 + \varepsilon)^{-N/H} h(t'_q)$$
$$= -h(t'_q) \left[\|f\|_{\mathcal{H}}^2 + (1 - \|f\|_{\mathcal{H}}^2)(1 + \varepsilon)^{-N/H} \right].$$

The term in square brackets is less than 1. It follows that

$$\sum_{q=0}^{\infty} \mathsf{P}\{\tilde{A}_{1q}(\varepsilon)\} = \infty. \tag{4.39}$$

If the events $\tilde{A}_{1q}(\varepsilon)$ were independent, the usage of the second Borel–Cantelli lemma would conclude the proof. However, they are dependent.

In order to create independence, we follow the lines of Li and Shao (2001) and Monrad and Rootzén (1995). We use another spectral expansion of the multiparameter fractional Brownian motion. Let $\overline{W}$ denote a complex-valued scattered Gaussian random measure on $\mathbb{R}^N$ with Lebesgue measure as its control measure.

Lemma 4.31 (Global spectral expansion, Ayache and Linde (2008)) *There exists a constant $C_4 = C_4(N, H)$ such that*

$$X(\mathbf{t}) = C_4 \int_{\mathbb{R}^N} \left(e^{i(\mathbf{p}, \mathbf{t})} - 1 \right) \|\mathbf{p}\|^{-(N/2)-H} \, d\overline{W}(\mathbf{p}). \tag{4.40}$$

We prove that

$$C_4^2(N, H) = \frac{2^{2H-1}\Gamma(H + N/2)\Gamma(H + 1)\sin(\pi H)}{\pi^{(N+2)/2}}. \tag{4.41}$$

For $N = 1$, our formula has the form

$$C_4^2(1, H) = \frac{2^{2H-1}\Gamma(H + 1/2)\Gamma(H + 1)\sin(\pi H)}{\pi^{3/2}},$$

or

$$C_4^2(1, H) = \frac{\Gamma(2H + 1)\sin(\pi H)}{2\pi}. \tag{4.42}$$

Here we used the duplication formula for gamma function (6.33). Formula (4.42) is known (Samorodnitsky and Taqqu, 1994). Therefore, in the rest of the proof we can and will suppose that $N \geq 2$.

It follows from (4.40) that the covariance function of the multiparameter fractional Brownian motion can be represented as

$$R(\mathbf{s}, \mathbf{t}) = C_4^2(N, H) \int_{\mathbf{R}^N} \left[e^{i(\mathbf{p}, \mathbf{t})} - 1\right]\left[e^{-i(\mathbf{p}, \mathbf{s})} - 1\right]\|\mathbf{p}\|^{-N-2H} \, d\mathbf{p}. \tag{4.43}$$

Rewrite (4.43) as

$$R(\mathbf{s}, \mathbf{t}) = C_4^2(N, H) \int_{\mathbf{R}^N} \left[1 - e^{i(\mathbf{p}, \mathbf{t})}\right]\|\mathbf{p}\|^{-N-2H} \, d\mathbf{p}$$

$$+ C_4^2(N, H) \int_{\mathbf{R}^N} \left[1 - e^{-i(\mathbf{p}, \mathbf{s})}\right]\|\mathbf{p}\|^{-N-2H} \, d\mathbf{p}$$

$$- C_4^2(N, H) \int_{\mathbf{R}^N} \left[1 - e^{i(\mathbf{p}, \mathbf{t}-\mathbf{s})}\right]\|\mathbf{p}\|^{-N-2H} \, d\mathbf{p}.$$

Consider the first term in the right-hand side of this formula. Using formula 3.3.2.3 from Prudnikov et al. (1986), we obtain

$$C_4^2(N, H) \int_{\mathbf{R}^N} \left[1 - e^{i(\mathbf{p}, \mathbf{t})}\right]\|\mathbf{p}\|^{-N-2H} \, d\mathbf{p}$$

$$= \frac{2\pi^{(N-1)/2}C_4^2(N, H)}{\Gamma((N-1)/2)} \int_0^\infty \lambda^{N-1} \, d\lambda \int_0^\pi \left[1 - e^{i\lambda\|\mathbf{t}\|\cos u}\right]\lambda^{-N-2H}\sin^{N-2}u \, du.$$

It is clear that the integral of the imaginary part is equal to 0. The integral of the real part may be rewritten as

$$C_4^2(N, H) \int_{\mathbf{R}^N} \left[1 - e^{i(\mathbf{p}, \mathbf{t})}\right]\|\mathbf{p}\|^{-N-2H} \, d\mathbf{p}$$

$$= \frac{2\pi^{(N-1)/2}C_4^2(N, H)}{\Gamma((N-1)/2)} \int_0^\infty \lambda^{-1-2H} \int_0^\pi \left[1 - \cos(\lambda\|\mathbf{x}\|\cos u)\right]\sin^{N-2}u \, du \, d\lambda. \tag{4.44}$$

To calculate the inner integral, we use formulae 2.5.3.1 and 2.5.55.7 from Prudnikov et al. (1986):

$$\int_0^\pi \sin^{N-2} u \, du = \frac{\sqrt{\pi}\,\Gamma((N-1)/2)}{\Gamma(N/2)},$$

$$\int_0^\pi \cos(\lambda\|\mathbf{x}\|\cos u)\sin^{N-2} u \, du = \sqrt{\pi}\,2^{(N-2)/2}\Gamma\big((N-1)/2\big)\frac{J_{(N-2)/2}(\lambda\|\mathbf{x}\|)}{(\lambda\|\mathbf{x}\|)^{(N-2)/2}}.$$

It follows that

$$\int_0^\pi \big[1-\cos(\lambda\|\mathbf{t}\|\cos u)\big]\sin^{N-2} u \, du = \frac{\sqrt{\pi}\,\Gamma((N-1)/2)}{\Gamma(N/2)}\big[1-g_0(\lambda\|\mathbf{x}\|)\big],$$

$$(4.45)$$

where the function $g_0(u)$ is given by (4.18). Substituting (4.45) in (4.44), we obtain

$$C_4^2(N,H)\int_{\mathbf{R}^N}\big[1-e^{i(\mathbf{p},\mathbf{t})}\big]\|\mathbf{p}\|^{-N-2H}\,d\mathbf{p}$$

$$= \frac{2\pi^{N/2}C_4^2(N,H)}{\Gamma(N/2)}\int_0^\infty \lambda^{-2H-1}\big[1-g_0(\lambda\|\mathbf{x}\|)\big]\,d\lambda. \qquad (4.46)$$

To calculate this integral, we use formula 2.2.3.1 from Prudnikov et al. (1986):

$$\int_0^1 (1-v^2)^{(N-3)/2}\,dv = \frac{\sqrt{\pi}\,\Gamma((N-1)/2)}{2\Gamma(N/2)}.$$

It follows that

$$1 = \frac{2\Gamma(N/2)}{\sqrt{\pi}\,\Gamma((N-1)/2)}\int_0^1 (1-v^2)^{(N-3)/2}\,dv. \qquad (4.47)$$

On the other hand, according to formula 2.5.6.1 from Prudnikov et al. (1986) we have

$$\int_0^1 (1-v^2)^{(N-3)/2}\cos(\lambda\|\mathbf{t}\|v)\,dv = \sqrt{\pi}\,2^{(N-4)/2}\Gamma\big((N-1)/2\big)\frac{J_{(N-2)/2}(\lambda\|\mathbf{t}\|)}{(\lambda\|\mathbf{t}\|)^{(N-2)/2}}.$$

It follows that

$$g_0(\lambda\|\mathbf{t}\|) = \frac{2\Gamma(N/2)}{\sqrt{\pi}\,\Gamma((N-1)/2)}\int_0^1 (1-v^2)^{(N-3)/2}\cos(\lambda\|\mathbf{t}\|v)\,dv. \qquad (4.48)$$

Subtracting (4.48) from (4.47), we obtain

$$1-g_0(\lambda\|\mathbf{t}\|) = \frac{4\Gamma(N/2)}{\sqrt{\pi}\,\Gamma((N-1)/2)}\int_0^1 (1-v^2)^{(N-3)/2}\sin^2\left(\frac{\lambda\|\mathbf{t}\|}{2}v\right)dv.$$

Substitute this formula into (4.46). We have

$$
\begin{aligned}
&C_4^2(N, H) \int_{\mathbf{R}^N} \left[1 - e^{i(\mathbf{p}, \mathbf{t})}\right] \|\mathbf{p}\|^{-N-2H} \, d\mathbf{p} \\
&= \frac{8\pi^{(N-1)/2} C_4^2(N, H)}{\Gamma((N-1)/2)} \int_0^\infty \int_0^1 \lambda^{-2H-1} (1 - v^2)^{(N-3)/2} \sin^2\left(\frac{\lambda \|\mathbf{t}\|}{2} v\right) \, dv \, d\lambda.
\end{aligned}
\tag{4.49}
$$

Consider the following integrals:

$$
\int_0^1 \lambda^{-2H-1} \sin^2\left(\frac{v \|\mathbf{t}\|}{2} \lambda\right) d\lambda \quad \text{and} \quad \int_1^\infty \lambda^{-2H-1} \sin^2\left(\frac{v \|\mathbf{t}\|}{2} \lambda\right) d\lambda.
$$

In the first integral, we bound the second multiplier by $\lambda^2/4$. In the second integral, we bound it by 1. It follows that the integral

$$
\int_0^\infty \lambda^{-2H-1} \sin^2\left(\frac{v \|\mathbf{x}\|}{2} \lambda\right) d\lambda
$$

converges uniformly, and we may change the order of integration in the right-hand side of (4.49). After that the inner integral is calculated using formula 2.5.3.13 from Prudnikov et al. (1986):

$$
\int_0^\infty \lambda^{-2H-1} \sin^2\left(\frac{v \|\mathbf{t}\|}{2} \lambda\right) d\lambda = \frac{\pi v^{2H}}{4\Gamma(2H+1) \sin(\pi H)} \|\mathbf{t}\|^{2H}.
$$

Now formula (4.46) may be rewritten as

$$
\begin{aligned}
&C_4^2(N, H) \int_{\mathbf{R}^N} \left[1 - e^{i(\mathbf{p}, \mathbf{t})}\right] \|\mathbf{p}\|^{-N-2H} \, d\mathbf{p} \\
&= \frac{2\pi^{(N-1)/2} C_4^2(N, H)}{\Gamma((N-1)/2) \Gamma(2H+1) \sin(\pi H)} \int_0^1 v^{2H} (1 - v^2)^{(N-3)/2} \, dv \cdot \|\mathbf{t}\|^{2H}.
\end{aligned}
$$

The integral in the right-hand side can be calculated by formula 2.2.4.8 from Prudnikov et al. (1986):

$$
\int_0^1 v^{2H} (1 - v^2)^{(N-3)/2} \, dv = \frac{\Gamma((N-1)/2) \Gamma(H+1/2)}{2\Gamma(H+N/2)}
$$

and (4.46) is rewritten once more as

$$
\begin{aligned}
&C_4^2(N, H) \int_{\mathbf{R}^N} \left[1 - e^{i(\mathbf{p}, \mathbf{t})}\right] \|\mathbf{p}\|^{-N-2H} \, d\mathbf{p} \\
&= \frac{\pi^{(N+2)/2} C_4^2(N, H)}{2^{2H} \Gamma(H+1) \Gamma(H+N/2) \sin(\pi H)} \|\mathbf{t}\|^{2H}.
\end{aligned}
\tag{4.50}
$$

On the other hand, the left-hand side of (4.50) is clearly equal to $\frac{1}{2}\|\mathbf{t}\|^{2H}$. Equation (4.41) follows.

Let $0 < a < b$ be two real numbers. Denote

$$X^{(a,b)}(\mathbf{t}) := C_4(N.H) \int_{\|\mathbf{p}\| \in (a,b]} \left(e^{i(\mathbf{p},\mathbf{t})} - 1\right) \|\mathbf{p}\|^{-(N/2)-H} \, d\overline{W}(\mathbf{p}),$$

$$\tilde{X}^{(a,b)}(\mathbf{t}) = X(\mathbf{t}) - X^{(a,b)}(\mathbf{t}).$$

The proof of the following Lemma is straightforward.

Lemma 4.32 *The random field $\tilde{X}^{(a,b)}(\mathbf{t})$ has the following properties.*

1. *It has homogeneous increments.*
2. *For any $u \in \mathbb{R}$*

$$\tilde{X}^{(a,b)}(u\mathbf{t}) \stackrel{d}{=} u^H \tilde{X}^{(ua,ub)}(\mathbf{t}). \tag{4.51}$$

3. *It is isotropic.*

Put

$$d_q := \left(u_q'\right)^{-1} \exp\{h(t_q')[\exp(h(t_q') + 1 - H)]\}$$

and consider the events

$$\tilde{A}_{2q}(\varepsilon) := \left\{ \left\| \frac{X^{(d_{q-1},d_q)}(s_q' + u_q'\mathbf{t}) - X^{(d_{q-1},d_q)}(s_q')}{(u_q')^H} - \sqrt{2h(t_q')}f \right\|_{C(\mathcal{B})} \right.$$

$$\left. \leq \beta(1+\varepsilon)[h(t_q')]^{-H/N} \right\},$$

$$\tilde{A}_{3q}(\varepsilon) := \left\{ \left\| \frac{\tilde{X}^{(d_{q-1},d_q)}(s_q' + u_q'\mathbf{t}) - \tilde{X}^{(d_{q-1},d_q)}(s_q')}{(u_q')^H} \right\|_{C(\mathcal{B})} \geq \varepsilon\beta[h(t_q')]^{-H/N} \right\}.$$

Lemma 4.33 *We have*

$$\sum_{q=0}^{\infty} P\{\tilde{A}_{3q}(\varepsilon)\} < \infty.$$

Proof Using Lemma 4.32, one can write

$$P\{\tilde{A}_{3q}(\varepsilon)\} = P\{\|\tilde{X}^{(u_q'd_{q-1}, u_q'd_q)}(\mathbf{t})\|_{C(\mathcal{B})} \geq \varepsilon\beta[h(t_q')]^{-H/N}\}.$$

Put

$$x_q := \exp\{-\exp[(1 - \|f\|_{\mathcal{H}}^2)h(t_q')]\}.$$

Using Lemma 4.32 once more, we have

$$P\big\{\tilde{A}_{3q}(\varepsilon)\big\} = P\bigg\{ \sup_{\|\mathbf{t}\|\le x_q} \big|\tilde{X}^{(x_q^{-1}u_q'd_{q-1},\,x_q^{-1}u_q'd_q)}(\mathbf{t})\big| \ge \varepsilon x_q^H \beta\big[h\big(t_q'\big)\big]^{-H/N}\bigg\}.$$

Denote

$$Z_q(\mathbf{t}) := \tilde{X}^{(x_q^{-1}u_q'd_{q-1},\,x_q^{-1}u_q'd_q)}(\mathbf{t}).$$

We estimate the variance of the random field $Z_q(\mathbf{t})$ for $\|\mathbf{t}\| \le x_q$. We have

$$E[Z_q(\mathbf{t})]^2 = 2C_4^2(N,H) \int_{\|\mathbf{p}\|\le x_q^{-1}u_q'd_{q-1}} (1 - \cos(\mathbf{p},\mathbf{t}))\,\|\mathbf{p}\|^{-N-2H}\,d\mathbf{p}$$

$$+\, 2C_4^2(N,H) \int_{\|\mathbf{p}\|> x_q^{-1}u_q'd_q} (1 - \cos(\mathbf{p},\mathbf{t}))\,\|\mathbf{p}\|^{-N-2H}\,d\mathbf{p}.$$

In the first integral, we bound $1 - \cos(\mathbf{p},\mathbf{t})$ by $\|\mathbf{p}\|^2 \cdot \|\mathbf{t}\|^2/2$. In the second integral, we bound it by 2. Then we have

$$E[Z_q(\mathbf{t})]^2 \le C_4^2(N,H)x_q^2 \int_{\|\mathbf{p}\|\le x_q^{-1}u_q'd_{q-1}} \|\mathbf{p}\|^{2-N-2H}\,d\mathbf{p}$$

$$+\, 4C_4^2(N,H) \int_{\|\mathbf{p}\|> x_q^{-1}u_q'd_q} \|\mathbf{p}\|^{-N-2H}\,d\mathbf{p}.$$

Now we pass to spherical coordinates and obtain

$$E\big[\tilde{X}^{(x_q^{-1}u_q'd_{q-1},\,x_q^{-1}u_q'd_q)}(\mathbf{t})\big]^2 \le Cx_q^2 \int_0^{x_q^{-1}u_q'd_{q-1}} p^{1-2H}\,dp + C\int_{x_q^{-1}u_q'd_q}^{\infty} p^{-1-2H}\,dp$$

$$= Cx_q^{2H}\big[\big(u_q'd_{q-1}\big)^{2-2H} + \big(u_q'd_q\big)^{-2H}\big].$$

Substituting the definitions of d_q and x_q into the last inequality, we obtain

$$E[Z_q(\mathbf{t})]^2 \le C \exp\big\{-2H\big\{\exp\big[\big(1 - \|f\|_{\mathcal{H}}^2\big)h\big(t_q'\big)\big] + (1 - H)h\big(t_q'\big)\big\}\big\},$$

or

$$E\big[\tilde{X}^{(x_q^{-1}u_q'd_{q-1},\,x_q^{-1}u_q'd_q)}(\mathbf{t}) - \tilde{X}^{(x_q^{-1}u_q'd_{q-1},\,x_q^{-1}u_q'd_q)}(\mathbf{s})\big]^2 \le \varphi_q^2(\|\mathbf{t}-\mathbf{s}\|)$$

for $\|\mathbf{t}-\mathbf{s}\| = \delta \le x_q$, where

$$\varphi_q^2(\delta) = C\min\big\{\delta^{2H}, \exp\big\{-2H\big\{\exp\big[\big(1 - \|f\|_{\mathcal{H}}^2\big)h\big(t_q'\big)\big] + (1 - H)h\big(t_q'\big)\big\}\big\}\big\}.$$

We need the following lemma of Fernique (1975).

Lemma 4.34 *Let $Z(\mathbf{t})$, $\mathbf{t} \in [0,t]^N$ be a separable centred Gaussian random field. Assume that*

$$\sup_{\substack{\mathbf{t},\mathbf{s}\in[0,t]^N \\ \|\mathbf{t}-\mathbf{s}\|\le\delta}} \mathsf{E}\big(Z(\mathbf{t}) - Z(\mathbf{s})\big)^2 \le \varphi^2(\delta).$$

Then, for any sequence of positive real numbers $y_0, y_1, \ldots, y_p, \ldots,$ and for any sequence of integer numbers $m_1, m_2, \ldots, m_p, \ldots,$ each divisible by the one before,

$$\mathsf{P}\left\{ \sup_{\mathbf{t}\in[0,t]^N} |Z(\mathbf{t})| \ge y_0\varphi(t) + \sum_{p=1}^{\infty} y_p\varphi(t/2m_p) \right\} \le \sqrt{2/\pi} \sum_{p=0}^{\infty} (m_{p+1})^N$$

$$\times \int_{y_p}^{\infty} \exp(-u^2/2)\,du.$$

Put $m_{p,q} := q^{2^p}$, $y_{0q} := 2\sqrt{(N+1)h(t'_q)}$, and

$$y_{p,q} := \varepsilon(p+1)^{-2}x_q^H \beta\big[h\big(t'_q\big)\big]^{-H/N} / \varphi_q\big(2x_q \cdot q^{-2^p}\big), \quad q \ge 1.$$

For large enough q,

$$y_{p,q} > 2\sqrt{(N+3)h\big(t'_q\big)}2^{p/2}$$

for all $p \ge 1$. Moreover,

$$y_{0,q}\varphi_q(x_q) + \sum_{p=1}^{\infty} y_{p,q}\varphi_q(x_q/2m_{p,q}) < \varepsilon x_q^H \beta\big[h\big(t'_q\big)\big]^{-H/N}.$$

We have

$$\sum_{q=1}^{\infty} q^2 \exp(-y_{0q}^2/2) + \sum_{q=1}^{\infty}\sum_{p=1}^{\infty} q^{N\cdot 2^{p+1}} \exp(-y_{pq}^2/2) < \infty,$$

and application of Lemma 4.34 finishes the proof. □

It follows from definition of the events $\tilde{A}_{1q}(\varepsilon)$, $\tilde{A}_{2q}(\varepsilon)$, and $\tilde{A}_{2q}(\varepsilon)$, that

$$\tilde{A}_{1q}(\varepsilon) \subset \tilde{A}_{2q}(2\varepsilon) \cup \tilde{A}_{3q}(\varepsilon) \subset \tilde{A}_{1q}(2\varepsilon) \cup \tilde{A}_{2q}(\varepsilon). \tag{4.52}$$

Combining (4.39), (4.52), and Lemma 4.33, we get

$$\sum_{q=0}^{\infty} \mathsf{P}\{\tilde{A}_{2q}(\varepsilon)\} = \infty. \tag{4.53}$$

Now we prove that the events $\tilde{A}_{2q}(\varepsilon)$ are independent. It is enough to prove that $d_{q-1} < d_q$. Using the definition of d_q, this inequality becomes

$$u'_q < e^{1-H} u'_{q-1}.$$

By our choice of u'_q, we have $\exp(p_q r) \leq u'_q \leq \exp(p_q+1)r)$. Since by construction of the parallelepipeds $R_{\mathbf{k}_q p_q}$ two adjacent parallelepipeds can lie in the same u-layer or in adjacent u-layers, we have

$$u'_{q-1} \leq e^{2r} u'_q.$$

We choose $r < (1 - H)/2$, and we are done.

It follows from the second Borel–Cantelli lemma that

$$\mathsf{P}\left\{\limsup_{q\to\infty} \tilde{A}_{2q}(\varepsilon)\right\} = 1. \tag{4.54}$$

Combining (4.52), (4.54), and Lemma 4.33, we get

$$\mathsf{P}\left\{\limsup_{q\to\infty} \tilde{A}_{1q}(3\varepsilon)\right\} = 1.$$

Since ε can be chosen arbitrarily close to 0, Lemma 4.28 is proved. This finishes proof of Theorem 4.26. $\square$

Example 4.35 (Local functional law of the iterated logarithm) Let $t_0 = 3$. Let $\mathcal{J}(t)$ contain only one element 0. Let $\mathbf{s}_0 := \mathbf{0}$ and $u_0 := t^{-1}$. Then we have

$$R_1(\mathbf{0}, u) = \left\{(\mathbf{s}, v)\colon \|\mathbf{s}\| \leq u, e^{-1}u \leq v \leq eu\right\}.$$

It is easy to see that $dA_1(u)$ is comparable to

$$du \int_{\|\mathbf{s}\|\leq u} \frac{d\mathbf{s}}{u^{N+1}} = C\frac{du}{u}.$$

The function $h(u) = \ln \ln u$ satisfies the conditions of Theorem 4.26. We obtain that, in the uniform topology, the set of a.s. limit points of the Strassen cloud

$$\frac{X(t\mathbf{t})}{\sqrt{2\ln\ln t^{-1}}t^H}$$

as $t \downarrow 0$ is Strassen ball $\mathcal{K}$. For the case of $N = 1$ and $H = 1/2$, this result is due to Gantert (1993).

Let $F(f) := \|f\|_{C(\mathcal{B})}$, $f \in C(\mathcal{B})$. On the one hand, we have

$$\limsup_{\|\mathbf{t}\|\downarrow 0} \frac{X(\mathbf{t})}{\sqrt{2\ln\ln\|\mathbf{t}\|^{-1}}\|\mathbf{t}\|^H} = \sup_{f\in\mathcal{K}} \|f\|_{C(\mathcal{B})} \quad \text{a.s.}$$

On the other hand, according to Benassi et al. (1997), we have

$$\limsup_{\|t\| \downarrow 0} \frac{X(t)}{\sqrt{2 \ln \ln \|t\|^{-1}} \|t\|^H} = 1 \quad \text{a.s.}$$

It follows that

$$\sup_{f \in \mathcal{K}} \|f\|_{C(\mathcal{B})} = 1. \tag{4.55}$$

Example 4.36 (Global functional law of the iterated logarithm) Let $t_0 := 3$. Let $\mathcal{J}(t)$ contain only one element 0. Let $s_0 := 0$ and $u_0 := t$. It is easy to check that $dA_1(t)$ is comparable to $t^{-1} dt$. It follows that, in the uniform topology, the set of a.s. limit points of the Strassen cloud

$$\frac{X(tt)}{\sqrt{2 \ln \ln tt^H}}$$

as $t \to \infty$ is Strassen ball $\mathcal{K}$. For the case of $N = 1$ and $H = 1/2$, this result is due to Strassen (1964). Using the continuous functional $F(f) = \|f\|_{C(\mathcal{B})}$, we obtain

$$\lim_{t \to \infty} \sup_{\|t\| \leq t} \frac{X(t)}{\sqrt{2 \ln \ln tt^H}} = \sup_{f \in \mathcal{K}} \|f\|_{C(\mathcal{B})} \quad \text{a.s.}$$

or, by (4.55),

$$\lim_{t \to \infty} \sup_{\|t\| \leq t} \frac{X(t)}{\sqrt{2 \ln \ln tt^H}} = 1 \quad \text{a.s.}$$

Example 4.37 (Functional Lévy modulus of continuity) Let $t_0 := 2$. Let $\mathcal{J}(t) := \{s \in \mathbb{R}^N : \|s\| \leq 1 - t^{-1}\}$ and $u := t^{-1}$ for any $s \in \mathcal{J}(t)$. Then we have

$$\mathcal{P}(t) = \left\{ (s, u) : \|s\| \leq 1 - t^{-1}, u = t^{-1} \right\}$$

and

$$\bigcup_{t \leq u} \mathcal{P}(t) = \left\{ (s, v) : \|s\| \leq 1 - u^{-1}, u^{-1} \leq v \leq 1 \right\}.$$

It is easy to see that $dA_1(u)$ is comparable to

$$\left(1 - u^{-1}\right)^N \int_1^{u^{-1}} v^{-N-1} \, dv \sim u^N.$$

The function $h(u) = N \ln u$ satisfies the conditions of Theorem 4.26. It follows that, in the uniform topology, the set of a.s. limit points of the family of Mueller clouds

$$\mathcal{S}(t) := \left\{ Y(t) = \frac{X(s + tt) - X(s)}{\sqrt{2N \ln t^{-1}} t^H} : \|s\| \leq 1 - t \right\}$$

as $t \downarrow 0$ is Strassen's ball $\mathcal{K}$. For the case of $N = 1$ and $H = 1/2$, this result is due to Mueller (1981). Using the continuous functional $F(f) = \|f\|_{C(\mathcal{B})}$ and (4.55), we obtain:

$$\limsup_{\|s\| \downarrow 0} \sup_{t \in \mathcal{B}} \frac{|X(t+s) - X(t)|}{\sqrt{2N \ln \|s\|^{-1}} \|s\|^H} = 1 \quad \text{a.s.,} \tag{4.56}$$

which coincides with the results by Benassi et al. (1997).

4.6 Bibliographical Remarks

Metric entropy was introduced by Kolmogorov (1955, 1956) under the name "ε-entropy". Lorentz (1966) introduced the term "metric entropy". The idea of using the metric entropy for investigation of local properties of sample paths of random fields is due to Dudley (1967) and Sudakov (1969, 1971). We would like to mention papers by Fernique (1975), Marcus (1973a,b) and Marcus and Shepp (1970, 1972).

Theorem 4.1 for the case of the sphere is due to Kozačenko and Yadrenko (1976a,b). Malyarenko (1999) proved the above theorem for all remaining compact two-point homogeneous spaces. Lemma 4.4 was proved by Yadrenko (1983) for the case of the space $\mathbb{R}^n$ with $C = 1/2$. Theorem 4.5 was proved by Kozačenko and Yadrenko (1976a) for the space $\mathbb{R}^n$. For the remaining noncompact two-point homogeneous spaces, these results are new.

The results of Sect. 4.3 were proved by Malyarenko (2005).

Reproducing kernel Hilbert spaces appeared in an implicit form in Cameron and Martin (1944). They were introduced independently by Aronszajn (1950) and Bergman (1950), see also the second, revised edition Bergman (1970). The above spaces are closely connected with the Bergman spaces A^p of the holomorphic functions in the unit disk that are pth-power integrable with respect to area, see Hedenmalm et al. (2000) and references cited there. The covariance operator C of a continuous Gaussian field $X(t)$ on a compact metric space T admits the *spectral factorisation* $C = SS^*$, where S is a linear continuous operator from the reproducing kernel Hilbert space H to the space $C(T)$. See the survey of spectral factorisation problem in Bingham (2012). The embedding $i: H \to C(T)$ is an example of an *abstract Wiener space* introduced by Gross (1967). See also classical books by Kuo (1975) and Bogachev (1998).

When $X(t)$ is a fractional Brownian motion on the interval, say, $[0, 1]$, S becomes the operator of fractional differentiation. Such operators are considered in the fundamental book by Samko et al. (1993). Applications of the above operators to stochastic calculus with respect to the fractional Brownian motion may be found in the books by Biagini et al. (2008) and Mishura (2008).

Approximation numbers were introduced in the framework of operator theory by Schatten and von Neumann (1946, 1948). The behaviour of approximation numbers for the Wiener process was determined by Maiorov and Wasilkowski (1996), for the fractional Brownian sheet by Kühn and Linde (2002), for the multiparameter fractional Brownian motion by Ayache and Linde (2008). Various optimal series representations for the fractional Brownian motion and fractional Brownian sheet were found. The expansions by Kühn and Linde (2002), Dzhaparidze and van Zanten (2005, 2006), and Malyarenko (2008) use classical special functions, while the expansions by Ayache and Taqqu (2003), Ayache and Linde (2008, 2009), and Schack (2009) use *wavelets* as basis functions.

Let $\{X_n : n \geq 1\}$ be a sequence of independent and identically distributed random variables with zero mean and positive finite variance σ^2. Put $S_0 := 0$, $S_n := X_1 + \cdots + X_n$, $n \geq 1$. Let $Y_n(t)$, $t \in [0, 1]$ be the following stochastic process:

$$Y_n(t) := \frac{1}{\sigma \sqrt{n}} \left(S_{[nt]} + (nt - [nt]) X_{[nt]+1} \right),$$

where the square brackets denote an integer part of a number. In other words, $Y_n(t)$ is the function defined by linear interpolation between its values $Y_n(j/n) = S_j/(\sigma \sqrt{n})$. Donsker (1951) proved that the sequence of probability measures μ_n in the space $C[0, 1]$ corresponding to the sequence $Y_n(t)$ weakly converges to the Wiener measure. Donsker called his theorem an *invariance principle* since the limiting distribution does not depend on the distribution of the summands. Another name is a *functional limit theorem*, because it describes weak convergence of probability measures in the functional space $C[0, 1]$. Donsker's result is the first functional limit theorem.

Various invariance principles for random fields were proved by Balan (2005), Bulinski and Shashkin (2006), Burton and Kim (1988), Berkes and Morrow (1981), El Machkouri and Ouchti (2006), Gorodetskiĭ (1982, 1984), Ivanov and Leonenko (1980, 1981a,b), Kim (1996), Kim and Seo (1995, 1996), Kim and Seok (1994, 1995), Lavancier (2007), Leonenko (1975, 1976, 1980, 1981), Leonenko and Yadrenko (1979a,b), Neaderhouser (1981), Poghosyan and Rœlly (1998), Poryvaĭ (2005), Shashkin (2003, 2004, 2008).

The results of Sect. 4.5 were proved by Malyarenko (2006). Large-deviation estimates used in the proof are part of a general theory. We mention the results by Schilder (1966) and the books by Dembo and Zeitouni (1998) and Deuschel and Stroock (1989).

References

N. Aronszajn. Theory of reproducing kernels. *Trans. Am. Math. Soc.*, 68(3):337–404, 1950. URL http://www.jstor.org/stable/1990404.

R. Askey. *Orthogonal polynomials and special functions*. Society for Industrial and Applied Mathematics, Philadelphia, 1975.

A. Ayache and W. Linde. Approximation of Gaussian random fields: general results and optimal wavelet representation of the Lévy fractional motion. *J. Theor. Probab.*, 21(1):69–96, 2008. doi:10.1007/s10959-007-0101-2.

A. Ayache and W. Linde. Series representations of fractional Gaussian processes by trigonometric and Haar systems. *Electron. J. Probab.*, 14(94):2691–2719, 2009 (electronic).

A. Ayache and M.S. Taqqu. Rate optimality of wavelet series approximations of fractional Brownian motion. *J. Fourier Anal. Appl.*, 9(5):451–471, 2003.

R.M. Balan. A strong invariance principle for associated random fields. *Ann. Probab.*, 33(2):823–840, 2005.

P. Baldi, G. Kerkyacharian, D. Marinucci, and D. Picard. Subsampling needlet coefficients on the sphere. *Bernoulli*, 15(2):438–463, 2009. URL http://projecteuclid.org/euclid.bj/1241444897.

A. Benassi, S. Jaffard, and D. Roux. Elliptic Gaussian random processes. *Rev. Mat. Iberoam.*, 13(1):19–90, 1997.

S. Bergman. *The kernel function and conformal mapping*, volume 5 of *Math. Surv.* Am. Math. Soc., New York, 1950.

S. Bergman. *The kernel function and conformal mapping*, volume 5 of *Math. Surv.* Am. Math. Soc., Providence, second, revised edition, 1970.

I. Berkes and G.J. Morrow. Strong invariance principles for mixing random fields. *Z. Wahrscheinlichkeitstheor. Verw. Geb.*, 57(1):15–37, 1981.

F. Biagini, Y. Hu, B. Øksendal, and T. Zhang. *Stochastic calculus for fractional Brownian motion and applications*. Probab. Appl. (N. Y.). Springer, London, 2008.

N.H. Bingham. Tauberian theorems for Jacobi series. *Proc. Lond. Math. Soc. (3)*, 36:285–309, 1978.

N.H. Bingham. Multivariate prediction and matrix Szegö theory. *Probab. Surv.*, 9:325–339, 2012. doi:10.1214/12-PS200.

N.H. Bingham, C.M. Goldie, and J.L. Teugels. *Regular variation*, volume 27 of *Encycl. Math. Appl.* Cambridge University Press, Cambridge, 1987.

V.I. Bogachev. *Gaussian measures*, volume 62 of *Math. Surveys Monogr.* Am. Math. Soc., Providence, 1998.

W.O. Bray and M.A. Pinsky. Growth properties of Fourier transforms via moduli of continuity. *J. Funct. Anal.*, 255:2265–2285, 2008.

A. Bulinski and A. Shashkin. Strong invariance principle for dependent random fields. In D. Denteneer, F. den Hollander and E. Verbitskiy, editors, *Dynamics & stochastics. Festschrift in honour of M.S. Keane*, volume 48 of *IMS Lect. Notes Monogr. Ser.*, pages 128–143. Inst. Math. Statist., Beachwood, 2006.

R.M. Burton and T.-S. Kim. An invariance principle for associated random fields. *Pac. J. Math.*, 132(1):11–19, 1988.

R.H. Cameron and W.T. Martin. Transformations of Wiener integrals under translations. *Ann. of Math. (2)*, 45:386–396, 1944.

O. Christensen. *An introduction to frames and Riesz bases*. Appl. Numer. Harmon. Anal. Birkhäuser, Basel, 2003.

A. Dembo and O. Zeitouni. *Large deviations techniques and applications*, volume 38 of *Appl. Math. (N. Y.)*. Springer, New York, second edition, 1998. Corrected reprint, 2010.

J.-D. Deuschel and D.W. Stroock. *Large deviations*, volume 137 of *Pure Appl. Math.* Academic Press, Boston, 1989.

M.D. Donsker. An invariance principle for certain probability limit theorems. *Mem. Am. Math. Soc.*, 1951(6):12 pp., 1951.

R.M. Dudley. The sizes of compact subsets of Hilbert space and continuity of Gaussian processes. *J. Funct. Anal.*, 1:290–330, 1967.

K. Dzhaparidze and H. van Zanten. Krein's spectral theory and the Paley–Wiener expansion for fractional Brownian motion. *Ann. Probab.*, 33(2):620–644, 2005.

K. Dzhaparidze and H. van Zanten. Optimality of an explicit series expansion of the fractional Brownian sheet. *Stat. Probab. Lett.*, 71(4):295–301, 2006.

M. El Machkouri and L. Ouchti. Invariance principles for standard-normalized and self-normalized random fields. *ALEA Lat. Am. J. Probab. Math. Stat.*, 2:177–194, 2006.

A. Erdélyi, W. Magnus, F. Oberhettinger, and F.G. Tricomi. *Higher transcendental functions*, volume II. McGraw-Hill, New York, 1953.

X. Fernique. Regularité des trajectoires des fonctions aléatoires gaussiennes. In *École d'Été de Probabilités de Saint-Flour, IV–1974*, volume 480 of *Lect. Notes Math.*, pages 1–96. Springer,

Berlin, 1975.

M. Flensted-Jensen and T.H. Koornwinder. Positive definite spherical functions on a noncompact, rank one symmetric space. In P. Eymard, J. Faraut, G. Schiffmann, and R. Takahashi, editors. *Analyse Harmonique sur les Groupes de Lie, II*, volume 739 of *Lect. Notes Math.*, pages 249–282. Springer, Berlin, 1979.

N. Gantert. An inversion of Strassen's law of the iterated logarithm for small time. *Ann. Probab.*, 21:1045–1049, 1993.

V.V. Gorodetskiĭ. The invariance principle for stationary random fields with strong mixing. *Teor. Veroâtn. Ee Primen.*, 27(2):358–364, 1982. In Russian.

V.V. Gorodetskiĭ. The central limit theorems and the invariance principle for weakly dependent random fields. *Dokl. Akad. Nauk SSSR*, 276(3):528–531, 1984. In Russian.

L. Gross. Abstract Wiener spaces. In *Proc. Fifth Berkeley Sympos. Math. Statist. and Probab.* (Berkeley, CA, 1965/66), volume II, pages 31–42. University of California Press, Berkeley, 1967.

H. Hedenmalm, B. Korenblum, and K. Zhu. *Theory of Bergman spaces*, volume 199 of *Grad. Texts in Math.* Springer, New York, 2000.

A.V. Ivanov and N.N. Leonenko. An invariance principle for an estimator of the correlation function of a homogeneous random field. *Ukr. Mat. Zh.*, 32(3):323–331, 1980. In Russian.

A.V. Ivanov and N.N. Leonenko. On the invariance principle for estimating the correlation function of a uniform isotropic random field. *Ukr. Mat. Zh.*, 33(3):313–323, 1981a. In Russian.

A.V. Ivanov and N.N. Leonenko. An invariance principle for an estimate of the correlation function of a homogeneous random field. *Dokl. Akad. Nauk Ukrain. SSR Ser. A*, 1:17–20, 1981b.

T.-S. Kim. The invariance principle for associated random fields. *Rocky Mt. J. Math.*, 26(4):1443–1454, 1996.

T.-S. Kim and H.-Y. Seo. An invariance principle for stationary strong mixing random fields. *J. Korean Stat. Soc.*, 24(2):281–292, 1995.

T.-S. Kim and H.-Y. Seo. The invariance principle for linearly positive quadrant dependent random fields. *J. Korean Math. Soc.*, 33(4):801–811, 1996.

T.-S. Kim and E.-Y. Seok. The invariance principle for associated random fields. *J. Korean Math. Soc.*, 31(4):679–689, 1994.

T.-S. Kim and E.-Y. Seok. The invariance principle for ρ-mixing random fields. *J. Korean Math. Soc.*, 32(2):321–328, 1995.

A.N. Kolmogorov. Evaluation of minimal number of elements of ε-nets in different functional classes and their application to the problem of representation of functions of several variables by superposition of functions of a smaller number of variables. *Usp. Mat. Nauk*, 10(1):192–194, 1955. In Russian.

A.N. Kolmogorov. On certain asymptotic characteristics of completely bounded metric spaces. *Dokl. Akad. Nauk SSSR (N.S.)*, 108:385–388, 1956. In Russian.

T.H. Koornwinder. Jacobi functions and analysis on noncompact semisimple Lie groups. In *Special functions: group theoretical aspects and applications*, Math. Appl. pages 1–85. Reidel, Dordrecht, 1984.

Yu.V. Kozačenko and M.Ĭ. Yadrenko. Local properties of sample functions of random fields. I. *Teor. Veroyatnost. i Mat. Statist.*, 14:53–66, 1976a. In Russian.

Yu.V. Kozačenko and M.Ĭ. Yadrenko. Local properties of sample functions of random fields. II. *Teor. Veroyatnost. i Mat. Statist.*, 15:82–98, 1976b. In Russian.

T. Kühn and W. Linde. Optimal series representation of fractional Brownian sheets. *Bernoulli*, 8:669–696, 2002. URL http://projecteuclid.org/euclid.bj/1078435223.

H.S. Kuo. *Gaussian measures in Banach spaces*, volume 463 of *Lect. Notes Math.* Springer, Berlin, 1975.

S. Kwapień and B. Szymański. Some remarks on Gaussian measures in Banach spaces. *Probab. Math. Stat.*, 1(1):59–65, 1980.

F. Lavancier. Invariance principles for non-isotropic long memory random fields. *Stat. Inference Stoch. Process.*, 10(3):255–282, 2007.

N. Leonenko. The invariance principle for homogeneous random fields. *Dopovīdī Akad. Nauk Ukraïn. RSR Ser. A*, 1975(6):497–499, 1975. In Ukrainian.

N. Leonenko. On an invariance principle for homogeneous random fields. *Teor. Veroyatnost. i Mat. Statist.*, 14:79–87, 1976. In Russian.

N. Leonenko. On the invariance principle for estimates of linear regression coefficients of a random field. *Teor. Veroyatnost. i Mat. Statist.*, 23:80–91, 1980. In Russian.

N. Leonenko. On the invariance principle for estimating regression coefficients of a random field. *Ukr. Mat. Zh.*, 33(6):771–778, 1981. In Russian.

N. Leonenko and M.Ĭ. Yadrenko. An invariance principle for homogeneous and isotropic random fields. *Teor. Veroâtn. Ee Primen.*, 24(1):175–181, 1979a. In Russian.

N. Leonenko and M.Ĭ. Yadrenko. The invariance principle for some classes of random fields. *Ukr. Mat. Zh.*, 31(5):559–565, 1979b. In Russian.

W.V. Li and Q.-M. Shao. Gaussian processes: inequalities, small probabilities, and applications. In D.N. Shanbhag and C.R. Rao, editors. *Handbook of Statistics, volume 19*, pages 533–597. North-Holland, Amsterdam, 2001.

M.A. Lifshits. *Gaussian random functions*, volume 322 of *Math. Appl.* Kluwer Academic, Dordrecht, 1995.

G.G. Lorentz. Metric entropy and approximation. *Bull. Am. Math. Soc.*, 72:903–937, 1966.

H. Luschgy and G. Pagès. Expansions for Gaussian processes and Parseval frames. *Electron. J. Probab.*, 14(42):1198–1221, 2009 (electronic).

V.E. Maiorov and G.W. Wasilkowski. Probabilistic and average linear widths in l_∞-norm with respect to r-fold Wiener measure. *J. Approx. Theory*, 84(1):31–40, 1996.

A. Malyarenko. Local properties of Gaussian random fields on compact symmetric spaces, and Jackson-type and Bernstein-type theorems. *Ukr. Math. J.*, 51(1):66–75, 1999.

A. Malyarenko. Abelian and Tauberian theorems for random fields on two-point homogeneous spaces. *Theory Probab. Math. Stat.*, 69:115–127, 2005.

A. Malyarenko. Functional limit theorems for multiparameter fractional Brownian motion. *J. Theor. Probab.*, 19(2):263–288, 2006.

A. Malyarenko. An optimal series expansion of the multiparameter fractional Brownian motion. *J. Theor. Probab.*, 21(2):459–475, 2008.

M.B. Marcus. A comparison of continuity conditions for Gaussian processes. *Ann. Probab.*, 1(1):123–130, 1973a.

M.B. Marcus. Continuity of Gaussian processes and random Fourier series. *Ann. Probab.*, 1:968–981, 1973b.

M.B. Marcus and L.A. Shepp. Continuity of Gaussian processes. *Trans. Am. Math. Soc.*, 151:377–391, 1970.

M.B. Marcus and L.A. Shepp. Sample behavior of Gaussian processes. In *Proceedings of the Sixth Berkeley Symposium on Mathematical Statistics and Probability* (Univ. California, Berkeley, CA, 1970/1971), volume II, pages 423–441. University of California Press, Berkeley, 1972.

Yu.S. Mishura. *Stochastic calculus for fractional Brownian motion and related processes*, volume 1929 of *Lect. Notes Math.* Springer, Berlin, 2008.

D. Monrad and H. Rootzén. Small values of Gaussian processes and functional laws of the iterated logarithm. *Probab. Theory Relat. Fields*, 101:173–192, 1995.

C. Mueller. A unification of Strassen's law and Lévy's modulus of continuity. *Z. Wahrscheinlichkeitstheor. Verw. Geb.*, 56:163–179, 1981.

C.C. Neaderhouser. An almost sure invariance principle for partial sums associated with a random field. *Stoch. Process. Appl.*, 11(1):1–10, 1981.

E.J.G. Pitman. On the behavior of the characteristic function of a probability distribution in the neighborhood of the origin. *J. Aust. Math. Soc.*, 8:423–443, 1968.

S. Poghosyan and S. Rœlly. Invariance principle for martingale-difference random fields. *Stat. Probab. Lett.*, 38(3):235–245, 1998.

D.V. Poryvaĭ. The invariance principle for a class of dependent random fields. *Theory Probab. Math. Stat.*, 70:123–134, 2005.

A.P. Prudnikov, Yu.A. Brychkov, and O.I. Marichev. *Integrals and series. Vol. 1. Elementary functions*. Gordon & Breach, New York, 1986.

S.G. Samko, A.A. Kilbas, and O.I. Marichev. *Fractional integrals and derivatives. Theory and applications*. Gordon & Breach, Yverdon, 1993.

G. Samorodnitsky and M.S. Taqqu. *Stable non-Gaussian random processes. Stochastic models with infinite variance*. Chapman & Hall, London, 1994.

H. Schack. An optimal wavelet series expansion of the Riemann–Liouville process. *J. Theor. Probab.*, 22(4):1030–1057, 2009.

R. Schatten and J. von Neumann. The cross-space of linear transformations. II. *Ann. of Math. (2)*, 47:608–630, 1946. URL http://www.jstor.org/stable/1969096.

R. Schatten and J. von Neumann. The cross-space of linear transformations. III. *Ann. of Math. (2)*, 49:557–582, 1948. URL http://www.jstor.org/stable/1969045.

M. Schilder. Some asymptotic formulas for Wiener integrals. *Trans. Am. Math. Soc.*, 125:63–85, 1966.

A.P. Shashkin. The invariance principle for a (bl, θ)-dependent random field. *Russ. Math. Surv.*, 58(3):617–618, 2003.

A.P. Shashkin. The invariance principle for a class of weakly dependent random fields. *Mosc. Univ. Math. Bull.*, 59(4):24–29, 2004.

A.P. Shashkin. A strong invariance principle for positively or negatively associated random fields. *Stat. Probab. Lett.*, 78(14):2121–2129, 2008.

V. Strassen. An invariance principle for the law of the iterated logarithm. *Z. Wahrscheinlichkeitstheor. Verw. Geb.*, 3:211–226, 1964.

V.N. Sudakov. Gauss and Cauchy measures and ε-entropy. *Dokl. Akad. Nauk SSSR*, 185:51–53, 1969. In Russian.

V.N. Sudakov. Gaussian random processes, and measures of solid angles in Hilbert space. *Dokl. Akad. Nauk SSSR*, 197:43–45, 1971. In Russian.

G. Szegö. *Orthogonal polynomials*, volume XXIII of *Colloquium Publ.* Am. Math. Soc., Providence, 1975.

V.I. Tarieladze. On nuclear covariance operators. In A. Weron, editor, *Proc. Second Internat. Conf., Błażejewko, 1979*, volume 828 of *Lect. Notes Math.*, pages 283–285. Springer, Berlin, 1980.

N.N. Vakhania, V.I. Tarieladze, and S.A. Chobanyan. *Probability distributions on Banach spaces*, volume 14 of *Math. Appl. (Sov. Ser.)*. Reidel, Dordrecht, 1987.

A.W. Van der Vaart and J.A. Wellner. *Weak convergence and empirical processes with applications to statistics*. Springer Ser. Stat. Springer, Berlin, 1996.

N.Ya. Vilenkin. *Special functions and the theory of group representations*, volume 22 of *Transl. Math. Monogr.* Am. Math. Soc., Providence, 1968. Translated from the Russian by V.N. Singh.

G.N. Watson. *A treatise on the theory of Bessel functions*. Cambridge University Press, Cambridge, 1944. Reprinted in 1995.

M.Ĭ. Yadrenko. *Spectral theory of random fields*. Translat. Ser. Math. Eng. Optimization Software, Publications Division, New York, 1983.

Chapter 5
Applications

Abstract Using the 0–1 law, we prove Bernstein-type approximation theorems on compact two-point homogeneous spaces. A short introduction to the deterministic theory of the Cosmic Microwave Background (CMB) for mathematicians is presented. Using the theory of invariant random fields in vector bundles, we prove a theorem about equivalence of two different groups of assumptions in cosmological theories. We present an efficient approach for the simulation of homogeneous and partially isotropic random fields based on their spectral expansion. Bibliographical remarks conclude.

5.1 Applications to Approximation Theory

One of the possible applications to approximation theory is based on the following theorem from the general theory of Gaussian measures. Let $\mathscr{X}$ be a Hausdorff locally convex topological space, and let P be a Gaussian measure on the Borel σ-field of $\mathscr{X}$. Let $\mathscr{H}$ be the corresponding reproducing-kernel Hilbert space. The 0–1 *law* (Lifshits, 1995) states that the P-measure of any measurable linear subset $\mathscr{L}$ of $\mathscr{X}$ is either 0 or 1. Moreover, if $\mathsf{P}(\mathscr{L}) = 1$, then $\mathscr{H} \subset \mathscr{L}$.

Let $T = (T, \varrho)$ be a compact metric space, and let $\mathscr{X} = C(T)$ be the Banach space of continuous functions on T equipped by the sup-norm. Let $X(t)$ be a centred Gaussian random field with continuous sample paths on the space T, and let P be the corresponding Gaussian measure on $\mathscr{X}$. Let

$$X(t) := \sum_{j \geq 1} X_j f_j(t)$$

be a uniformly a.s. convergent series expansion of the random field $X(t)$ with independent standard normal X_j. By results of Sect. 4.4, the system $\{f_j(t) \colon j \geq 1\}$ is a Parseval frame for $\mathscr{H}$. Moreover, in many cases it is also an orthonormal basis of $\mathscr{H}$. Therefore, the elements of $\mathscr{H}$ have the following description:

$$\mathscr{H} = \left\{ \sum_{j \geq 1} c_j f_j(t) \colon \sum_{j \geq 1} c_j^2 < \infty \right\}. \tag{5.1}$$

A. Malyarenko, *Invariant Random Fields on Spaces with a Group Action*,
Probability and Its Applications, DOI 10.1007/978-3-642-33406-1_5,
© Springer-Verlag Berlin Heidelberg 2013

In the terms of approximation theory, the space $\mathcal{H}$ consists of functions that have the *constructive property* (5.1).

On the other hand, let $f(\varepsilon)$ be a uniform modulus of continuity of the random field $X(t)$, i.e.

$$\limsup_{\substack{\varepsilon \downarrow 0}} \sup_{\substack{s,t \in T \\ \varrho(s,t) \le \varepsilon}} \frac{|X(s) - X(t)|}{f(\varepsilon)} < \infty \quad \text{a.s.,} \tag{5.2}$$

and let $\mathcal{L}$ be the measurable linear set of functions satisfying (5.2). In the terms of approximation theory, the space $\mathcal{L}$ consists of functions that have the *descriptive property* (5.2). Clearly $\mathsf{P}(\mathcal{L}) = 1$. The 0–1 law states that the descriptive property (5.2) follows from the constructive property (5.1). In other words, we have proved a *Bernstein-type approximation theorem*, see Achieser (1992).

Example 5.1 Let $\mathcal{B}$ be the centred ball of radius 1 in the space $\mathcal{R}^N$. Let P be the Gaussian measure in the space $C(\mathcal{B})$ corresponding to the multiparameter fractional Brownian motion. On the one hand, the reproducing-kernel Hilbert space of the measure P is described by (4.31) and (4.32). On the other hand, it follows from the Lévy modulus of continuity (4.56) that $\mathsf{P}(\mathcal{L}) = 1$, where $\mathcal{L}$ is the set of functions $f(\mathbf{t})$ satisfying

$$\limsup_{\|\mathbf{s}\| \downarrow 0} \sup_{\mathbf{t} \in \mathcal{B}} \frac{|f(\mathbf{t} + \mathbf{s}) - f(\mathbf{t})|}{\sqrt{2N \ln \|\mathbf{s}\|^{-1}} \|\mathbf{s}\|^H} < \infty. \tag{5.3}$$

The corresponding Bernstein-type theorem reads: any function satisfying (4.31) and (4.32), satisfies (5.3).

Example 5.2 Let $T = G/K$ be a compact two-point homogeneous space, and let μ be the G-invariant probability measure on T. Let $\delta > 0$, let $\gamma(\lambda) := \ln^{1+2\delta}(1 + \lambda)$, and let $\{b_\ell : \ell \ge 0\}$ be a sequence of positive real numbers satisfying

$$\sum_{\ell=0}^{\infty} h(T, \ell) b_\ell \gamma(\ell) < \infty. \tag{5.4}$$

Let $\{X_\ell^m : \ell \ge 0, 1 \le m \le h(T, \ell)\}$ be a set of independent standard normal random variables. Put

$$X(t) := \sum_{\ell=0}^{\infty} \sum_{m=1}^{h(T,\ell)} X_\ell^m \sqrt{b_\ell} S_\ell^m(t),$$

where $S_\ell^m(t)$ are real-valued harmonics on T (Subsection 6.4.3). Let P be the corresponding Gaussian measure in the space $C(T)$.

The reproducing-kernel Hilbert space of the field $X(t)$ is the set of all functions $f(t)$ whose Fourier coefficients

$$c_{\ell m}(f) := \int_T f(t) S_\ell^m(t) \, d\mu(t) \tag{5.5}$$

satisfy the following condition:

$$\sum_{\ell=0}^{\infty} \sum_{m=1}^{h(T,\ell)} \frac{c_{\ell m}^2}{b_\ell} < \infty. \tag{5.6}$$

By the result of Example 4.3, $P(\mathscr{L}) = 1$, where $\mathscr{L}$ is the set of functions $f(t)$ satisfying

$$\limsup_{\varepsilon \downarrow 0} \sup_{\substack{s,t \in T \\ \theta(s,t) \le \varepsilon}} |f(s) - f(t)| \ln^\delta (\varepsilon^{-1}) < \infty. \tag{5.7}$$

The corresponding Bernstein-type theorem reads: any function whose Fourier coefficients satisfy (5.6) satisfies (5.7).

Using the second pair of functions in Example 4.3, we obtain the following result. Let $\{b_\ell : \ell \ge 0\}$ be a sequence of positive real numbers satisfying

$$\sum_{\ell=0}^{\infty} h(T,\ell) b_\ell \ell^{2\delta} \ln^{2\beta}(1+\ell) < \infty$$

with $\delta \in (0,1)$ and $\beta \in \mathbb{R}$. Any function $f(t)$ whose Fourier coefficients (5.5) satisfy (5.6) satisfies the following condition:

$$\limsup_{\varepsilon \downarrow 0} \sup_{\substack{s,t \in T \\ \theta(s,t) \le \varepsilon}} \frac{|f(s) - f(t)|}{\varepsilon^\delta \ln^{1/2-\beta}(\varepsilon^{-1})} < \infty.$$

By applying Abelian theorems of Subsection 4.3.1, the reader may obtain many Bernstein-type theorems. We prove yet another Bernstein-type theorem instead.

Example 5.3 In Example 5.2, let $\mathscr{P}_\ell$ be the direct sum of the spaces of harmonics up to degree ℓ. Let $f \in L^2(T, d\mu)$, and let $E_\ell(f)$ be the error of the optimal approximation of the function f with elements of the space $\mathscr{P}_\ell$:

$$E_\ell(f) := \inf_{g \in \mathscr{P}_\ell} \|f - g\|_{L^2(T,d\mu)}.$$

Theorem 5.4 *Let $\{a_\ell : \ell \ge 0\}$ be a sequence of positive numbers that monotonically converges to 0 and satisfies the following condition: for some $\varepsilon > 0$ and $\delta > 0$,*

$$a_{\ell-1}^2 - a_\ell^2 = O\left(\frac{1}{h(T,\ell)(1+\ell)^{2+2\varepsilon} \ln^{1+2\delta}(1+\ell)}\right).$$

If $E_\ell(f) = O(a_\ell)$, then f satisfies (5.7).

Proof Put

$$b_\ell := \frac{1}{h(T,\ell)(1+\ell)^{1+\varepsilon} \ln^{1+2\delta}(1+\ell)}.$$

This sequence satisfies condition (5.4). By the results of Example 5.2, it is enough to prove that f satisfies (5.6). We have

$$E_\ell(f) = \left(\sum_{k=\ell+1}^\infty \sum_{m=1}^{h(T,k)} c_{km}^2(f) \right)^{1/2}.$$

It follows that

$$\sum_{m=1}^{h(T,k)} c_{\ell m}^2(f) = E_{\ell-1}^2(f) - E_\ell^2(f)$$

$$= O\left(a_{\ell-1}^2 - a_\ell^2\right)$$

$$= O\left(\frac{1}{h(T,\ell)(1+\ell)^{2+2\varepsilon} \ln^{1+2\delta}(1+\ell)} \right).$$

Then

$$\sum_{\ell=0}^\infty \sum_{m=1}^{h(T,\ell)} \frac{c_{\ell m}^2}{b_\ell} \le C \sum_{\ell=0}^\infty \frac{1}{(1+\ell)^{1+\varepsilon}} < \infty. \qquad \square$$

Similar Bernstein-type theorems can be proved using the technique of the previous example and Abelian theorems of Subsection 4.3.1.

5.2 Applications to Cosmology

After reading several physical books and papers the author has found a jungle of various choices of coordinates, phase conventions etc. Therefore, Subsection 5.2.1 contains a short introduction to the *deterministic* model of the cosmic microwave background (CMB) for mathematicians. The *probabilistic* model of the CMB is discussed in Subsection 5.2.2.

5.2.1 The Cosmic Microwave Background (CMB)

According to the Standard Cosmological Model, also known as ΛCDM (Lambda Cold Dark Matter) model (Weinberg, 2008), our Universe started in a "Big Bang". This term refers to the idea that the Universe has expanded from a hot and dense initial condition at some *finite* time in the past, and continues to expand now.

As the Universe expanded, both the plasma and the radiation grew cooler. When the Universe cooled enough, it became transparent. The photons that were around at that time are observable now as the *relic radiation*. Their glow is strongest in the microwave region of the electromagnetic spectrum, hence another name *cosmic microwave background radiation*, or just CMB.

In cosmological models, it is usually assumed that the CMB is a single realisation of a random field. A CMB detector measures an electric field $\mathbf{E}$ perpendicular to the direction of observation (or line of sight) $\mathbf{n}$. Mathematically, $\mathbf{n}$ is a point on the sphere S^2. The vector $\mathbf{E}(\mathbf{n})$ lies in the tangent plane, $T_{\mathbf{n}}S^2$. In other words, $\mathbf{E}(\mathbf{n})$ is a section of the tangent bundle $\xi = (TS^2, \pi, S^2)$ with

$$\pi(\mathbf{n}, \mathbf{x}) = \mathbf{n}, \quad \mathbf{n} \in S^2, \ \mathbf{x} \in T_{\mathbf{n}}S^2.$$

From the observations, we define the intensity tensor. In physical terms, the intensity tensor is

$$\mathsf{P} := C \langle \mathbf{E}(\mathbf{n}) \otimes \mathbf{E}^*(\mathbf{n}) \rangle,$$

where $\langle \cdot \rangle$ denotes time average over the historical accidents that produced a particular pattern of fluctuations. Assuming ergodicity, the time average is equal to the space average, i.e., average over the possible positions from which the radiation could be observed. The constant C is chosen so that P is measured in brightness temperature units (in these units, the intensity tensor is independent of radiation frequency). This will be ignored in what follows.

Introduce a basis in each tangent plane $T_{\mathbf{n}}S^2$. Realise S^2 as $\{(x, y, z) \in \mathbb{R}^3 : x^2 + y^2 + z^2 = 1\}$ and define the chart $(U_I, \mathbf{h}_I)$ as $U_I = S^2 \setminus \{(0, 0, 1), (0, 0, -1)\}$ and $\mathbf{h}_I(\mathbf{n}) = (\theta(\mathbf{n}), \varphi(\mathbf{n})) \in \mathbb{R}^2$, the spherical coordinates. Let SO(3) be the rotation group in $\mathbb{R}^3$. For any rotation g, define the chart $(U_g, \mathbf{h}_g)$ as

$$U_g := gU_I, \qquad \mathbf{h}_g(\mathbf{n}) = \mathbf{h}_I(g^{-1}\mathbf{n}).$$

The sphere S^2, equipped with the atlas $\{(U_g, \mathbf{h}_g): g \in \text{SO}(3)\}$, becomes a real-analytic manifold. The local θ-axis in each tangent plane is along the direction of decreasing the inclination θ:

$$\mathbf{e}_\theta := -\frac{\partial}{\partial \theta}.$$

The local φ-axis is along the direction of increasing the azimuth φ:

$$\mathbf{e}_\varphi := (1/\sin\theta)\frac{\partial}{\partial \varphi}.$$

With this convention, $\mathbf{e}_\theta$, $\mathbf{e}_\varphi$, and the direction of radiation propagation $-\mathbf{n}$ form a right-hand basis. This convention is in accordance with the International Astronomic Union standard. The orthonormal basis $(\mathbf{e}_\theta, \mathbf{e}_\varphi)$ turns S^2 into a Riemannian manifold and each tangent plane $T_{\mathbf{n}}S^2$ can be identified with the space $\mathbb{R}^2$.

In this basis, the intensity tensor becomes the intensity matrix:

$$P_{ab} = \langle E_a(\mathbf{n}) \otimes E_b^*(\mathbf{n}) \rangle, \quad a, b \in \{\theta, \varphi\}.$$

The rotations about the line of sight together with the *parity transformation* $\mathbf{n} \to -\mathbf{n}$ generate the group $O(2)$ of orthogonal matrices in $\mathbb{R}^2$. The action of $O(2)$ on the intensity matrix extends to the representation $g \mapsto g \mathscr{A} g^{-1}$ of $O(2)$ in the real 4-dimensional space of Hermitian 2×2 matrices $\mathscr{A}$ with inner product

$$(\mathscr{A}, \mathscr{B}) := \operatorname{tr}(\mathscr{A} \mathscr{B}).$$

This representation is reducible and may be decomposed into the direct sum of three irreducible representations.

The standard choice of an orthonormal basis in the spaces of the irreducible components is as follows. The space of the first irreducible component is generated by the matrix

$$\frac{1}{2}\sigma_0 := \frac{1}{2}\begin{pmatrix} 1 & 0 \\ 0 & 1 \end{pmatrix}.$$

The representation in this space is the trivial representation of the group $O(2)$. Physicists call the elements of this space *scalars*.

The space of the second irreducible component is generated by the matrices

$$\frac{1}{2}\sigma_1 := \frac{1}{2}\begin{pmatrix} 0 & 1 \\ 1 & 0 \end{pmatrix}, \qquad \frac{1}{2}\sigma_3 := \frac{1}{2}\begin{pmatrix} 1 & 0 \\ 0 & -1 \end{pmatrix}.$$

Let $g_\alpha \in SO(2)$ with

$$g_\alpha := \begin{pmatrix} \cos\alpha & \sin\alpha \\ -\sin\alpha & \cos\alpha \end{pmatrix}. \tag{5.8}$$

It is easy to check that

$$g_\alpha \sigma_1 g_\alpha^{-1} = \cos(2\alpha)\sigma_1 + \sin(2\alpha)\sigma_3,$$
$$g_\alpha \sigma_3 g_\alpha^{-1} = -\sin(2\alpha)\sigma_1 + \cos(2\alpha)\sigma_3. \tag{5.9}$$

The elements of this space are *symmetric trace-free tensors*.

Finally, the space of the third irreducible component is generated by the matrix

$$\frac{1}{2}\sigma_2 := \frac{1}{2}\begin{pmatrix} 0 & -i \\ i & 0 \end{pmatrix}.$$

The representation in this space is the representation $g \mapsto \det g$ of the group $O(2)$. Physicists call the elements of this space *pseudo-scalars* (they do not change under rotation but change sign under reflection). The matrices σ_1, σ_2, and σ_3 are known as *Pauli matrices*.

The standard physical notation for the components of the intensity matrix in the above basis is as follows:

$$P := \frac{1}{2}(I\sigma_0 + U\sigma_1 + V\sigma_2 + Q\sigma_3),$$

or

$$P = \frac{1}{2} \begin{pmatrix} I + Q & U - iV \\ U + iV & I - Q \end{pmatrix}.$$

The real numbers I, Q, U, and V are called *Stokes parameters*. Their physical sense is as follows. I is the total intensity of the radiation (which is directly proportional to the fourth power of the absolute temperature T by the Stefan–Boltzmann law). On the tangent plane $T_{\mathbf{n}} S^2$, the tip of the electric vector $\mathbf{E}(\mathbf{n})$ traces out an ellipse as a function of time. The parameters U and Q measure the orientation of the above ellipse relative to the local θ-axis, $\mathbf{e}_\theta$. The *polarisation angle* between the major axis of the ellipse and $\mathbf{e}_\theta$ is

$$\chi = \frac{1}{2} \tan^{-1} \frac{U}{Q},$$

and the length of the major semi-axis is $(Q^2 + U^2)^{1/2}$. The last parameter, V, measures circular polarisation.

According to modern cosmological theories, the polarisation of the CMB was introduced while scattering off the photons by charged particles. This process cannot induce circular polarisation in the scattered light. Therefore, in what follows we put $V = 0$.

The physics of the CMB polarisation is described in Cabella and Kamionkowski (2005), Challinor (2005), Challinor and Peiris (2008), Durrer (2008), Lin and Wandelt (2006), among others. Of these, Challinor and Peiris use the right-hand basis, while the remaining authors use the left-hand basis, in which $\mathbf{e}_\theta = \partial/\partial\theta$. In what follows, we use the left-hand basis $\mathbf{e}_\theta$, $\mathbf{e}_\varphi$, $-\mathbf{n}$ with

$$\mathbf{e}_\theta := \frac{\partial}{\partial\theta}, \qquad \mathbf{e}_\varphi := (1/\sin\theta)\frac{\partial}{\partial\varphi}. \tag{5.10}$$

5.2.2 The Probabilistic Model of the CMB

The absolute temperature, $T(\mathbf{n})$, is a section of the homogeneous vector bundle $\xi_0 = (E_0, \pi, S^2)$, where the representation of the rotation group $G = \mathrm{SO}(3)$ induced by the representation $\mu(g_\alpha) = 1$ of the massive subgroup $K = \mathrm{SO}(2)$ is realised.

The representations λ of the group G are enumerated by nonnegative integers $\ell = 0, 1, \ldots$. The restriction of the representation λ_ℓ to K is the direct sum of the representations $e^{im\alpha}$, $m = -\ell, -\ell + 1, \ldots, \ell$. Therefore we have $\dim \lambda_\ell = 2\ell + 1$ and $|m| \le \ell$. By Frobenius reciprocity (Subsection 6.3.6), $\hat{G}_K(\mu) = \{\lambda_0, \lambda_1, \ldots, \lambda_\ell, \ldots\}$.

The representations $e^{im\alpha}$ of K act in one-dimensional complex spaces H_m. Choose the Gel'fand–Tsetlin basis in the space $H^{(\ell)}$ of the representation V_ℓ. Each vector $\mathbf{e}_m$ of a basis can be multiplied by a *phase* $\exp(i\alpha_m)$.

Any rotation $g \in \mathrm{SO}(3)$ is defined by the Euler angles $g = (\varphi, \theta, \psi)$ with $\varphi, \psi \in [0, 2\pi]$ and $\theta \in [0, \pi]$. The order in which the angles are given and the axes about which they are applied are not standard. We adopt the so-called *zxz convention*: the first rotation is about the z-axis by ψ, the second rotation is about the x-axis by θ, and the third rotation is about the z-axis by φ. Note that the chart defined by the Euler angles has the property mentioned in Subsection 2.3.2: the first two local coordinates (φ, θ) are spherical coordinates in S^2 (up to order) with dense domain U_I.

The matrix elements of the representation λ_ℓ are traditionally denoted by

$$D_{mn}^{(\ell)}(\varphi, \theta, \psi) := \left(\lambda_\ell(\varphi, \theta, \psi) \mathbf{e}_m, \mathbf{e}_n \right)_{H^{(\ell)}}$$

and called *Wigner D-functions*. The explicit formula for the Wigner D-function depends on the phase convention. Choose the basis $\{\mathbf{e}_m : -\ell \le m \le \ell\}$ in every space $H^{(\ell)}$ to obtain (6.16) and (6.17).

The following symmetry relation follows:

$$d_{-m,-n}^{(\ell)}(\theta) = (-1)^{n-m} d_{mn}^{(\ell)}(\theta). \tag{5.11}$$

In this particular case, formula (2.50) takes the form

$$\mu Y_{\ell m}(\theta, \varphi) = \sqrt{2\ell + 1} \overline{D_{m0}^{(\ell)}(\varphi, \theta, 0)}.$$

The functions in the left-hand side form an orthonormal basis in the space of the square-integrable functions on S^2 with respect to the probability SO(3)-invariant measure. It is conventional to form a basis with respect to the Lebesgue measure induced by the embedding $S^2 \subset \mathbb{R}^3$ which is 4π times the invariant probability measure, and omit the first subscript:

$$Y_{\ell m}(\theta, \varphi) := \sqrt{\frac{2\ell + 1}{4\pi}} \overline{D_{m0}^{(\ell)}(\varphi, \theta, 0)}.$$

This is formula (A4.40) from Durrer (2008) defining the *spherical harmonics*.

In cosmological models, one assumes that $T(\mathbf{n})$ is a single realisation of the mean-square continuous strict-sense SO(3)-invariant random field in the homogeneous vector bundle $\xi = (E_0, \pi, S^2)$. It is customary to use the term *"isotropic"* instead of "SO(3)-invariant". By Theorem 2.26, we have

$$T(\mathbf{n}) = \sum_{\ell=0}^{\infty} \sum_{m=-\ell}^{\ell} Z_{\ell m} Y_{\ell m}(\mathbf{n}),$$

where $\mathsf{E}[Z_{\ell m}] = 0$ unless $\ell = 0$ and $\mathsf{E}[Z_{\ell m} \overline{Z_{\ell' m'}}] = \delta_{\ell \ell'} \delta_{m m'} R^{(\ell)}$ with

$$\sum_{\ell=0}^{\infty} (2\ell + 1) R^{(\ell)} < \infty.$$

Physicists call $Z_{\ell m}$s the *expansion coefficients*, and $R^{(\ell)}$ the *power spectrum* of the CMB. Different notations for the expansion coefficients and power spectrum may be found in the literature. Some of them are shown in Table 5.1.

Table 5.1 Examples of different notation for temperature expansion coefficients and power spectrum

Source	$Z_{\ell m}$	$R^{(\ell)}$
Cabella and Kamionkowski (2005)	$a_{\ell m}^T$	C_ℓ^{TT}
Challinor (2005), Challinor and Peiris (2008)	$T_{\ell m}$	C_ℓ^T
Durrer (2008), Weinberg (2008)	$a_{\ell m}$	C_ℓ
Lin and Wandelt (2006), Zaldarriaga and Seljak (1997)	$a_{T,\ell m}$	$C_{T\ell}$
Kamionkowski et al. (1997)	$a_{\ell m}^T$	C_ℓ^T

In what follows, we use the notation of Lin and Wandelt (2006). In this notation, the expansion for the temperature has·the form

$$T(\mathbf{n}) = \sum_{\ell=0}^{\infty} \sum_{m=-\ell}^{\ell} a_{T,\ell m} Y_{\ell m}(\mathbf{n}). \tag{5.12}$$

Since $T(\mathbf{n})$ is real, the coefficients $a_{T,\ell m}$ must satisfy the *reality condition* which depends on the phase convention. For our current convention, when the Wigner d-function is determined by (6.17), we have

$$Y_{\ell -m}(\theta, \varphi) = \sqrt{\frac{2\ell+1}{4\pi}} e^{-im\varphi} d_{-m,0}^{(\ell)}(\theta)$$

$$= \sqrt{\frac{2\ell+1}{4\pi}} e^{-im\varphi} (-1)^m d_{m0}^{(\ell)}(\theta)$$

$$= (-1)^m \overline{Y_{\ell m}(\theta, \varphi)}.$$

Here we used the symmetry relation (5.11). The reality condition is

$$a_{T,\ell -m} = (-1)^m \overline{a_{T,\ell m}}. \tag{5.13}$$

This form of the reality condition is used by Cabella and Kamionkowski (2005), Challinor (2005), Challinor and Peiris (2008), Challinor (2009), Durrer (2008), Kamionkowski et al. (1997), Lin and Wandelt (2006), among others.

Introduce the following notation: $m^+ = \max\{m, 0\}$, $m^- = \max\{-m, 0\}$. We have $(-m)^- = m^+$ and $m^+ - m = m^-$. If we choose another basis, $\{(-1)^m\, \mathbf{e}_m : -\ell \le m \le \ell\}$, then the Wigner D-function, $D_{mn}^{(\ell)}(\varphi, \theta, \psi)$, is multiplied by $(-1)^{m^-+n^-}$, and we obtain

$$Y_{\ell -m}(\theta, \varphi) = \sqrt{\frac{2\ell+1}{4\pi}} (-1)^{(-m)^-} e^{-im\varphi} d_{-m,0}^{(\ell)}(\theta)$$

$$= \sqrt{\frac{2\ell+1}{4\pi}} (-1)^{m+m^-} e^{-im\varphi} (-1)^m d_{m0}^{(\ell)}(\theta)$$

$$= \overline{Y_{\ell m}(\theta, \varphi)}.$$

The modified reality condition is

$$a_{T,\ell-m} = \overline{a_{T,\ell m}}.$$ (5.14)

This form of reality condition is used by Geller and Marinucci (2010), Weinberg (2008), Zaldarriaga and Seljak (1997), among others.

Let $T_0 := \mathsf{E}[T(\mathbf{n})]$. The temperature fluctuation, $\Delta T(\mathbf{n}) = T(\mathbf{n}) - T_0$, expands as

$$\Delta T(\mathbf{n}) = \sum_{\ell=1}^{\infty} \sum_{m=-\ell}^{\ell} a_{T,\ell m} Y_{\ell m}(\mathbf{n}).$$

The part of this sum corresponding to $\ell = 1$ is called a *dipole*. When analysing data, the dipole is usually removed since it linearly depends on the velocity of the observer's motion relative to the surface of last scattering.

The *complex polarisation* is defined as $Q + iU$. It follows easily from (5.9) that any rotation (5.8) maps $Q + iU$ to $e^{2i\alpha}(Q + iU)$. Then, by (6.22), $(Q + iU)(\mathbf{n})$ is a section of the homogeneous vector bundle $\xi_{-2} = (E_{-2}, \pi, S^2)$, where the representation of the rotation group $G = \mathrm{SO}(3)$ induced by the representation $\mu(g_\alpha) = e^{-2i\alpha}$ of the massive subgroup $K = \mathrm{SO}(2)$ is realised. By Frobenius reciprocity, $\hat{G}_K(\mu) = \{\lambda_2, \lambda_3, \ldots, \lambda_\ell, \ldots\}$.

In general, let $s \in \mathbb{Z}$, and let $\xi_{-s} := (E_{-s}, \pi, S^2)$ be the homogeneous vector bundle where the representation of the rotation group $\mathrm{SO}(3)$ induced by the representation $\mu(g_\alpha) = e^{-is\alpha}$ of the massive subgroup $\mathrm{SO}(2)$ is realised. In the physical literature, the sections of these bundle are called

- quantities of spin s by Challinor and Peiris (2008), Challinor (2009), Geller and Marinucci (2010), Newman and Penrose (1966), Weinberg (2008) among others;
- quantities of spin $-s$ by Cabella and Kamionkowski (2005), Lin and Wandelt (2006), Zaldarriaga and Seljak (1997), among others;
- quantities of spin $|s|$ and helicity s by Durrer (2008) among others.

Let $g = (\theta, \varphi, \psi)$ be the Euler angles in $\mathrm{SO}(3)$. Put $\psi = 0$. Then, $\mathbf{n} = (\theta, \varphi, 0)$ are spherical coordinates in S^2. By (2.51) we obtain

$$(Q + iU)(\mathbf{n}) = \sum_{\ell=2}^{\infty} \sum_{m=-\ell}^{\ell} a_{-2,\ell m} {}_{-2}Y_{\ell m}(\mathbf{n}),$$

where

$$a_{-2,\ell m} := \int_{S^2} (Q + iU)(\mathbf{n}) \overline{{}_{-2}Y_{\ell m}(\mathbf{n})} \, d\mathbf{n}$$

and, by (2.50),

$${}_{-2}Y_{\ell m}(\theta, \varphi) = \sqrt{2\ell + 1} \overline{D^{(\ell)}_{m,-2}(\varphi, \theta, 0)}.$$

The functions in the left-hand side form an orthonormal basis in the space of the square-integrable sections of the homogeneous vector bundle ξ_{-2} with respect to the SO(3)-invariant probability measure.

There exist different conventions. The first convention is used by Durrer (2008) among others. In this convention, a basis is formed with respect to the Lebesgue measure induced by the embedding $S^2 \subset \mathbb{R}^3$ which is 4π times the invariant probability measure and the sign of the second index of the Wigner D-function is changed (because we would like to expand $Q + iU$ with respect to $_2Y_{\ell m}$):

$$_{-2}Y_{\ell m}(\theta, \varphi) = \sqrt{\frac{2\ell + 1}{4\pi}} \overline{D^{(\ell)}_{m,2}(\varphi, \theta, 0)}.$$

In the general case, for any $s \in \mathbb{Z}$, this convention reads (Durrer, 2008, formula (A4.51))

$$_sY_{\ell m}(\theta, \varphi) := \sqrt{\frac{2\ell + 1}{4\pi}} \overline{D^{(\ell)}_{m,-s}(\varphi, \theta, 0)}. \tag{5.15}$$

These functions are called *spherical harmonics of spin s* or the *spin-weighted spherical harmonics*. They appeared in Gel'fand and Šapiro (1952) under the name *generalised spherical harmonics*. The current name goes back to Newman and Penrose (1966). Note that the spin-weighted spherical harmonics are defined for $\ell \geq |s|$ and $|m| \leq \ell$.

The second harmonic convention is used by Lin and Wandelt (2006), Newman and Penrose (1966), among others. It reads as

$$_sY_{\ell m}(\theta, \varphi) = (-1)^m \sqrt{\frac{2\ell + 1}{4\pi}} \overline{D^{(\ell)}_{m,-s}(\varphi, \theta, 0)}.$$

Both conventions are coherent with the following phase convention:

$$\overline{_sY_{\ell m}} := (-1)^{m+s} {}_{-s}Y_{\ell-m}. \tag{5.16}$$

In particular, for $s = 0$ we return back to the convention $\overline{Y_{\ell m}} = (-1)^m Y_{\ell-m}$ corresponding to reality condition (5.13).

To produce the harmonic convention coherent with the phase convention

$$\overline{_sY_{\ell m}} = (-1)^s {}_{-s}Y_{\ell-m}$$

corresponding to reality condition (5.14), one must multiply the right-hand side of the convention equation by $(-1)^{m^-}$. Thus, the modified first convention, used by Weinberg (2008) among others, is

$$_sY_{\ell m}(\theta, \varphi) = (-1)^{m^-} \sqrt{\frac{2\ell + 1}{4\pi}} \overline{D^{(\ell)}_{m,-s}(\varphi, \theta, 0)},$$

while the modified second convention, used by Geller and Marinucci (2010), among others, is

$$_sY_{\ell m}(\theta, \varphi) = (-1)^{m^+}\sqrt{\frac{2\ell+1}{4\pi}}\overline{D^{(\ell)}_{m,-s}(\varphi, \theta, 0)}.$$

In what follows, we use the convention (5.15). The explicit expression for the spherical harmonics of spin s in the chart determined by spherical coordinates follows from (6.17) and (5.15):

$$_sY_{\ell m}(\theta, \varphi) = (-1)^m\sqrt{\frac{(2\ell+1)(\ell+m)!(\ell-m)!}{4\pi(\ell+s)!(\ell-s)!}}\sin^{2\ell}(\theta/2)e^{im\varphi}$$

$$\times \sum_{r=\max\{0,m-s\}}^{\min\{\ell+m,\ell-s\}}\binom{\ell-s}{r}\binom{\ell+s}{r-m+s}(-1)^{\ell-r-s}\cot^{2r-m+s}(\theta/2).$$

$$(5.17)$$

The decomposition of the complex polarisation takes the form

$$(Q+iU)(\mathbf{n}) = \sum_{\ell=2}^{\infty}\sum_{m=-\ell}^{\ell}a_{2,\ell m}\,_2Y_{\ell m}(\mathbf{n}),$$

where

$$a_{2,\ell m} := \int_{S^2}(Q+iU)(\mathbf{n})\overline{_2Y_{\ell m}(\mathbf{n})}\,d\mathbf{n},$$

while the decomposition of the conjugate complex polarisation is

$$(Q-iU)(\mathbf{n}) = \sum_{\ell=2}^{\infty}\sum_{m=-\ell}^{\ell}a_{-2,\ell m}\,_{-2}Y_{\ell m}(\mathbf{n}),$$

where

$$a_{-2,\ell m} := \int_{S^2}(Q-iU)(\mathbf{n})\overline{_{-2}Y_{\ell m}(\mathbf{n})}\,d\mathbf{n}.$$

In cosmological models, one assumes that $(Q+iU)(\mathbf{n})$ is a single realisation of the mean-square continuous strict-sense isotropic random field in the homogeneous vector bundle ξ_2. Isotropic random fields in vector bundles ξ_s, $s \in \mathbb{Z}$ were defined by Geller and Marinucci (2010). By Theorem 2.27, we have

$$(Q+iU)(\mathbf{n}) = \sum_{\ell=2}^{\infty}\sum_{m=-\ell}^{\ell}a_{2,\ell m}\,_2Y_{\ell m}(\mathbf{n}), \qquad (5.18)$$

where $\mathsf{E}[a_{2,\ell m}] = 0$ and $\mathsf{E}[a_{2,\ell m}\overline{a_{2,\ell'm'}}] = \delta_{\ell\ell'}\delta_{mm'}C_{2\ell}$ with

$$\sum_{\ell=2}^{\infty}(2\ell+1)C_{2\ell} < \infty.$$

Table 5.2 Examples of different notation for complex polarisation expansion coefficients

Source	Expansion coefficients
Durrer (2008)	$a_{\ell m}^{(\pm 2)}$
Lin and Wandelt (2006), Zaldarriaga and Seljak (1997)	$a_{\pm 2, \ell m}$
Weinberg (2008)	$a_{P, \ell m}$

Different notations for the complex polarisation expansion coefficients $a_{\pm 2, \ell m}$ may be found in the literature. Some of them are shown in Table 5.2.

In what follows we use the notation by Lin and Wandelt (2006). The expansion for the conjugate complex polarisation has the form

$$(Q - iU)(\mathbf{n}) = \sum_{\ell=2}^{\infty} \sum_{m=-\ell}^{\ell} a_{-2, \ell m} \, {}_{-2}Y_{\ell m}(\mathbf{n}). \tag{5.19}$$

Since $Q(\mathbf{n})$ and $U(\mathbf{n})$ are real, the coefficients $a_{2, \ell m}$ and $a_{-2, \ell m}$ must satisfy the reality condition which depends on the phase convention. We agreed to use the first harmonic convention (5.15). Therefore, our phase convention is (5.16), and the reality condition is

$$\overline{a_{-2, lm}} = (-1)^m a_{2, l-m}. \tag{5.20}$$

Along with the standard basis (5.10), it is useful to use the so-called *helicity basis*. Again, there exist different names and conventions. Durrer (2008) defines the helicity basis as

$$\mathbf{e}_{+} := \frac{1}{\sqrt{2}}(\mathbf{e}_{\theta} - i\mathbf{e}_{\varphi}), \qquad \mathbf{e}_{-} := \frac{1}{\sqrt{2}}(\mathbf{e}_{\theta} + i\mathbf{e}_{\varphi}),$$

while Weinberg (2008) uses the opposite definition

$$\mathbf{e}_{+} := \frac{1}{\sqrt{2}}(\mathbf{e}_{\theta} + i\mathbf{e}_{\varphi}), \qquad \mathbf{e}_{-} := \frac{1}{\sqrt{2}}(\mathbf{e}_{\theta} - i\mathbf{e}_{\varphi}).$$

Challinor (2005) and Thorne (1980) use the notation

$$\mathbf{m} := \frac{1}{\sqrt{2}}(\mathbf{e}_{\theta} + i\mathbf{e}_{\varphi}), \qquad \mathbf{m}^{*} := \frac{1}{\sqrt{2}}(\mathbf{e}_{\theta} - i\mathbf{e}_{\varphi}),$$

while Challinor (2009) uses the notation

$$\mathbf{m}_{+} := \frac{1}{\sqrt{2}}(\mathbf{e}_{\theta} + i\mathbf{e}_{\varphi}), \qquad \mathbf{m}_{-} := \frac{1}{\sqrt{2}}(\mathbf{e}_{\theta} - i\mathbf{e}_{\varphi})$$

and calls these the *null basis*. We will use the definition and notation by Durrer (2008).

The helicity basis is useful for the following reason. Let $\eth$ be a covariant derivative in direction $-\sqrt{2}\mathbf{e}_-$:

$$\eth := \nabla_{-\sqrt{2}\mathbf{e}_-}.$$

Let $C^\infty(\xi_s)$ be the space of infinitely differentiable sections of the vector bundle ξ_s. Durrer (2008) proves that for any $_sf \in C^\infty(\xi_s)$ we have

$$\eth\,_sf = \left(s\cot\theta - \frac{\partial}{\partial\theta} - \frac{i}{\sin\theta}\frac{\partial}{\partial\varphi} \right)\,_sf.$$

In particular, put $_sf := {}_sY_{\ell m}$. Using (5.17), we obtain

$$\eth\,_sY_{\ell m} = \sqrt{(\ell - s)(\ell + s + 1)}\,_{s+1}Y_{\ell m}.$$

For $s \geq 0$ and $\ell = s$, the spherical harmonic $_{s+1}Y_{\ell m}$ is not defined and we use the convention $\sqrt{(\ell - \ell)(2\ell + 1)}\,_{\ell+1}Y_{\ell m} = 0$. Then, $\eth\colon C^\infty(\xi_s) \to C^\infty(\xi_{s+1})$. Therefore, $\eth$ is called the *spin-raising operator*. Moreover, the last display shows that the restriction of $\eth$ to the space $H^{(\ell)}$, $\ell > s$, is an intertwining operator between equivalent representations λ_ℓ.

The adjoint operator, $\eth^*$, is the covariant derivative in direction $-\sqrt{2}\mathbf{e}_+$:

$$\eth^* := \nabla_{-\sqrt{2}\mathbf{e}_+}.$$

For any $_sf \in C^\infty(\xi_s)$ we have

$$\eth^*\,_sf = \left(s\cot\theta - \frac{\partial}{\partial\theta} + \frac{i}{\sin\theta}\frac{\partial}{\partial\varphi} \right)\,_sf.$$

In particular,

$$\eth^*\,_sY_{\ell m} = -\sqrt{(\ell + s)(\ell - s + 1)}\,_{s-1}Y_{\ell m}.$$

For $s \leq 0$ and $\ell = -s$, the spherical harmonic $_{s-1}Y_{\ell m}$ is not defined and we use the convention $\sqrt{(\ell - \ell)(2\ell + 1)}\,_{-\ell-1}Y_{\ell m} = 0$. Then, $\eth^*\colon C^\infty(\xi_s) \to C^\infty(\xi_{s-1})$. Therefore, $\eth^*$ is called the *spin-lowering operator*. Moreover, the last display shows that the restriction of $\eth^*$ to the space $H^{(\ell)}$, $\ell > -s$, is an intertwining operator between equivalent representations λ_ℓ.

Zaldarriaga and Seljak (1997) introduced the following idea. Assume for a moment that

$$\sum_{\ell=2}^{\infty} \frac{(2\ell + 1)(\ell + 2)!}{(l - 2)!} C_{2\ell} < \infty. \tag{5.21}$$

Then, it is possible to act twice with $\eth$ on both sides of (5.19) and to interchange differentiation and summation:

$$\eth^2(Q-iU)(\mathbf{n}) = \eth^2 \sum_{\ell=2}^{\infty}\sum_{m=-\ell}^{\ell} a_{-2,\ell m}\,_{-2}Y_{\ell m}(\mathbf{n})$$

$$= \sum_{\ell=2}^{\infty}\sum_{m=-\ell}^{\ell} a_{-2,\ell m}\eth^2\,_{-2}Y_{\ell m}(\mathbf{n})$$

$$= \sum_{\ell=2}^{\infty}\sum_{m=-\ell}^{\ell} \sqrt{\frac{(\ell+2)!}{(\ell-2)!}}a_{-2,\ell m}Y_{\ell m}(\mathbf{n}).$$

By the same argument, we have

$$\left(\eth^*\right)^2(Q+iU)(\mathbf{n}) = \sum_{\ell=2}^{\infty}\sum_{m=-\ell}^{\ell} \sqrt{\frac{(\ell+2)!}{(\ell-2)!}}a_{2,\ell m}Y_{\ell m}(\mathbf{n}).$$

Unlike complex polarisation, the new random fields are rotationally invariant and no ambiguities connected with rotations (5.9) arise. However, they have complex behaviour under parity transformation, because $Q(\mathbf{n})$ and $U(\mathbf{n})$ behave differently (Lin and Wandelt, 2006): Q has even parity: $Q(-\mathbf{n}) = Q(\mathbf{n})$ while U has odd parity: $U(-\mathbf{n}) = -U(\mathbf{n})$.

Therefore, it is customary to group together quantities of the same parity:

$$\tilde{E}(\mathbf{n}) := -\frac{1}{2}\left((\eth^*)^2(Q+iU)(\mathbf{n}) + \eth^2(Q-iU)(\mathbf{n})\right),$$

$$\tilde{B}(\mathbf{n}) := -\frac{1}{2i}\left((\eth^*)^2(Q+iU)(\mathbf{n}) - \eth^2(Q-iU)(\mathbf{n})\right).$$

The random fields $\tilde{E}(\mathbf{n})$ and $\tilde{B}(\mathbf{n})$ are scalar (spin 0), real-valued, and isotropic. To find their behaviour under parity transformation, follow Lin and Wandelt (2006). Notice that if $\mathbf{n}$ has spherical coordinates (θ, φ), then $-\mathbf{n}$ has spherical coordinates $\theta' = \pi - \theta$ and $\varphi' = \varphi + \pi$. Therefore,

$$\frac{\partial}{\partial\theta'} = -\frac{\partial}{\partial\theta}, \qquad \frac{\partial}{\partial\varphi'} = \frac{\partial}{\partial\varphi}.$$

Because $(Q+iU)(-\mathbf{n}) = (Q-iU)(\mathbf{n})$, we obtain

$$\left(\eth^*\right)'(Q+iU)(-\mathbf{n}) = \left(2\cot\theta' - \frac{\partial}{\partial\theta'} + \frac{i}{\sin\theta'}\frac{\partial}{\partial\varphi'}\right)(Q+iU)(-\mathbf{n})$$

$$= \left(-2\cot\theta + \frac{\partial}{\partial\theta} - \frac{i}{\sin\theta}\frac{\partial}{\partial\varphi}\right)(Q-iU)(\mathbf{n})$$

$$= -\eth(Q-iU)(\mathbf{n})$$

and

$$((\eth^*)')^2(Q+iU)(-\mathbf{n}) = \left(2\cot\theta' - \frac{\partial}{\partial\theta'} + \frac{i}{\sin\theta'}\frac{\partial}{\partial\varphi'}\right)(-\eth(Q-iU)(\mathbf{n}))$$

$$= \eth^2(Q-iU)(\mathbf{n}).$$

Similarly, we have $(\eth')^2(Q+iU)(-\mathbf{n}) = (\eth^*)^2(Q-iU)(\mathbf{n})$. Therefore,

$$\tilde{E}(-\mathbf{n}) = -\frac{1}{2}\left((\eth^*)^2(Q+iU)(-\mathbf{n}) + \eth^2(Q-iU)(-\mathbf{n})\right)$$

$$= -\frac{1}{2}\left((\eth^*)^2(Q-iU)(\mathbf{n}) + \eth^2(Q+iU)(\mathbf{n})\right)$$

$$= \tilde{E}(\mathbf{n})$$

and

$$\tilde{B}(-\mathbf{n}) = -\frac{1}{2i}\left((\eth^*)^2(Q+iU)(-\mathbf{n}) - \eth^2(Q-iU)(-\mathbf{n})\right)$$

$$= -\frac{1}{2i}\left((\eth^*)^2(Q-iU)(\mathbf{n}) - \eth^2(Q+iU)(\mathbf{n})\right)$$

$$= -\tilde{B}(\mathbf{n}).$$

This means that $\tilde{E}(\mathbf{n})$ has even parity, like the electric field, while $\tilde{B}(\mathbf{n})$ has odd parity, like the magnetic field.

The spectral expansion of the fields $\tilde{E}(\mathbf{n})$ and $\tilde{B}(\mathbf{n})$ has the form

$$\tilde{E}(\mathbf{n}) = \sum_{\ell=2}^{\infty}\sum_{m=-\ell}^{\ell} a_{\tilde{E},\ell m} Y_{\ell m}(\mathbf{n}),$$

$$\tilde{B}(\mathbf{n}) = \sum_{\ell=2}^{\infty}\sum_{m=-\ell}^{\ell} a_{\tilde{B},\ell m} Y_{\ell m}(\mathbf{n}),$$

where

$$a_{\tilde{E},\ell m} := -\frac{1}{2}\sqrt{\frac{(\ell+2)!}{(\ell-2)!}}(a_{2,\ell m} + a_{-2,\ell m}),$$

$$a_{\tilde{B},\ell m} := -\frac{1}{2i}\sqrt{\frac{(\ell+2)!}{(\ell-2)!}}(a_{2,\ell m} - a_{-2,\ell m}).$$

Table 5.3 Examples of different notation for the fields $E(\mathbf{n})$ and $B(\mathbf{n})$ and their expansion coefficients

Source	Fields	Multipoles
Challinor (2005), Challinor (2009)	–	$E_{\ell m}, B_{\ell m}$
Durrer (2008)	$\mathscr{E}(\mathbf{n}), \mathscr{B}(\mathbf{n})$	$e_{\ell m}, b_{\ell m}$
Geller and Marinucci (2010)	$f_{\mathrm{E}}, f_{\mathrm{M}}$	$A_{\ell m \mathrm{E}}, A_{\ell m \mathrm{M}}$
Lin and Wandelt (2006)	$E(\mathbf{n}), B(\mathbf{n})$	$a_{E,\ell m}, a_{B,\ell m}$
Weinberg (2008), Zaldarriaga and Seljak (1997)	–	$a_{E,\ell m}, a_{B,\ell m}$

It is convenient to introduce the fields $E(\mathbf{n})$ and $B(\mathbf{n})$ as

$$E(\mathbf{n}) := \sum_{\ell=2}^{\infty} \sum_{m=-\ell}^{\ell} a_{E,\ell m} Y_{\ell m}(\mathbf{n}),$$

$$B(\mathbf{n}) := \sum_{\ell=2}^{\infty} \sum_{m=-\ell}^{\ell} a_{B,\ell m} Y_{\ell m}(\mathbf{n}),$$

(5.22)

with

$$a_{E,\ell m} := -\frac{1}{2}(a_{2,\ell m} + a_{-2,\ell m}),$$

$$a_{B,\ell m} := -\frac{1}{2\mathrm{i}}(a_{2,\ell m} - a_{-2,\ell m}).$$

(5.23)

The random fields $E(\mathbf{n})$ and $B(\mathbf{n})$ are scalar (spin 0), real-valued, and isotropic. Moreover, $E(\mathbf{n})$ has even parity, while $B(\mathbf{n})$ has odd parity. The advantage of $E(\mathbf{n})$ and $B(\mathbf{n})$ is that their definition does not use assumption (5.21). The expansion coefficients $a_{E,\ell m}$ are called *electric multipoles*, while the expansion coefficients $a_{B,\ell m}$ are called *magnetic multipoles*.

Different notations for the fields $E(\mathbf{n})$ and $B(\mathbf{n})$ and electric and magnetic multipoles may be found in the literature. Some of them are shown in Table 5.3. In what follows, we use notation by Lin and Wandelt (2006).

We prove the following theorem.

Theorem 5.5 *Let $T(\mathbf{n})$ be a real-valued random field defined by (5.12). Let $(Q \pm \mathrm{i}U)(\mathbf{n})$ be random fields defined by (5.18) and (5.19). Let $E(\mathbf{n})$ and $B(\mathbf{n})$ be random fields (5.22) whose expansion coefficients are determined by (5.23). The following statements are equivalent.*

1. *$((Q - \mathrm{i}U)(\mathbf{n}), T(\mathbf{n}), (Q + \mathrm{i}U)(\mathbf{n}))$ is an isotropic random field in $\xi_{-2} \oplus \xi_0 \oplus \xi_2$. The fields $Q(\mathbf{n})$ and $U(\mathbf{n})$ are real-valued.*
2. *$(T(\mathbf{n}), E(\mathbf{n}), B(\mathbf{n}))$ is an isotropic random field in $\xi_0 \oplus \xi_0 \oplus \xi_0$ with real-valued components. The components $T(\mathbf{n})$ and $B(\mathbf{n})$ are uncorrelated. The components $E(\mathbf{n})$ and $B(\mathbf{n})$ are uncorrelated.*

Proof Let $((Q - iU)(\mathbf{n}), T(\mathbf{n}), (Q + iU)(\mathbf{n}))$ be an isotropic random field in $\xi_{-2} \oplus \xi_0 \oplus \xi_2$, and let $Q(\mathbf{n})$ and $U(\mathbf{n})$ be real-valued. By Theorem 2.28 and reality conditions (5.13) and (5.20), we have $\mathsf{E}[a_{T,\ell m}] = 0$ for $\ell \neq 0$, $\mathsf{E}[a_{\pm 2,\ell m}] = 0$ and

$$
\begin{aligned}
\mathsf{E}[a_{T,\ell m}\overline{a_{T,\ell'm'}}] &= \delta_{\ell\ell'}\delta_{mm'}C_{T,\ell}, \\
\mathsf{E}[a_{\pm 2,\ell m}\overline{a_{\pm 2,\ell'm'}}] &= \delta_{\ell\ell'}\delta_{mm'}C_{2,\ell}, \\
\mathsf{E}[a_{T,\ell m}\overline{a_{\pm 2,\ell'm'}}] &= \delta_{\ell\ell'}\delta_{mm'}C_{T,\pm 2,\ell}, \\
\mathsf{E}[a_{-2,\ell m}\overline{a_{2,\ell'm'}}] &= \delta_{\ell\ell'}\delta_{mm'}C_{-2,2,\ell}
\end{aligned}
\tag{5.24}
$$

with

$$
\sum_{\ell=0}^{\infty}(2\ell + 1)C_{T,\ell} + 2\sum_{\ell=2}^{\infty}(2\ell + 1)C_{2,\ell} < \infty.
\tag{5.25}
$$

It is enough to prove that $\mathsf{E}[a_{E,\ell m}] = \mathsf{E}[a_{B,\ell m}] = 0$ and

$$
\mathsf{E}[a_{X,\ell m}\overline{a_{Y,\ell'm'}}] = \delta_{\ell\ell'}\delta_{mm'}C_{XY,\ell}
\tag{5.26}
$$

with

$$
\sum_{\ell}(2\ell + 1)C_{X,\ell} < \infty
\tag{5.27}
$$

for all $X, Y \in \{T, E, B\}$. Then, the second statement of the theorem follows from Theorem 2.28.

The first condition trivially follows from (5.23). Condition (5.26) with $X = Y = T$ is obvious. We prove condition (5.26) with $X = Y = E$. Indeed, by (5.23) and (5.20),

$$
\begin{aligned}
\mathsf{E}[a_{E,\ell m}\overline{a_{E,\ell'm'}}] &= \frac{1}{4}\Big(\mathsf{E}\big[(a_{2,\ell m} + a_{-2,\ell m})(\overline{a_{2,\ell'm'}} + \overline{a_{-2,\ell'm'}})\big]\Big) \\
&= \frac{1}{4}\Big(\mathsf{E}[a_{2,\ell m}\overline{a_{2,\ell'm'}}] + \mathsf{E}[a_{2,\ell m}\overline{a_{-2,\ell'm'}}] \\
&\quad + \mathsf{E}[a_{-2,\ell m}\overline{a_{2,\ell'm'}}] + \mathsf{E}[a_{-2,\ell m}\overline{a_{-2,\ell'm'}}]\Big) \\
&= \frac{1}{2}\delta_{\ell\ell'}\delta_{mm'}(C_{2,\ell} + \operatorname{Re}C_{-2,2,\ell}).
\end{aligned}
$$

Condition (5.26) with $X = Y = B$ can be proved similarly.

Next, we prove condition (5.26) with $X = T$ and $Y = B$. Indeed,

$$
\begin{aligned}
\mathsf{E}[a_{T,\ell m}\overline{a_{B,\ell'm'}}] &= \frac{1}{2i}\Big(\mathsf{E}[a_{T,\ell m}\overline{a_{2,\ell'm'}}] - \mathsf{E}[a_{T,\ell m}\overline{a_{-2,\ell'm'}}]\Big) \\
&= -\frac{1}{2}\delta_{\ell\ell'}\delta_{mm'}(C_{T,2,\ell} - C_{T,-2,\ell}) \\
&= 0
\end{aligned}
$$

by (5.20), which also proves that $T(\mathbf{n})$ and $B(\mathbf{n})$ are uncorrelated. Condition (5.26) for other cross-correlations can be proved similarly.

Next, we prove (5.27) with $X = E$. Indeed,

$$\sum_{\ell=2}^{\infty}(2\ell+1)C_{E,\ell} = \frac{1}{2}\sum_{\ell=2}^{\infty}(2\ell+1)(C_{2,\ell} + \operatorname{Re}C_{-2,2,\ell}) < \infty.$$

Condition (5.27) for $X = B$ can be proved similarly.

Next, we prove that $E(\mathbf{n})$ is real-valued. It is enough to prove the reality condition $a_{E,\ell-m} = (-1)^m\overline{a_{E,\ell m}}$. We have

$$a_{E,\ell-m} = -\frac{1}{2}(a_{2,\ell-m} + a_{-2,\ell-m})$$

$$= -\frac{1}{2}\left[(-1)^m\overline{a_{-2,\ell m}} + (-1)^{-m}\overline{a_{2,\ell m}}\right]$$

$$= (-1)^m\overline{a_{E,\ell m}}.$$

$B(\mathbf{n})$ is real-valued for similar reasons.

Finally, we prove that $E(\mathbf{n})$ and $B(\mathbf{n})$ are uncorrelated. Indeed, $\mathsf{E}[E(\mathbf{n}_1)B(\mathbf{n}_2)] = C_{EB}(\mathbf{n}_1 \cdot \mathbf{n}_2)$, because $(T(\mathbf{n}), E(\mathbf{n}), B(\mathbf{n}))$ is an isotropic random field in $\xi_0 \oplus \xi_0 \oplus \xi_0$. So, $C_{EB}((-\mathbf{n}_1)\cdot(-\mathbf{n}_2)) = C_{EB}(\mathbf{n}_1 \cdot \mathbf{n}_2)$. On the other hand,

$$C_{EB}\big((-\mathbf{n}_1)\cdot(-\mathbf{n}_2)\big) = \mathsf{E}\big[E(-\mathbf{n}_1)B(-\mathbf{n}_2)\big]$$

$$= \mathsf{E}\big[E(\mathbf{n}_1)\big(-B(-\mathbf{n}_2)\big)\big]$$

$$= -C_{EB}(\mathbf{n}_1 \cdot \mathbf{n}_2),$$

because $E(-\mathbf{n}_1) = E(\mathbf{n}_1)$ and $B(-\mathbf{n}_1) = -B(\mathbf{n}_1)$. Therefore, $C_{EB}(\mathbf{n}_1 \cdot \mathbf{n}_2) = 0$.

Conversely, let $(T(\mathbf{n}), E(\mathbf{n}), B(\mathbf{n}))$ be an isotropic random field in $\xi_0 \oplus \xi_0 \oplus \xi_0$ with real-valued components, let the components $T(\mathbf{n})$ and $B(\mathbf{n})$ be uncorrelated, and let the components $E(\mathbf{n})$ and $B(\mathbf{n})$ be also uncorrelated. Solving system of equations (5.23), we obtain

$$\begin{aligned} a_{2,\ell m} &= -a_{E,\ell m} + a_{B,\ell m}\mathrm{i}, \\ a_{-2,\ell m} &= -a_{E,\ell m} - a_{B,\ell m}\mathrm{i}. \end{aligned} \tag{5.28}$$

It is obvious that $\mathsf{E}[a_{\pm 2,\ell m}] = 0$. We have to prove (5.24), (5.25), and (5.20). The first equation in (5.24) is obvious. The second equation is proved as follows.

$$\mathsf{E}[a_{2,\ell m}\overline{a_{2,\ell'm'}}] = \mathsf{E}\big[(-a_{E,\ell m} + a_{B,\ell m}\mathrm{i})(-a_{E,\ell'm'} - a_{B,\ell'm'}\mathrm{i})\big]$$

$$= \delta_{\ell\ell'}\delta_{mm'}(C_{E,\ell} + C_{B,\ell}),$$

because $E(\mathbf{n})$ and $B(\mathbf{n})$ are uncorrelated. The proof for negative coefficients is similar.

The third equation in (5.24) is proved as follows.

$$\mathsf{E}[a_{T,\ell m}\overline{a_{2,\ell'm'}}] = \mathsf{E}\big[a_{T,\ell m}(-a_{E,\ell'm'} - a_{B,\ell'm'}\mathrm{i})\big]$$
$$= -\delta_{\ell\ell'}\delta_{mm'}C_{TE,\ell},$$

because $T(\mathbf{n})$ and $B(\mathbf{n})$ are uncorrelated. The proof for negative coefficients is similar.

The fourth equation in (5.24) is proved as follows.

$$\mathsf{E}[a_{-2,\ell m}\overline{a_{2,\ell'm'}}] = \mathsf{E}\big[(-a_{E,\ell m} - a_{B,\ell m}\mathrm{i})(-a_{E,\ell'm'} - a_{B,\ell m}\mathrm{i})\big]$$
$$= \delta_{\ell\ell'}\delta_{mm'}(C_{E,\ell} - C_{B,\ell}),$$

because $E(\mathbf{n})$ and $B(\mathbf{n})$ are uncorrelated.

Because $C_{2,\ell} = C_{E,\ell} + C_{B,\ell}$, we have

$$\sum_{\ell=0}^{\infty}(2\ell+1)C_{T,\ell} + 2\sum_{\ell=2}^{\infty}(2\ell+1)C_{2,\ell}$$

$$= \sum_{\ell=0}^{\infty}(2\ell+1)C_{T,\ell} + 2\sum_{\ell=2}^{\infty}(2\ell+1)(C_{E,\ell} + C_{B,\ell})$$

$$< \infty,$$

which proves (5.25). The reality condition (5.20) is proved as

$$\overline{a_{-2,\ell m}} = \overline{-a_{E,\ell m} - a_{B,\ell m}\mathrm{i}}$$
$$= -\overline{a_{E,\ell m}} + \overline{a_{B,\ell m}}\mathrm{i}$$
$$= -(-1)^m a_{E,\ell-m} + (-1)^m a_{B,\ell-m}\mathrm{i}$$
$$= (-1)^m a_{2,\ell-m}. \qquad\qquad \square$$

In *Gaussian* cosmological theories, the random field $(T(\mathbf{n}), E(\mathbf{n}), B(\mathbf{n}))$ is supposed to be Gaussian and isotropic with real-valued components. Let $\eta_{\ell 0 j}$, $\ell \geq 0$, $1 \leq j \leq 3$ and $\eta_{\ell m j}$, $\ell \geq 1$, $1 \leq m \leq \ell$, $1 \leq j \leq 6$ be independent standard normal random variables. Put

$$\zeta_{\ell m j} := \begin{cases} \eta_{\ell 0 j}, & m = 0, \\ \frac{1}{\sqrt{2}}(\eta_{\ell m 2j-1} + \eta_{\ell m 2j}\mathrm{i}), & m > 0, \end{cases}$$

where $\ell \geq 0$, $0 \leq m \leq \ell$, and $1 \leq j \leq 3$. Now put

$$a_{T,\ell m} := (C_{T,\ell})^{1/2}\zeta_{\ell m 1},$$

$$a_{E,\ell m} := \frac{C_{TE,\ell}}{(C_{T,\ell})^{1/2}}\zeta_{\ell m 1} + \left(C_{E,\ell} - \frac{(C_{TE,\ell})^2}{C_{T,\ell}}\right)^{1/2}\zeta_{\ell m 2},$$

$$a_{B,\ell m} := (C_{B,\ell m})^{1/2}\zeta_{\ell m 3}$$

for $m \geq 0$ and $a_{X,\ell-m} = (-1)^m \overline{a_{X,\ell m}}$ for $m < 0$ and $X \in \{T, E, B\}$. The random fields

$$T(\mathbf{n}) := \sum_{\ell=0}^{\infty} \sum_{m=-\ell}^{\ell} a_{T,\ell m} Y_{\ell m}(\mathbf{n}),$$

$$E(\mathbf{n}) := \sum_{\ell=2}^{\infty} \sum_{m=-\ell}^{\ell} a_{E,\ell m} Y_{\ell m}(\mathbf{n}),$$

$$B(\mathbf{n}) := \sum_{\ell=2}^{\infty} \sum_{m=-\ell}^{\ell} a_{B,\ell m} Y_{\ell m}(\mathbf{n})$$

satisfy all conditions of the second statement of Theorem 5.5. The random fields $(Q \pm iU)(\mathbf{n})$ can be reconstructed by (5.28), (5.18), and (5.19). By Theorem 5.5, $((Q - iU)(\mathbf{n}), T(\mathbf{n}), (Q + iU)(\mathbf{n}))$ is an isotropic Gaussian random field in $\xi_{-2} \oplus \xi_0 \oplus \xi_2$. The fields $Q(\mathbf{n})$ and $U(\mathbf{n})$ are real-valued.

Finally, we note that Kamionkowski et al. (1997) proposed a different formalism for computations of the polarisation field on the whole sky. Instead of spin-weighted harmonics $_sY_{\ell m}$, they use *tensor harmonics* $Y_{\ell m}^E$ and $Y_{\ell m}^B$ which are related to the spin-weighted harmonics as follows.

$$Y_{\ell m}^E = \frac{1}{\sqrt{2}} (_{-2}Y_{\ell m} \mathbf{e}_- \otimes \mathbf{e}_- + {_2}Y_{\ell m} \mathbf{e}_+ \otimes \mathbf{e}_+),$$

$$Y_{\ell m}^B = \frac{1}{i\sqrt{2}} (_{-2}Y_{\ell m} \mathbf{e}_- \otimes \mathbf{e}_- - {_2}Y_{\ell m} \mathbf{e}_+ \otimes \mathbf{e}_+).$$

This formalism is also used by Cabella and Kamionkowski (2005), Challinor (2005), Challinor and Peiris (2008), Challinor (2009), among others. An excellent survey of different types of spherical harmonics may be found in Thorne (1980).

5.3 Applications to Earthquake Engineering

Let $X(\mathbf{x}, t)$ be the ground acceleration at the point with coordinates $\mathbf{x} = (x, y)$ at time t during a strong earthquake. We assume that $X(\mathbf{x}, t)$ is a centred Gaussian homogeneous random field in the space-time domain $\mathbb{R}^3$. By (2.14a) the covariance function of the random field $X(\mathbf{x}, t)$ has the form

$$R_{XX}(\mathbf{x}, t_1; \mathbf{y}, t_2) = \int_{\mathbb{R}^3} e^{i(\kappa, \mathbf{x} - \mathbf{y}) + i\omega(t_1 - t_2)} \, d\mu(\kappa, \omega),$$

where $\kappa = (\kappa_1, \kappa_2)^\top$ is a vector in the wavenumber domain, and where ω is the frequency.

Assume that the measure μ is absolutely continuous with respect to Lebesgue measure. The corresponding Radon–Nikodym derivative, $f(\kappa, \omega)$, is called the *spectral density* of the random field $X(\mathbf{x}, t)$.

Suppose that the spectral density is even with respect to the second argument: $f(\kappa, -\omega) = f(\kappa, \omega)$. Denote $S(\kappa, \omega) := 2f(\kappa, \omega)$, and call $S(\kappa, \omega)$ the *one-sided spectral density* of the random field $X(\mathbf{x}, t)$. Then we have

$$R(\mathbf{x}, t_1; \mathbf{y}, t_2) = \int_0^\infty \int_{\mathbb{R}^2} \cos\big((\kappa, \mathbf{x} - \mathbf{y}) + \omega(t_1 - t_2)\big) S(\kappa, \omega) \, d\kappa \, d\omega. \qquad (5.29)$$

To simulate the sample paths of the random field $X(\mathbf{x}, t)$, we may do the following. Choose an upper cutoff frequency ω_u, above which the values of the spectral density are insignificant for practical purposes. Choose also upper cutoff wavenumbers κ_{1u} and κ_{2u}. Select three positive integers N, N_1, and N_2 and define $\Delta\omega := \omega_u/N$, $\Delta\kappa_1 := \kappa_{1u}/N_1$, $\Delta\kappa_2 := \kappa_{2u}/N_2$. Choose three real numbers ω_1, κ_{1N_1}, and κ_{2N_2} with $0 \le \omega_1 \le \Delta\omega$, $0 \le \kappa_{1,N_1+1} \le \Delta\kappa_1$, and $0 \le \kappa_{2,N_2+1} \le \Delta\kappa_2$. Put

$$\omega_k := \omega_1 + (k-1)\Delta\omega, \quad 1 \le k \le N,$$

$$\kappa_{1,k_1} := \kappa_{1N_1} + (-N_1 + k_1 - 1)\Delta\kappa_1, \quad 1 \le k_1 \le 2N_1,$$

$$\kappa_{2,k_2} := \kappa_{2N_2} + (-N_2 + k_2 - 1)\Delta\kappa_2, \quad 1 \le k_2 \le 2N_2.$$

The sample paths are simulated by the following formula:

$$\tilde{X}(x_1, x_2, t) = \sqrt{\Delta\omega\Delta\kappa_1\Delta\kappa_2} \sum_{k=1}^{N} \sum_{k_1=1}^{2N_1} \sum_{k_2=1}^{2N_2} \sqrt{f(\kappa_{1,k_1}, \kappa_{2,k_2}, \omega_k)}$$

$$\times \big[\cos(\kappa_{1,k_1}x_1 + \kappa_{2,k_2}x_2 + \omega_k t) X_{k_1 k_2 k}$$

$$+ \sin(\kappa_{1,k_1}x_1 + \kappa_{2,k_2}x_2 + \omega_k t) Y_{k_1 k_2 k}\big], \qquad (5.30)$$

where $X_{k_1 k_2 k}$ and $Y_{k_1 k_2 k}$ are independent standard normal random variables. In other words, we apply Karhunen's theorem to (5.29):

$$X(x_1, x_2, t) = \int_0^\infty \int_{\mathbb{R}^2} \cos\big((\kappa, \mathbf{x}) + \omega t\big) \sqrt{S(\kappa, \omega)} \, dZ_1(\kappa, \omega)$$

$$+ \int_0^\infty \int_{\mathbb{R}^2} \sin\big((\kappa, \mathbf{x}) + \omega t\big) \sqrt{S(\kappa, \omega)} \, dZ_2(\kappa, \omega),$$

truncate the region of integration, and use a cubature formula. For the case of the stationary random process, this approach goes back to Rice (1944).

An alternative approach has been proposed by Shinozuka (1972) for stationary stochastic processes and by Shinozuka (1987) for homogeneous random fields. The sample paths are simulated as follows:

$$\tilde{X}(x_1, x_2, t) = \sqrt{2\Delta\omega\Delta\kappa_1\Delta\kappa_2} \sum_{k=1}^{N} \sum_{k_1=1}^{2N_1} \sum_{k_2=1}^{2N_2} \sqrt{f(\kappa_{1,k_1}, \kappa_{2,k_2}, \omega_k)}$$

$$\times \cos(\kappa_{1,k_1} x_1 + \kappa_{2,k_2} x_2 + \omega_k t + Z_{k_1 k_2 k}), \tag{5.31}$$

where $Z_{k_1 k_2 k}$ are independent random variables having uniform distribution on the interval $[0, 2\pi]$. In what follows we refer to Rice's approach as Method 1, while Shinozuka's approach will be referred to as Method 2.

The sample paths simulated by the above methods have the following properties.

- The random function (5.30) is Gaussian, while the random function (5.31) is not Gaussian. As $\min\{N, N_1, N_2\} \to \infty$, the finite-dimensional distributions of the random function (5.31) converge to those of a Gaussian random field, by a suitable version of the Central Limit Theorem.
- The random function (5.30) is the sum of trigonometric functions with random *amplitudes*, while the random function (5.31) is the sum of trigonometric functions with random *phases*.
- Assume that

$$\frac{\omega_1}{\Delta\omega} = \frac{p_0}{q_0}, \qquad \frac{\kappa_{1,N_1}}{\Delta\kappa_1} = \frac{p_1}{q_1}, \qquad \frac{\kappa_{2,N_2}}{\Delta\kappa_2} = \frac{p_2}{q_2},$$

where p_j and q_j, $0 \le j \le 2$, are mutually prime numbers. Then, the vector

$$\mathbf{T} = (2\pi q_0/\Delta\omega, 2\pi q_1/\Delta\kappa_1, 2\pi q_2/\Delta\kappa_2)^{\top}$$

is the period of both random functions (5.30) and (5.31).

- Both random functions are centred and have the same covariance function

$$R_{\tilde{X}\tilde{X}}(x_1, x_2, t) = \Delta\omega\Delta\kappa_1\Delta\kappa_2 \sum_{k=1}^{N} \sum_{k_1=1}^{2N_1} \sum_{k_2=1}^{2N_2} \cos(\kappa_{1,k_1} x_1 + \kappa_{2,k_2} x_2 + \omega_k t)$$

$$\times f(\omega_k, \kappa_{1,k_1}, \kappa_{2,k_2}).$$

As simultaneously $\min\{N, N_1, N_2\} \to \infty$ and $\min\{\omega_u, \kappa_{1u}, \kappa_{2u}\} \to \infty$, the above covariance function converges to that of the random field $X(x_1, x_2, t)$.

- The simulation of the random field $X(x_1, x_2, t)$ over a grid of points in the space-time domain, assuming M points in each direction, requires $O(M^6)$ operations for both methods. The use of the Fast Fourier Transform (FFT) reduces the amount of computations to $O((M \ln M)^3)$ operations. However, it requires the storage of three-dimensional arrays in memory.

Assume that the random field $X(\mathbf{x}, t)$ is *isotropic with respect to the space variable*, i.e., for any rotation g of the space domain we have $R(g\mathbf{x}, s; g\mathbf{y}, t) =$

$R(\mathbf{x}, s; \mathbf{y}, t)$. On the one hand,

$$R(g\mathbf{x}, t_1; g\mathbf{y}, t_2) = \int_0^\infty \int_{\mathbb{R}^2} e^{i(\kappa, g\mathbf{x} - g\mathbf{y}) + i\omega(t_1 - t_2)} f(\kappa, \omega)\, d\kappa\, d\omega$$

$$= \int_0^\infty \int_{\mathbb{R}^2} \exp\big(i(g^{-1}\kappa, \mathbf{x} - \mathbf{y}) + i\omega(t_1 - t_2)\big) f(\kappa, \omega)\, d\kappa\, d\omega$$

$$= \int_0^\infty \int_{\mathbb{R}^2} \exp\big(i(\kappa, \mathbf{x} - \mathbf{y}) + i\omega(t_1 - t_2)\big) f(g\kappa, \omega)\, d\kappa\, d\omega,$$

where in the last line we denote $g^{-1}\kappa$ again by κ and use the fact that Lebesgue measure $d\kappa$ is rotationally invariant. On the other hand, the left-hand side of the above expression is clearly equal to the right-hand side of (5.29). By the uniqueness property of the Fourier transform, we obtain $f(g\kappa, \omega) = f(\kappa, \omega)$. In other words, $f(\kappa, \omega) = f(\kappa, \omega)$, where κ is the length of the vector κ.

Using formulae 3.3.2.3 and 2.5.40.6 in Prudnikov (1988), we obtain

$$R(\mathbf{x}, t_1; \mathbf{y}, t_2) = 2\pi \int_0^\infty \int_0^\infty J_0\big(\kappa \|\mathbf{x} - \mathbf{y}\|\big) \cos\big(\omega(t_1 - t_2)\big)\kappa f(\kappa, \omega)\, d\kappa\, d\omega. \quad (5.32)$$

Pass to polar coordinates and use the general addition theorem for Bessel functions Watson (1944, Chap. XI). We obtain:

$$R(r_1, \varphi_1, t_1; r_2, \varphi_2, t_2)$$

$$= (2\pi)^2 \sum_{\ell=0}^\infty \sum_{m=1}^{h(S^2,\ell)} Y_{\ell m}(\varphi_1)\overline{Y_{\ell m}(\varphi_2)} \int_0^\infty \int_0^\infty J_\ell(\kappa r_1) J_\ell(\kappa r_2)$$

$$\times \big[\cos(\omega t_1)\cos(\omega t_2) + \sin(\omega t_1)\sin(\omega t_2)\big]\kappa f(\kappa, \omega)\, d\kappa\, d\omega.$$

Using the values $H(S^2, 0) = 1$, $h(S^2, \ell) = 2$ for $\ell \geq 0$, and

$$Y_{\ell 1}(\varphi) = \frac{1}{\sqrt{2\pi}} e^{im\varphi}, \quad \ell \geq 0; \qquad Y_{\ell 2}(\varphi) = \frac{1}{\sqrt{2\pi}} e^{-im\varphi}, \quad \ell \geq 1$$

and making use of the symmetry property (6.52), we obtain

$$R(r_1, \varphi_1, t_1; r_2, \varphi_2, t_2) = 2\pi \sum_{\ell=-\infty}^\infty \big[\cos(\ell\varphi_1)\cos(\ell\varphi_2) + \sin(\ell\varphi_1)\sin(\ell\varphi_2)\big]$$

$$\times \int_0^\infty \int_0^\infty J_\ell(\kappa r_1) J_\ell(\kappa r_2)\big[\cos(\omega t_1)\cos(\omega t_2)$$

$$+ \sin(\omega t_1)\sin(\omega t_2)\big]\kappa f(\kappa, \omega)\, d\kappa\, d\omega.$$

For simulation purposes, the sum in the above formula can be truncated to include a relatively small number of terms. We choose a positive integer L, an upper cutoff frequency ω_u, and an upper cutoff wave-number κ_u, above which the contribution of the spectral density $f(\kappa, \omega)$ to the simulations is insignificant for practical purposes. We choose positive integers N and N_1, define the frequency step $\Delta\omega := \omega_u/N$ and the wave-number step $\Delta\kappa := \kappa_u/N_1$. The frequency nodes ω_k and the wave-number nodes κ_k are based on the classical centroid rule:

$$\omega_k := (k - 1/2)\Delta\omega, \quad 1 \le k \le N,$$

$$\kappa_k := (k_1 - 1/2)\Delta\kappa, \quad 1 \le k_1 \le M_1.$$

In particular, using Method 1 to simulate the random field, we obtain

$$\tilde{X}(r, \varphi, t) = \sqrt{2\pi\,\Delta\kappa\,\Delta\omega} \sum_{\ell=-L}^{L} \sum_{k=1}^{N} \sum_{k_1=1}^{N_1} \sqrt{\kappa_{k_1} f(\kappa_{k_1}, \omega_k)} J_\ell(\kappa_{k_1} r)$$

$$\times \Big[\cos(\ell\varphi)\cos(\omega_k t)X_{k_1 k 1} + \cos(\ell\varphi)\sin(\omega_k t)X_{k_1 k 2}$$

$$+ \sin(\ell\varphi)\cos(\omega_k t)X_{k_1 k 3} + \sin(\ell\varphi)\sin(\omega_k t)X_{k_1 k 4}\Big], \quad (5.33)$$

where $X_{k_1 k j}$, $1 \le j \le 4$, are independent standard normal random variables. On the other hand, Method 2 leads to the following simulation formula:

$$\tilde{X}(r, \varphi, t) = 2\sqrt{\pi\,\Delta\kappa\,\Delta\omega} \sum_{\ell=-L}^{L} \sum_{k=1}^{N} \sum_{k_1=1}^{N_1} \sqrt{\kappa_{k_1} f(\kappa_{k_1}, \omega_k)} J_\ell(\kappa_{k_1} r)$$

$$\times \Big[\cos(\ell\varphi)\cos(\omega_k t + Z_{k_1 k 1}) + \sin(\ell\varphi)\cos(\omega_k t + Z_{k_1 k 2})\Big], \quad (5.34)$$

where $Z_{k_1 k 1}$ and $Z_{k_1 k 2}$ are independent random variables having uniform distribution on the interval $[0, 2\pi]$.

For Method 1 the algorithm of simulation based on (5.33) and using the Fast Fourier Transform consists of the following steps.

1. Define two arrays

$$A^\ell_{kk_1} := \begin{cases} 2N\sqrt{2\pi\,\Delta\kappa\,\Delta\omega\kappa_{k_1} f(\kappa_{k_1}, \omega_k)}(X_{k_1 k 1} - iX_{k_1 k 2}), & 1 \le k \le N, \\ 0, & N+1 \le k \le 2N, \end{cases}$$

$$B^\ell_{kk_1} := \begin{cases} 2N\sqrt{2\pi\,\Delta\kappa\,\Delta\omega\kappa_{k_1} f(\kappa_{k_1}, \omega_k)}(X_{k_1 k 3} - iX_{k_1 k 4}), & 1 \le k \le N, \\ 0, & N+1 \le k \le 2N, \end{cases}$$

where $X_{k_1 k j}$, $1 \le j \le 4$, are independent standard normal random variables.

2. Calculate and save the values of $\text{Re}[a^\ell_{jk_1} c_j]$ and $\text{Re}[b^\ell_{jk_1} c_j]$, where $a^\ell_{jk_1}$ and $b^\ell_{jk_1}$ are inverse Fast Fourier Transforms of the columns of the arrays $A^\ell_{kk_1}$ and $B^\ell_{kk_1}$ defined in step 1, and $c_j = \exp[i\pi(j - 1)/(2N)]$.

3. Calculate the values of the time histories $\tilde{X}_j(r, \varphi) = \tilde{X}(r, \varphi, t_j)$ at any location (r, φ) and at discrete time moments $t_j = (j - 1)\pi/\omega_u$, $1 \le j \le 2N$, using the

formula

$$\tilde{X}_j(r, \varphi) = \sum_{\ell=-L}^{L} \sum_{k_1=1}^{N_1} J_\ell(\kappa_{k_1} r) \left(\cos(\ell\varphi) \operatorname{Re}\left[a_{jk_1}^\ell c_j\right] + \sin(\ell\varphi) \operatorname{Re}\left[b_{jk_1}^\ell c_j\right] \right).$$

(5.35)

Equation (5.35) produces simulations of time histories over half a period $T/2 = 2\pi N/\omega_u$. Simulations over the second half of a period are obtained as $\tilde{X}_{j+2N}(r, \varphi) = -\tilde{X}_j(r, \varphi)$.

For Method 2 the algorithm of simulation based on (5.34) consists of the same three steps as that for Method 1, except that the arrays $A_{kk_1}^\ell$ and $B_{kk_1}^\ell$ in step 1 are defined as

$$A_{kk_1}^\ell := \begin{cases} 4N\sqrt{\pi \Delta\kappa \Delta\omega\kappa_{k_1} f(\kappa_{k_1}, \omega_k)} \exp(iZ_{k_1 k1}), & 1 \le k \le N, \\ 0, & N+1 \le k \le 2N, \end{cases}$$

$$B_{kk_1}^\ell := \begin{cases} 4N\sqrt{\pi \Delta\kappa \Delta\omega\kappa_{k_1} f(\kappa_{k_1}, \omega_k)} \exp(iZ_{k_1 k2}), & 1 \le k \le N, \\ 0, & N+1 \le k \le 2N, \end{cases}$$

where $Z_{k_1 k1}$ and $Z_{k_1 k2}$ are independent random variables having uniform distribution on the interval $[0, 2\pi]$.

Assume for simplicity, that $N = N_1 = N_2$ in either (5.30) or (5.31), the simulation of the field as homogeneous only, would require the generation of sample functions at a two-dimensional space grid and a computational effort of $O((N \ln N)^3)$.

On the other hand, assume for simplicity that $N = N_1$ in either (5.33) or (5.34). The computational effort for simulation of a time history at one location is $O(N^2)$. The simulation of time histories at more than one location in space requires minimal additional computational efforts, namely, only the repetition of step 3. Such an advantage is of particular importance in the simulation of spatially variable seismic ground motions for the seismic response of bridges. For this case, simulated ground motions are required only at the supports of the structure, but not over an entire two-dimensional grid of locations.

The description of the spatial variability of seismic ground motions is based on the statistical analysis of data recorded at dense instrument arrays. Assume that the ground acceleration field $X(r, \varphi, t)$ is the realisation of a centred Gaussian homogeneous and spatially isotropic random field. Consider two points $\mathbf{x} = (r_1, \varphi_1)$ and $\mathbf{y} = (r_2, \varphi_2)$ on the ground surface at a distance ϱ apart from one another. The two-dimensional stochastic process $(X(r_1, \varphi_1, t), X(r_2, \varphi_2, t))^\top$ is stationary. Moreover, its one-sided cross-spectral density, $f_a(r_1, \varphi_1, r_2, \varphi_2, \omega)$, defined by

$$\mathrm{E}\left[X(r_1, \varphi_1, s) X(r_2, \varphi_2, s+t)\right] = \int_0^\infty \cos(\omega t) f_a(r_1, \varphi_1, r_2, \varphi_2, \omega)\, d\omega$$

depends only on ϱ and therefore will be denoted by $f_a(\varrho, \omega)$.

In applications, the above cross-spectral density is usually described by the product of two terms:

$$f_a(\varrho, \omega) = f_a(\omega)\gamma(\varrho, \omega),$$

where the first term, $f_a(\omega)$, describes the apparent propagation of the motions, while
the second term, $\gamma(\varrho, \omega)$, is called the *coherency* and represents the loss of similarity
in the seismic motions at various locations on the ground surface.

Assume that both terms are known. How to calculate the spectral density $f(\kappa, \omega)$
of the random field $X(r, \varphi, t)$? On the one hand, by definition of the one-sided cross-
spectral density, the cross-covariance of the motions between the above two locations should be

$$R(\mathbf{x}, t_1; \mathbf{y}, t_2) = \int_0^\infty \cos\big(\omega(t_2 - t_1)\big) f_a(\omega)\gamma(\varrho, \omega)\, d\omega.$$

On the other hand, according to (5.32), the above cross-covariance should be

$$R(\mathbf{x}, t_1; \mathbf{y}, t_2) = 2\pi \int_0^\infty \int_0^\infty J_0(\kappa\varrho) \cos\big(\omega(t_2 - t_1)\big)\kappa f(\kappa, \omega)\, d\kappa\, d\omega.$$

The Fourier transform is a one-to-one mapping. Therefore, we have

$$f_a(\omega)\gamma(\varrho, \omega) = 2\pi \int_0^\infty J_0(\kappa\varrho)\kappa f(\kappa, \omega)\, d\kappa.$$

In other words, the right-hand side is 2π times the Hankel transform of order 0 of
the unknown spectral density. By the inversion formula for Hankel transform (6.54),
we obtain

$$f(\kappa, \omega) = \frac{1}{2\pi} f_a(\omega) \int_0^\infty \gamma(\varrho, \omega) J_0(\kappa\varrho)\varrho\, d\varrho.$$

The right-hand side of the last display should be nonnegative. It follows that *the
Hankel transform of order 0 of the coherency $\gamma(\varrho, \omega)$ of a homogeneous and
isotropic random field in the space $\mathbb{R}^3$ with respect to the variable ϱ is a nonnegative
function.*

Luco and Wong (1986) proposed the following coherency model:

$$\gamma(\varrho, \omega) := \exp\big(-\alpha^2\varrho^2\omega^2\big),$$

where α is the incoherence parameter. This model may serve as a coherency model
of a homogeneous and spatially isotropic random field. Indeed, using formula
2.12.9.3 from Prudnikov (1988), we obtain

$$f(\kappa, \omega) = \frac{1}{4\pi\alpha^2\omega^2} f_a(\omega) \exp\left(-\frac{\kappa^2}{4\alpha^2\omega^2}\right) \geq 0.$$

Kanai (1957) and Tajimi (1960) proposed constructing a model for the spectral
density of the acceleration of apparent propagation of ground motions in the following way. We start from the bedrock excitation process, which is assumed to be

a Gaussian white noise with constant spectral density S_0. This process is passed through a soil filter with frequency ω_g and damping ζ_g. The spectral density of the filtered process becomes

$$f_a(\omega) = S_0 \frac{1 + 4\zeta_g^2(\omega/\omega_g)^2}{[1 - (\omega/\omega_g)^2]^2 + 4\zeta_g^2(\omega/\omega_g)^2}.$$

The above spectral density, the *Kanai–Tajimi spectrum*, has the following drawback. The spectral density of the velocity

$$f_v(\omega) = \frac{f_a(\omega)}{\omega^2}$$

and the spectral density of the displacement

$$f_d(\omega) = \frac{f_a(\omega)}{\omega^4}$$

are not integrable. To overcome this difficulty, Clough and Penzien (1975) proposed passing the Kanai–Tajimi spectrum through an additional filter with parameters ω_f and ζ_f. The spectral density becomes

$$f_a(\omega) = S_0 \frac{1 + 4\zeta_g^2(\omega/\omega_g)^2}{[1 - (\omega/\omega_g)^2]^2 + 4\zeta_g^2(\omega/\omega_g)^2} \frac{(\omega/\omega_f)^4}{[1 - (\omega/\omega_f)^2]^2 + 4\zeta_f^2(\omega/\omega_f)^2}.$$

The resulting spectral density, the *Clough–Penzien spectrum*, yields integrable spectral densities for both velocity and displacement.

The results of simulation of a Gaussian homogeneous and spatially isotropic random field with Clough–Penzien spectrum of the acceleration of apparent propagation of ground motions and Luco–Wong coherency model may be found in Katafygiotis et al. (1999).

5.4 Bibliographical Remarks

Applications to approximation theory were considered by Malyarenko (1999, 2008).

The *Copernican principle* states that our place in the Solar system is not a special one. Its extension, the *cosmological principle*, states that our Universe is to a good approximation homogeneous and isotropic on sufficiently large scales.

The theory of an expanding and cooling Universe which started in a "Big Bang" was confirmed by the discovery of the cosmic microwave background (CMB) by Penzias and Wilson in 1964 (Nobel prize 1978). The precise measurements of the CMB spectrum and the discovery of its anisotropy by the team led by George Smoot and John Mather were awarded another Nobel prize in 2006 and established what we now call the Standard Cosmological Model.

The anisotropy of the CMB is characterised by a perturbation in its intensity tensor. As we have seen in Subsection 5.2.1, the intensity tensor is a section of a tensor bundle over the sphere S^2. The above perturbation is random, having its origin in quantum uncertainty. By the cosmological principle, the perturbation must be a homogeneous and isotropic random field. Therefore, the intensity tensor is an isotropic *random* section of the above tensor bundle. A variant of the rigorous mathematical theory of invariant random sections of vector and tensor bundles was published by Malyarenko (2011) and is described in Subsection 2.4.2.

Subsection 5.2.1 is a short introduction to the deterministic model of the CMB for mathematicians. While studying physical literature, we have found that there exist various definitions of both ordinary and spin-weighted spherical harmonics. The choice of a definition is called the *phase convention*. In terms of the representation theory, the phase convention is the choice of a basis in the space of the group representation. We made an attempt to describe different phase conventions in order to help the mathematicians to read the physics literature. We also describe different notations for power spectra. The material of this subsection is based on Cabella and Kamionkowski (2005), Challinor (2005, 2009), Challinor and Peiris (2008), Durrer (2008), and Lin and Wandelt (2006).

Our new result in Subsection 5.2.2 is Theorem 5.5. It states that the standard assumptions of cosmological theories (the random fields $T(\mathbf{n})$, $E(\mathbf{n})$, and $B(\mathbf{n})$ are jointly isotropic) is equivalent to the assumption that $((Q - iU)(\mathbf{n}), T(\mathbf{n}), (Q + iU)(\mathbf{n}))$ is an isotropic random field in $\xi_{-2} \oplus \xi_0 \oplus \xi_2$. The exposition is based on Malyarenko (2011). Yet another, more group-theoretic approach to studying tensor random fields on the two-dimensional sphere was recently developed by Leonenko and Sakhno (2012).

We would like to mention the books by Lyth and Liddle (2009) and Marinucci and Peccati (2011), papers by Baldi et al. (2007, 2009a,b), Baldi and Marinucci (2007), Cabella and Marinucci (2009), Durastanti et al. (2012), Geller et al. (2009), Lan and Marinucci (2008, 2009), Marinucci (2004, 2005, 2006, 2008), Marinucci and Peccati (2010a,b,c), Marinucci and Piccioni (2004), and Marinucci and Wigman (2011a,b). Note that we do not consider questions connected with statistical analysis of the observation data of the recent and forthcoming experiments. For an introduction to this field of research, see Geller et al. (2009), Jaffe (2009) and the references therein.

Scalar, vector, and tensor spherical harmonics are considered by Freeden and Schreiner (2009).

Seismic time histories of the same earthquake at various locations on the ground surface are different. The simulation of the spatial variation of the seismic ground motions becomes extremely important for the probabilistic engineering community. An excellent survey of the above subject may be found in Zerva (2009). Our treatment is based on Katafygiotis et al. (1999).

References

N.I. Achieser. *Theory of approximation.* Dover, New York, 1992.

P. Baldi and D. Marinucci. Some characterizations of the spherical harmonics coefficients for isotropic random fields. *Stat. Probab. Lett.*, 77(5):490–496, 2007.

P. Baldi, D. Marinucci, and V.S. Varadarajan. On the characterization of isotropic Gaussian fields on homogeneous spaces of compact groups. *Electron. Commun. Probab.*, 12:291–302, 2007.

P. Baldi, G. Kerkyacharian, D. Marinucci, and D. Picard. Subsampling needlet coefficients on the sphere. *Bernoulli*, 15(2):438–463, 2009a. URL http://projecteuclid.org/euclid.bj/1241444897.

P. Baldi, G. Kerkyacharian, D. Marinucci, and D. Picard. Asymptotics for spherical needlets. *Ann. Stat.*, 37(3):1150–1171, 2009b. URL http://projecteuclid.org/euclid.aos/1239369018.

P. Cabella and M. Kamionkowski. Theory of cosmic microwave background polarization. Preprint. Available at http://arxiv.org/abs/astro-ph/0403392, 2005.

P. Cabella and D. Marinucci. Statistical challenges in the analysis of cosmic microwave background radiation. *Ann. Appl. Stat.*, 3(1):61–95, 2009. doi:10.1214/08-AOAS190.

A. Challinor. Cosmic microwave background anisotropies. In E. Papantonopoulos, editor. *The physics of the early Universe*, volume 653 of *Lect. Notes Phys.*, pages 71–103. Springer, Berlin, 2005. doi:10.1007/b99562.

A. Challinor. Cosmic microwave background polarization analysis. In V. Martinez, E. Saar, E. Martínez-González, and M. Pons-Borderia, editors. *Data analysis in cosmology*, volume 665 of *Lect. Notes Phys.*, pages 121–158. Springer, Berlin, 2009.

A. Challinor and H. Peiris. Lecture notes on the physics of cosmic microwave background anisotropies. In M. Novello and S. Perez Bergliaffa, editors. *Cosmology and gravitation: XIII Brazilian school on cosmology and gravitation (XIII BSCG)*, Rio de Janeiro (Brazil), 20 July–2 August 2008, volume 1132 of *AIP Conf. Proc.*, pages 86–140, Am. Inst. Phys., Melville, 2008.

R.W. Clough and J. Penzien. *Dynamics of structures*. McGraw–Hill, New York, 1975.

C. Durastanti, D. Geller, and D. Marinucci. Adaptive nonparametric regression on spin fiber bundles. *J. Multivar. Anal.*, 104(1):16–38, 2012.

R. Durrer. *The cosmic microwave background*. Cambridge University Press, Cambridge, 2008.

W. Freeden and M. Schreiner. *Spherical functions of mathematical geosciences: a scalar, vectorial and tensorial setup*. Adv. Geophys. Environ. Mech. Math. Springer, Berlin, 2009.

I.M. Gel'fand and Z.Ya. Šapiro. Representations of the group of rotations in three-dimensional space and their applications. Usp. Mat. Nauk (N.S.), 7(1(47)):3–117, 1952. In Russian.

D. Geller and D. Marinucci. Spin wavelets on the sphere. *J. Fourier Anal. Appl.*, 16:840–884, 2010.

D. Geller, X. Lan, and D. Marinucci. Spin needlets spectral estimation. *Electron. J. Stat.*, 3:1497–1530, 2009. doi:10.1214/09-EJS448. URL http://projecteuclid.org/euclid.ejs/1262617416.

A.H. Jaffe. Bayesian analysis of cosmic microwave background data. In M. Hobson, A. Jaffe, A. Liddle, P. Mukherjee, and D. Parkinson, editors. *Bayesian methods in cosmology*, Lect. Notes Phys., pages 229–244. Cambridge University Press, Cambridge, 2009.

M. Kamionkowski, A. Kosowsky, and A. Stebbins. Statistics of cosmic microwave background polarization. *Phys. Rev. D*, 55(12):7368–7388, 1997.

K. Kanai. Semi-empirical formula for the seismic characteristics of the ground. *Bull. Earthq. Res. Inst. Univ. Tokyo*, 35:309–325, 1957.

L.S. Katafygiotis, A. Zerva, and A. Malyarenko. Simulation of homogeneous and partially isotropic random fields. *J. Eng. Mech.*, 125:1180–1189, 1999.

X. Lan and D. Marinucci. The needlets bispectrum. *Electron. J. Stat.*, 2:332–367, 2008. doi:10.1214/08-EJS197. URL http://projecteuclid.org/euclid.ejs/1211317529.

X. Lan and D. Marinucci. On the dependence structure of wavelet coefficients for spherical random fields. *Stoch. Process. Appl.*, 119(10):3749–3766, 2009. doi:10.1016/j.spa.2009.07.005.

N. Leonenko and L. Sakhno. On spectral representations of tensor random fields on the sphere. *Stoch. Anal. Appl.*, 30:44–66, 2012. doi:10.1080/07362994.2012.628912.

M.A. Lifshits. *Gaussian random functions*, volume 322 of *Math. Appl.* Kluwer Academic, Dordrecht, 1995.

Y.-T. Lin and B.D. Wandelt. A beginner's guide to the theory of CMB temperature and polarization power spectra in the line-of-sight formalism. *Astropart. Phys.*, 25:151–166, 2006.

J.E. Luco and H.L. Wong. Response of a rigid foundation to a spatially random ground motion. *Earthquake Eng. Struct. Dyn.*, 14:891–908, 1986.

D.H. Lyth and A.R. Liddle. *The primordial density perturbation*. Cambridge University Press, Cambridge, 2009.

A. Malyarenko. Local properties of Gaussian random fields on compact symmetric spaces, and Jackson-type and Bernstein-type theorems. *Ukr. Math. J.*, 51(1):66–75, 1999.

A. Malyarenko. An optimal series expansion of the multiparameter fractional Brownian motion. *J. Theor. Probab.*, 21(2):459–475, 2008.

A. Malyarenko. Invariant random fields in vector bundles and application to cosmology. *Ann. Inst. H. Poincaré Probab. Stat.*, 47(4):1068–1095, 2011. doi:10.1214/10-AIHP409. URL http://projecteuclid.org/euclid.aihp/1317906502.

D. Marinucci. Testing for non-Gaussianity on cosmic microwave background radiation: a review. *Stat. Sci.*, 19(2):204–307, 2004. doi:10.1214/088342304000000783.

D. Marinucci. Regression methods for testing Gaussianity on a spherical random field. *Random Oper. Stoch. Equ.*, 13(4):313–324, 2005. doi:10.1515/156939705775992411.

D. Marinucci. High-resolution asymptotics for the angular bispectrum of spherical random fields. *Ann. Stat.*, 34(1):1–41, 2006. URL http://projecteuclid.org/euclid.aos/1146576254.

D. Marinucci. A central limit theorem and higher order results for the angular bispectrum. *Probab. Theory Relat. Fields*, 141:389–409, 2008. doi:10.1007/s00440-007-0088-8.

D. Marinucci and G. Peccati. Ergodicity and Gaussianity for spherical random fields. *J. Math. Phys.*, 51:23 p., 2010a. doi:10.1063/1.3329423.

D. Marinucci and G. Peccati. Group representations and high-resolution central limit theorems for subordinated spherical random fields. *Bernoulli*, 16(3):798–824, 2010b. doi:10.3150/09-BEJ230.

D. Marinucci and G. Peccati. Representations of SO(3) and angular polyspectra. *J. Multivar. Anal.*, 101(1):77–100, 2010c. doi:10.1016/j.jmva.2009.04.017.

D. Marinucci and G. Peccati. *Random fields on the sphere. Representation, limit theorems and cosmological applications*, volume 389 of *London Math. Soc. Lect. Note Ser.* Cambridge University Press, Cambridge, 2011.

D. Marinucci and M. Piccioni. The empirical process on Gaussian spherical harmonics. *Ann. Stat.*, 32(3):1261–1288, 2004. URL http://projecteuclid.org/euclid.aos/1085408502.

D. Marinucci and I. Wigman. On the area of excursion sets of spherical Gaussian eigenfunctions. *J. Math. Phys.*, 52(9):093301, 2011a. doi:10.1063/1.3624746.

D. Marinucci and I. Wigman. The defect variance of random spherical harmonics. *J. Phys. A, Math. Theor.*, 44(35):355206, 2011b. doi:10.1088/1751-8113/44/35/355206.

E.T. Newman and R. Penrose. Note on the Bondi–Metzner–Sachs group. *J. Math. Phys.*, 7(5):863–870, 1966.

A.P. Prudnikov, Yu.A. Brychkov, and O.I. Marichev. *Integrals and series. Vol. 2. Special functions*. Gordon & Breach, New York, second edition, 1988.

S.O. Rice. Mathematical analysis of random noise. *Bell Syst. Tech. J.*, 23:282–332, 1944.

M. Shinozuka. Monte Carlo solution of structural dynamics. *Comput. Struct.*, 2:855–874, 1972.

M. Shinozuka. Stochastic fields and their digital simulation. In G. Schuëller and M. Shinozuka, editors. *Stochastic methods in structural dynamics*, volume 10 of *Mech. Dyn. Syst.*, pages 93–133. Nijhoff, Dordrecht, 1987.

H. Tajimi. A statistical method of determining the maximum response of a building structure during an earthquake. In *Proceedings of the Second World Conference on Earthquake Engineering*, volume 2, pages 781–797, Tokyo–Kyoto, 1960.

K.S. Thorne. Multipole expansions of gravitational radiation. *Rev. Mod. Phys.*, 52(2):299–339, 1980.

G.N. Watson. *A treatise on the theory of Bessel functions*. Cambridge University Press, Cambridge, 1944. Reprinted in 1995.

S. Weinberg. *Cosmology*. Oxford University Press, London, 2008.

M. Zaldarriaga and U. Seljak. An all-sky analysis of polarisation in the microwave background. *Phys. Rev. D*, 55(4):1830–1840, 1997.

A. Zerva. *Spatial variation of seismic ground motion. Modeling and engineering applications*. Adv. Eng. CRC Press, Boca Raton, 2009.

Chapter 6
Appendix A: Mathematical Background

Abstract The theory of invariant random fields on spaces with a group action requires good knowledge of various parts of mathematics other than Probability and Statistics. We discuss differentiable manifolds, vector bundles, Lie groups and Lie algebras, group actions and group representations, special functions, rigged Hilbert spaces, Abelian and Tauberian theorems. Bibliographical remarks conclude.

6.1 Differential Geometry

In this section, we summarise necessary definitions and theorems from Differential Geometry.

6.1.1 Manifolds

Let $\mathbb{K}$ be either the field of real numbers $\mathbb{R}$ or the field of complex numbers $\mathbb{C}$. Let m be a positive integer, and let U and V be two subsets in $\mathbb{K}^m$. Let $\varphi : U \to V$. Then φ may be written as

$$y_1 = \varphi_1(x_1, \ldots, x_m), \ldots, y_m = \varphi_m(x_1, \ldots, x_m). \tag{6.1}$$

Let r be either a nonnegative integer or a symbol ∞ or a symbol ω. A mapping φ is called a *mapping of class C^r* if one of the following conditions holds.

- $\mathbb{K} = \mathbb{R}$, $r \notin \{\infty, \omega\}$, and functions (6.1) have continuous partial derivatives up to order r;
- $\mathbb{K} = \mathbb{R}$, $r = \infty$, and functions (6.1) have partial derivatives of all orders;
- $r = \omega$ and functions (6.1) are $\mathbb{K}$-analytic.

Let T be a set. A *chart* in T is a pair (U, φ), where $U \subseteq T$, and φ is a one-to-one map from U to an open set $\varphi(U) \subseteq \mathbb{K}^m$. For any charts $(U_\alpha, \varphi_\alpha)$ and (U_β, φ_β) with $U_\alpha \cap U_\beta \neq \varnothing$, consider the mapping

$$\varphi_\beta \circ \varphi_\alpha^{-1} : \varphi_\alpha(U_\alpha \cap U_\beta) \to \varphi_\beta(U_\alpha \cap U_\beta),$$

A. Malyarenko, *Invariant Random Fields on Spaces with a Group Action*,
Probability and Its Applications, DOI 10.1007/978-3-642-33406-1_6,
© Springer-Verlag Berlin Heidelberg 2013

which is called a *chart change*. A family $\{(U_\alpha, \varphi_\alpha): \alpha \in A\}$ of charts on T is called an *atlas of class C^r* if

- U_α form a covering of T;
- both the domain and the range of any chart change are open subsets of $\mathbb{K}^n$;
- any chart change is of class C^r.

The set of all atlases of class C^r is partially ordered by inclusion. Any atlas $\mathscr{A}$ is contained in a unique *maximal* atlas $\mathscr{A}_{\max}$ with respect to the above partial ordering. Any maximal atlas $\mathscr{A}_{\max}$ is called a C^r-*structure* on T. A pair $(T, \mathscr{A}_{\max})$ is called a *manifold of class C^r*.

Introduce a topology in T by declaring a subset $O \subseteq T$ open, if for any chart (U, h) of the manifold T the set $h(O \cap U)$ is open in $\mathbb{K}^n$. In what follows we assume that the above topology satisfies the Hausdorff separation axiom: any two different points in T can be separated by disjoint open sets.

6.1.2 Vector Bundles

Let π be a continuous map from a topological space E on a topological space T, and let n be a positive integer. A triple $\xi = (E, \pi, T)$ is called a *vector bundle* of rank n over $\mathbb{K}$, if

1. for any $t \in T$ the set $E_t = \pi^{-1}(t)$ is the n-dimensional $\mathbb{K}$-vector space;
2. for every $t \in T$ there exist a neighbourhood U of t and a homeomorphism $\psi: \pi^{-1}(U) \to U \times \mathbb{K}^n$ such that

$$\psi(E_t) \subset \{t\} \times \mathbb{K}^n$$

and the composition $\mathrm{proj}_2 \circ \psi: E_t \to \mathbb{K}^n$ is a $\mathbb{K}$-vector space isomorphism, where proj_2 is the projection from $\{t\} \times \mathbb{K}^n$ on $\mathbb{K}^n$.

The space T is called the *base* of the vector bundle ξ, while the space E is called its *total space*. The set E_t is called the *fibre* of the vector bundle ξ over the point $t \in T$. The pair (U, ψ) is called a *local trivialisation*.

All operations over vector spaces can be naturally extended to vector bundles. In particular, it is possible to consider direct sums, tensor products of vector bundles, conjugate vector bundles, etc.

A *section* of the vector bundle ξ, or a *vector field* is a continuous map $s: T \to E$ such that πs maps T identically to itself.

The standard example of a vector bundle is as follows. Let f be a real-valued function on a manifold of class C^∞. The function f is called *differentiable* at a point $t \in T$ if there exists a chart (U, φ) with $t \in U$ such that the composite function $f \circ \varphi^{-1}$ is an infinitely differentiable function on $\varphi(U)$. The function f is called differentiable if it is differentiable at each point $t \in T$.

Let $C^\infty(T)$ denote the set of all differentiable functions on T, and let $C^\infty(t)$ denote the set of functions on T which are differentiable at t. The operations

$$(\lambda f)(t) := \lambda f(t),$$

$$(f + g)(t) := f(t) + g(t),$$

$$(fg)(t) := f(t)g(t)$$

for $\lambda \in \mathbb{R}$, $t \in T$, $f, g \in C^\infty(T)$ turn $C^\infty(T)$ into an algebra over $\mathbb{R}$. A *vector field* on T is a *derivation* of the algebra $C^\infty(T)$, i.e., a mapping $X \colon C^\infty(T) \to C^\infty(T)$ such that

$$X(\alpha f + \beta g) = \alpha X f + \beta X g,$$

$$X(fg) = f(Xg) + (Xf)g$$

for $\alpha, \beta \in \mathbb{R}$, $f, g \in C^\infty(T)$. Let $D(T)$ be the set of all vector fields on T. The operations

$$fX \colon g \to f(Xg),$$

$$X + Y \colon g \to Xg + Yg$$

for $g \in C^\infty(T)$ turn $D(T)$ into a module over the ring $C^\infty(T)$.

Let (U, φ) be a chart in T, and let t be an arbitrary point in U. For a point $s \in U$, put $\varphi(s) := (x_1(s), \dots, x_m(s))$. For a function $f \in C^\infty(T)$ put $f^* := f \circ \varphi^{-1}$. For an arbitrary vector field X we have

$$(Xf)(t) = \sum_{j=1}^m \left(\frac{\partial f^*}{\partial x_j} \right)_{\varphi(t)} (Xx_j)(t), \quad t \in U. \tag{6.2}$$

The mapping $f \mapsto (\partial f^*/\partial x_j) \circ \varphi$ is a vector field on U and is denoted by $\partial/\partial x_j$. By (6.2),

$$X = \sum_{j=1}^m (Xx_j) \frac{\partial}{\partial x_j}.$$

Therefore, $\partial/\partial x_j$ $(1 \le j \le m)$ is a basis of the module $D(U)$.

Let $t \in T$ and $X \in D(T)$. Let X_t be the linear mapping $X_t(f) = (Xf)(t)$ from $C^\infty(T)$ into $\mathbb{R}$. The set $\{X_t \colon X \in D(T)\}$ is called the *tangent space* to T at t. It is denoted by $D(t)$; its elements are called the *tangent vectors* to T at t. Equation (6.2) shows that $D(t)$ is a vector space over $\mathbb{R}$ spanned by the m linearly independent vectors

$$\mathbf{e}_j \colon f \to \left(\frac{\partial f^*}{\partial x_j} \right)_{\varphi(t)}, \quad f \in C^\infty(T).$$

Now put $E := \bigcup_{t \in T} \{t\} \times D(t)$, $\pi(t, X_t) := t$. Let $\{(U_\alpha, \varphi_\alpha) \colon \alpha \in A\}$ be an atlas of T. Define a map $\psi_\alpha \colon \pi^{-1}(U_\alpha) \to \mathbb{R}^{2m}$ by

$$\psi_\alpha(t, x_1 \mathbf{e}_1 + \cdots + x_m \mathbf{e}_m) := (\varphi_\alpha(t), x_1, \dots, x_m).$$

By definition, a subset $G \subset E$ is open iff $\psi_\alpha(G \cap \pi^{-1}(U_\alpha))$ is open in $\mathbb{R}^{2m}$ for each $\alpha \in A$. Then the triple (E, π, T) is the vector bundle with fibres $D(t)$ and local trivialisations (U_α, ψ_α). It is called the *tangent bundle* of the manifold T.

To simplify formulations, in what follows we consider only manifolds over $\mathbb{R}$ with $r = \infty$. All mappings will be *smooth*, i.e., of class C^∞.

Let $F : T \to S$ be a map between two manifolds of class C^∞. It is called *smooth*, if for every $t \in T$ there is a chart (U, φ) containing t and a chart (V, ψ) in S containing $f(t)$ with $f(U) \subset V$ such that $\psi \circ F \circ \varphi^{-1}$ is a smooth map from $\varphi(U)$ to $\psi(V)$. The *tangent map* of F at a point $t \in T$ is the map $dF_t : D(t) \to D(f(t))$ acting by $dF_t(X)(f) = X(f \circ F)$, where $f \in C^\infty(T)$ and $X \in D(t)$.

A *Riemannian metric* on T is a family

$$g_t : D(t) \times D(t) \to \mathbb{R}$$

of positive-definite inner products such that for all vector fields X, Y on T the mapping

$$T \to \mathbb{R}, \quad t \mapsto g_t\big(X(t), Y(t)\big)$$

is smooth. A pair (T, g_t) is called a *Riemannian manifold*.

Let T be a connected manifold and let $\gamma : [a, b] \to T$ be a smooth curve. The length of γ is

$$L(\gamma) = \int_a^b \sqrt{g\big(\gamma'(s), \gamma'(s)\big)}\, ds.$$

The distance between points p and q in T is defined as $d(p, q) := \inf L(\gamma)$, where the infimum is taken over all smooth curves beginning at p and ending at q. With this distance, (T, d) is a metric space.

A one-to one mapping $f : T \to T$ is called the *isometry* if for any points s, t from T, $d(f(s), f(t)) = d(s, t)$. A Riemannian manifold T is called *two-point homogeneous* if for any four points s_1, s_2, t_1, and t_2 of T with $d(s_1, s_2) = d(t_1, t_2)$ there exist an isometry f of the manifold T such that $g(s_1) = t_1$ and $g(s_2) = t_2$.

6.1.3 Differential Operators

Let T be a manifold of class C^∞ or C^ω. Let $C_0^\infty(T)$ be the set of all infinitely differentiable functions φ on T with compact support $\mathrm{supp}(\varphi)$. The *differential operator* D on T is a linear mapping of $C_0^\infty(T)$ on itself such that $\mathrm{supp}(D\varphi) \subset \mathrm{supp}(\varphi)$, $\varphi \in C_0^\infty(T)$. Formula

$$(Df)(t) = (D\varphi)(t),$$

where $\varphi(t) \in C_0^\infty(T)$ is any function equal to $f \in C^\infty(T)$ in some neighbourhood of a point $t \in M$ correctly extends the definition of D to the space $C^\infty(T)$.

6.1.4 Invariant Differential Operators on Manifolds

Let φ be a diffeomorphism of a manifold T of class C^∞ on itself. For any $f \in C^\infty(T)$, put $f^\varphi := f \circ \varphi^{-1}$. For any differential operator D on T, define another differential operator, D^φ, as

$$D^\varphi : f \to \left(D(f \circ \varphi)\right) \circ \varphi^{-1}, \quad f \in C^\infty(T).$$

Operator D is called *invariant* with respect to φ if $D^\varphi = D$.

6.2 Lie Groups and Lie Algebras

Almost all groups under consideration in this book are Lie groups. In this section, we summarise all the necessary definitions and theorems from the theory of Lie groups.

6.2.1 Basic Definitions

A set G is a *Lie group* if it is both a group and a manifold of class C^ω over $\mathbb{K}$ and the mapping

$$G \times G \to G, \quad (g_1, g_2) \mapsto g_1 g_2^{-1}$$

is also of class C^ω.

Let $L_g : G \to G$ be the left translation: $L_g(h) := gh$, $g, h \in G$. The corresponding tangent map dL_g acts from $D(h)$ to $D(gh)$. A vector field $X(h)$ in the tangent bundle (E, π, G) is called *left-invariant* if $dL_g X(h) = X(gh)$ for every h in G.

Let X be a tangent vector at the identity element $e \in G$. Then $dL_g(X)$ is a left-invariant vector field on G. This correspondence is one-to-one. In what follows, we identify a tangent vector with the corresponding vector field and denote both by capital letters $X, Y, \ldots$.

The *Lie bracket* $[X, Y]$ of two left-invariant vector fields (tangent vectors) X and Y is

$$[X, Y](f) := X\big(Y(f)\big) - Y\big(X(f)\big), \quad f \in C^\omega(G). \tag{6.3}$$

It has the following properties.

1. It is bilinear:

$$[\alpha X_1 + \beta X_2, Y] = \alpha[X_1, Y] + \beta[X_2, Y],$$
$$[X, \alpha Y_1 + \beta Y_2] = \alpha[X, Y_1] + \beta[X, Y_2]$$

for all $X_1, X_2, Y_1, Y_2 \in D(e)$, and for all $\alpha, \beta \in \mathbb{K}$.

2. It is skew symmetric:

$$[X, Y] = -[Y, X], \quad X, Y \in D(e).$$

3. It satisfies the Jacobi identity:

$$[[X, Y], Z] + [[Y, Z], X] + [[Z, X], Y] = 0, \quad X, Y, Z \in D(e).$$

Any finite-dimensional $\mathbb{K}$-linear space with an operation satisfying the above properties is called a *Lie algebra*. The tangent space $D(e)$ with Lie bracket (6.3) is called the Lie algebra of the Lie group G.

Conversely, for any Lie algebra $\mathfrak{g}$ there exists a (not always unique) Lie group G such that $\mathfrak{g}$ is the Lie algebra of the group G. Among all Lie groups with Lie algebra $\mathfrak{g}$ there is a unique (connected) simply-connected group $\tilde{G}$. Any connected Lie group with Lie algebra $\mathfrak{g}$ has the form $\tilde{G}/D$, where D is a discrete normal subgroup lying in the centre of $\tilde{G}$.

Example 6.1 Let $\mathfrak{g}$ be the space $\mathbb{R}^3$ with Lie bracket the vector product in $\mathbb{R}^3$. Then $\tilde{G} = SU(2)$, the group of all unitary unimodular 2×2 matrices. The centre of $\tilde{G}$ is $\pm I$, where I is the identity matrix. The only remaining connected Lie group with Lie algebra $\mathfrak{g}$ is the quotient group $SU(2)/\pm I$, which is isomorphic to $SO(3)$.

The *exponential map* $\exp: \mathfrak{g} \to G$ is given by $\exp(X) = \gamma(1)$, where $\gamma: \mathbb{R} \to G$ is the unique analytic group homomorphism whose tangent vector at the identity is equal to X.

Let $\mathfrak{g}$ be a Lie algebra. For any $X \in \mathfrak{g}$, the mapping

$$\operatorname{ad} X: Y \mapsto [X, Y], \quad Y \in \mathfrak{g}$$

is a linear operator on $\mathfrak{g}$. Moreover, ad respects the Lie algebra structure:

$$\operatorname{ad}[X, Y] = \operatorname{ad} X \operatorname{ad} Y - \operatorname{ad} Y \operatorname{ad} X.$$

Therefore, ad is called the *adjoint representation* of the algebra $\mathfrak{g}$.

The bilinear form

$$B(X, Y) := \operatorname{tr}(\operatorname{ad} X \operatorname{ad} Y)$$

is called the *Killing form* on $\mathfrak{g}$. A Lie algebra $\mathfrak{g}$ is called *semisimple* if its Killing form is non-degenerate. From now on, we consider only real semisimple Lie algebras.

An automorphism θ of a Lie algebra $\mathfrak{g}$ is called a *Cartan involution* if it is involutive, i.e., $\theta^2 = 1$, and the bilinear form

$$B_\theta(X, Y) := -B(X, \theta Y)$$

is positive-definite. Put

$$\mathfrak{k} := \{X \in \mathfrak{g}: \theta X = X\},$$

$$\mathfrak{p} := \{X \in \mathfrak{g}: \theta X = -X\}.$$

Then B is positive-definite on $\mathfrak{p}$ and negative-definite on $\mathfrak{k}$.

Let $\mathfrak{g}$ be a real semisimple Lie algebra, and let G be a connected Lie group with Lie algebra $\mathfrak{g}$. Then G is called a *semisimple Lie group*. The Cartan involution θ is the differential of a unique involutive automorphism of G, denoted also by θ. The subgroup

$$K := \{g \in G: \theta(g) = g\}$$

is the connected subgroup of G with Lie algebra $\mathfrak{k}$.

Among all connected Lie groups having the same semisimple Lie algebra $\mathfrak{g}$, there always exists at least one Lie group G with finite centre. In this case, the subgroup K is compact. Moreover, K is also a *maximal* compact subgroup of G.

Let $\mathfrak{a}$ be a maximal abelian subspace of $\mathfrak{p}$, and let $\mathfrak{m}$ be the centraliser of $\mathfrak{a}$ in $\mathfrak{k}$. Let $\mathfrak{a}^*$ be the real linear dual space of $\mathfrak{a}$. For any $\alpha \in \mathfrak{a}^*$, put

$$\mathfrak{g}_\alpha := \{X \in \mathfrak{g}: [H, X] = \alpha(H)X, H \in \mathfrak{a}\}.$$

Then α is called a *root* or a *restricted root* of a pair $(\mathfrak{g}, \mathfrak{a})$ if $\alpha \neq 0$ and $\mathfrak{g}_\alpha \neq \{0\}$. The linear space $\mathfrak{g}$ has direct sum decomposition

$$\mathfrak{g} = \mathfrak{g}_0 \oplus \sum_{\alpha \in \Sigma} \oplus \mathfrak{g}_\alpha, \quad \mathfrak{g}_0 = \mathfrak{a} \oplus \mathfrak{m},$$

where Σ is the set of restricted roots. The spaces $\mathfrak{g}_\alpha$ are called *root subspaces*. A point $H \in \mathfrak{a}$ is called *regular* if $\alpha(H) \neq 0$ for all $H \in \Sigma$ and *singular* otherwise. The subset $\mathfrak{a}' \subset \mathfrak{a}$ of regular elements is the complement of a union of finitely many hyperplanes. The connected components of $\mathfrak{a}'$ are called the *Weyl chambers*. Fix a Weyl chamber $\mathfrak{a}^+$. A root α is called *positive* if α has positive values on $\mathfrak{a}^+$. The direct sum of the root subspaces of all positive roots is a subalgebra $\mathfrak{n}$ of Lie algebra $\mathfrak{g}$. A positive root is called *simple* if it is not the sum of two positive roots. The set of simple roots forms a basis in the linear space $\mathfrak{a}^*$.

Let $A = \exp(\mathfrak{a})$ and let $N = \exp(\mathfrak{n})$. Let M be the centraliser of A in K, and let M' be the normaliser of A in K. The quotient group M'/M is called the *Weyl group* and is denoted by W. Let $A^+ = \exp(\mathfrak{a}^+)$, and let $\overline{A^+}$ be the closure of A^+ in G. The dimension $\dim \mathfrak{a}$ is called the *real rank* of $\mathfrak{g}$ and of G.

The *Cartan decomposition* states that $G = K\overline{A^+}K$. In other words, each $g \in G$ can be written as $g = k_1 a k_2$ with $k_1, k_2 \in K$ and $a \in \overline{A^+}$. Moreover, a is unique.

The *Iwasawa decomposition* states that $G = KAN$. In other words, each $g \in G$ can be written as $g = kan$ with $k \in K$, $a \in A$, and $n \in N$. All elements of the Iwasawa decomposition are unique.

6.2.2 Spherical Functions

Let G be a connected Lie group, K a compact subgroup, and $T = G/K$. Let $\mathbf{D}(T)$ be the algebra of differential operators on T invariant under all the translations $\tau(g): tK \to gtK$ of T.

Definition 6.2 A complex-valued function $\varphi(t)$ on T of class C^∞ with $\varphi(o) = 1$ is called a *zonal spherical function* if $\varphi(kt) = \varphi(t)$ for all $k \in K$ and φ is an eigenfunction of each operator in $\mathbf{D}(T)$.

Let H be a closed subgroup of G. Let $\mathfrak{h} \subset \mathfrak{g}$ be their respective Lie algebras and $\mathfrak{m}$ be a complementary subspace, $\mathfrak{g} = \mathfrak{m} \oplus \mathfrak{h}$. The coset space $T = G/H$ is called *reductive* if the subspace $\mathfrak{m} \subset \mathfrak{g}$ can be chosen in such a way that

$$\mathrm{Ad}_G(h)\mathfrak{m} \subset \mathfrak{m}, \quad h \in H.$$

Let $S(\mathfrak{m})$ be the *symmetric algebra* over $\mathfrak{m}$, i.e. the algebra of complex-valued polynomial functions on the dual space $\mathfrak{m}^*$. Let $(X_1, \ldots, X_r)$ and $(X_{r+1}, \ldots, X_n)$ be bases of $\mathfrak{m}$ and $\mathfrak{h}$ respectively. Let $\pi: G \to G/H$ be the natural projection. If f is a function on G/H, put $\tilde{f} := f \circ \pi$. Let $I(\mathfrak{m})$ be the set of $\mathrm{Ad}_G(H)$-invariants in $S(\mathfrak{m})$. The algebra $\mathbf{D}(T)$ consists of all operators of the form

$$(L_Q f)(g \cdot o) := \left[Q\left(\frac{\partial}{\partial x_1}, \ldots, \frac{\partial}{\partial x_r}\right) \tilde{f}\left(g \exp(x_1 X_1 + \cdots + x_r X_r)\right)\right](0), \quad (6.4)$$

where $Q \in I(\mathfrak{m})$.

Let T be a two-point homogeneous manifold. In this case, $r = 1$, and $\mathbf{D}(T)$ consists of the polynomials in the *Laplace–Beltrami operator* $L_{X_1^2}$.

6.3 Group Actions and Group Representations

Spectral expansions of G-invariant random fields are based on the theory of irreducible unitary representations of the group G. In this section, we summarise the above theory.

6.3.1 Group Actions

Let T be a set, and let G be a group with identity element e. A *left group action* of G on T is a function

$$G \times T \to T, \quad (g, t) \mapsto gt, \quad g \in G, t \in T, \quad (6.5)$$

which satisfies the following condition: for all g, h in G, and for any $t \in T$ we have

$$(gh)t = g(ht), \quad et = t.$$

On the other hand, a *right group action* of G on T is a function

$$T \times G \to T, \quad (t, g) \mapsto tg, \quad g \in G, \ t \in T,$$

such that for all g, h in G, and for any $t \in T$ we have

$$t(gh) = (tg)h, \quad te = t.$$

In what follows, we consider only left group actions.

The *orbit Gt* of a point $t \in T$ is the set

$$Gt := \{gt : g \in G\}.$$

The *quotient* of the group action is the set of all orbits of T under the action of G. It is denoted by T/G.

A left action of a group G on a space T is called *transitive* if for any $t_1, t_2 \in T$, there exists at least one element $g \in G$ with $gt_1 = t_2$. In this and only in this case, the quotient space T/G contains only one orbit.

A left action of a topological group G on a topological space T is called *continuous* if (6.5) maps the topological product $G \times T$ into T continuously. The quotient T/G is equipped with the quotient topology.

For every $t \in T$, the *stabiliser subgroup* of G is

$$G_t := \{g \in G : g \cdot t = t\}.$$

6.3.2 Positive-Definite Functions on Groups and Group Representations

Let G be a set carrying both the structure of a group and the structure of a topological space. If the mapping

$$G \times G \to G, \quad (g, h) \mapsto gh^{-1}$$

is continuous, then G is called a *topological group*.

A continuous function $R \colon G \to \mathbb{C}$ is called *positive-definite* if for any positive integer n, for any elements $g_1, g_2, \ldots, g_n$ in G, and for any complex numbers $\lambda_1, \lambda_2, \ldots, \lambda_n$ we have

$$\sum_{j,k=1}^{n} \lambda_j \overline{\lambda_k} R(g_k^{-1} g_j) \geq 0.$$

To describe the class of positive-definite functions, we need some preparation.

Let $H \neq \{0\}$ be a complex separable Hilbert space with the inner product $(\cdot, \cdot)$, and let $\mathbf{x} \in H$. Let $\mathbf{x}^*$ be the element of the conjugate space H^* acting on a vector $\mathbf{y} \in H$ by

$$\mathbf{x}^*(\mathbf{y}) := (\mathbf{y}, \mathbf{x}).$$

Note that the mapping

$$H \to H^*, \quad \mathbf{x} \mapsto \mathbf{x}^*, \quad \mathbf{x} \in H$$

is semi-linear:

$$\lambda \mathbf{x} + \mu \mathbf{y} \mapsto \overline{\lambda} \mathbf{x} + \overline{\mu} \mathbf{y}, \quad \lambda, \mu \in \mathbb{C}, \ \mathbf{x}, \mathbf{y} \in H.$$

Let H_1 and H_2 be two complex separable Hilbert spaces with inner products $(\cdot, \cdot)_1$ and $(\cdot, \cdot)_2$, and let $\mathbf{x}_1 \in H_1$ and $\mathbf{x}_2 \in H_2$. The *tensor product* $\mathbf{x}_1 \otimes \mathbf{x}_2$ is the rank-one linear operator from H_1^* to H_2 acting by

$$\mathbf{x}_1 \otimes \mathbf{x}_2(\mathbf{x}^*) := \mathbf{x}^*(\mathbf{x}_1)\mathbf{x}_2, \quad \mathbf{x}^* \in H_1^*. \tag{6.6}$$

Define the inner product of two linear operators $\mathbf{x}_1 \otimes \mathbf{x}_2$ and $\mathbf{y}_1 \otimes \mathbf{y}_2$ by

$$(\mathbf{x}_1 \otimes \mathbf{x}_2, \mathbf{y}_1 \otimes \mathbf{y}_2) := (\mathbf{x}_1, \mathbf{y}_1)_1 (\mathbf{x}_2, \mathbf{y}_2)_2.$$

This inner product can be extended by linearity to the space of finite rank operators from H_1^* to H_2. The completion of that space under the above inner product is called the *Hilbertian tensor product* of the Hilbert spaces H_1 and H_2 and is denoted by $H_1 \otimes H_2$. The elements of $H_1 \otimes H_2$ are *Hilbert–Schmidt* operators from H_1^* to H_2.

Let $H_1 = H^*$ and $H_2 = H$. A product AB of two Hilbert–Schmidt operators $A, B: H \to H$ is called a *trace-class operator*. The inner product (A, B) is called the *trace* of the operator AB and is denoted by $\mathrm{tr}(AB)$.

Let G be a topological group. Let $L(H)$ be the topological group of all invertible operators on H equipped with the strong operator topology. A continuous homomorphism U from G into $L(H)$ is called a *representation* of the topological group G. A representation U is called *unitary* if $U(G)$ is a subset in the group $U(H)$ of all unitary operators on H.

For a unitary representation U, the mapping

$$G \times H \to H, \quad (g, \mathbf{x}) \mapsto U(g)\mathbf{x}, \quad g \in G, \ \mathbf{x} \in H,$$

defines the left action of G on H which preserves the inner product in H.

A closed subspace $H_1 \subseteq H$ is called *invariant* under the representation $U(g)$ if

$$U(g)H_1 \subseteq H_1, \quad g \in G.$$

In any case, there exist at least two invariant subspaces of $U(g)$, $\{0\}$ and H. The representation $U(g)$ is called *irreducible* if there exist no other invariant subspaces.

In what follows, generic representations of G will be denoted by capital Latin letters U, V, W, ..., while irreducible unitary representations will be denoted by small Greek letters λ, μ, ν, ...

Let $V(g)$ be a representation of G in H. The representation $\tilde{V}(g) = V^*(g^{-1})$ is called the *conjugate-dual* representation. The representation $V^+(g) = V^\top(g^{-1})$ is called the *dual* representation.

For some topological groups, any unitary representation can be uniquely decomposed into irreducible unitary representations. To describe this class, we need to introduce more definitions.

A unitary representation U_1 in the space H_1 is called *equivalent* to a unitary representation U_2 in the space H_2 if there exists a unitary isomorphism U from H_1 on H_2 such that

$$U_2(g) = U U_1(g) U^{-1}, \quad g \in G.$$

The operator U is called the *intertwining operator* between the equivalent representations U_1 and U_2.

The set of all equivalence classes of unitary irreducible representations of a group G is denoted by $\hat{G}$. Introduce a topology on $\hat{G}$ in the following way. Note that equivalent representations have the same set of *matrix elements*

$$u_{\lambda \mathbf{x}}(g) := \big(\lambda(g)\mathbf{x}, \mathbf{x}\big), \quad \lambda \in \hat{G}, \ \mathbf{x} \in H^{(\lambda)}, \ g \in G,$$

where $H^{(\lambda)}$ is the space of the representation λ. A point $\lambda \in \hat{G}$ is a *limit point* of a set $M \subseteq \hat{G}$ if any matrix element $u_{\lambda \mathbf{x}}$ ($\mathbf{x} \in H^{(\lambda)}$) can be approximated by matrix elements $u_{\mu \mathbf{y}}$ ($\mu \in M$, $\mathbf{y} \in H^{(\mu)}$) uniformly on any compact subset $K \subseteq G$. A set $M \subseteq \hat{G}$ is *closed* if it contains all its limit points, and *open* if $\hat{G} \setminus M$ is closed.

Let X be a locally compact topological space, let $\mathfrak{B}$ be the Borel σ-algebra on X, and let μ be a locally finite measure ($\mu(K) < \infty$ for any compact $K \in \mathfrak{B}$) on the measurable space $(X, \mathfrak{B})$. For any $x \in X$, let $H^{(x)}$ be a separable Hilbert space, and let $\dim H^{(x)}$ be a measurable function on X. Then X may be represented as a union of Borel sets

$$X_n := \{x \in X \colon \dim X = n\}, \quad n = 0, 1, \dots, \infty.$$

Moreover, all spaces $H^{(x)}$, $x \in X_n$ may be identified with the *standard space*

$$H_n := \begin{cases} \mathbb{C}^n, & n < \infty, \\ \ell^2, & n = \infty. \end{cases}$$

A vector function $\mathbf{f}(x)$ that maps any point $x \in X$ into $H^{(x)}$ is called *measurable* if for all n, and for any $\mathbf{x} \in H_n$, all complex-valued functions $(\mathbf{f}(x), \mathbf{x})$ are measurable on X_n. Put

$$\|\mathbf{f}\|^2 := \int_X \|\mathbf{f}(x)\|^2 \, d\mu(x).$$

The Hilbert space of all vector-functions $\mathbf{f}(x)$ for which this integral converges, is called a *direct integral of the Hilbert spaces* $H^{(x)}$ and is denoted by

$$\int_X \oplus H^{(x)} \, d\mu(x).$$

A function $A^{(x)}$ on X that maps any point $x \in X$ into the set of all linear continuous operators on $H^{(x)}$ is called *measurable*, if for any measurable vector-function $\mathbf{f}(x)$ the vector-function

$$(\mathbf{A}\mathbf{f})(x) := A^{(x)}\mathbf{f}(x), \quad x \in X$$

is again measurable. The linear operator A is continuous iff the real-valued function $\|A^{(x)}\|$ is essentially bounded. In this case, A is called a *direct integral of operators* $A^{(x)}$ and is denoted by

$$A = \int_X \oplus A^{(x)} \, d\mu(x).$$

A representation U is called a *direct integral of representations* $U^{(x)}$ if

$$U(g) = \int_X \oplus U^{(x)}(g) \, d\mu(x), \quad g \in G. \tag{6.7}$$

When the space X is discrete, a direct integral of representations is called a *direct sum* of representations.

Any unitary continuous representation U of a locally compact topological group G is equivalent to a representation of type (6.7), where μ-almost all representations are continuous, unitary, and irreducible. The representation (6.7) is not unique.

The representation (6.7) has a *simple spectrum* if the representations $U^{(x)}$ are pairwise non-equivalent. A direct sum $n \oplus U$ on n copies of a representation U with simple spectrum is called a *representation with spectrum of multiplicity* n, where $n = 0, 1, \ldots, \infty$. A representation U is called the *representation of type I* if it can be represented as a direct sum of representations with spectra of multiplicity n.

A locally compact topological group is called the *group of type I* if all its unitary representations are of type I. For a group of type I with countable base, any unitary representation has the form

$$U(g) = \int_{\hat{G}} \oplus \left(n(\lambda) \oplus \lambda \right) d\mu(\lambda), \tag{6.8}$$

where $\lambda \in \hat{G}$ is an irreducible unitary representation, and $n(\lambda)$ is a measurable function. Moreover, the measure μ is uniquely determined by the representation U up to equivalence of measures, and the function $n(\lambda)$ is uniquely determined by the representation U μ-almost everywhere.

Let G be the *Mautner group*, i.e. the product $\mathbb{C}^2 \times \mathbb{R}$ in which multiplication is given by

$$(z_1, w_1, t_1)(z_2, w_2, t_2) := \left(z_1 + \exp(\mathrm{i}t_1)z_2, \, w_1 + \exp(2\pi \mathrm{i}t_1)w_1, \, t_1 + t_2 \right).$$

The Mautner group is not of type I.

Any *compact* topological group G is of type I. Moreover, the space $\hat{G}$ is finite or countable and discrete, and the direct integral in (6.8) becomes a direct sum.

Let G be a topological group of type I with countable base, and let

$$\hat{G}_n := \{\lambda \in \hat{G} : \dim \lambda = n\}, \qquad n = 1, 2, \ldots, \infty.$$

Let F_n be a measure on the measurable space $(\hat{G}_n, \mathfrak{B}(\hat{G}_n))$ with values in the set of self-adjoint nonnegative-definite trace-class operators on the standard space H_n. A continuous function $R: G \to \mathbb{C}$ is positive-definite iff

$$R(g) = \sum_{n \geq 1} \int_{\hat{G}_n} \mathrm{tr}\big[\lambda(g)\,\mathrm{d}F_n(\lambda)\big], \tag{6.9}$$

where $\lambda \in \hat{G}_n$ acts on the standard space H_n, and the measures F_n satisfy the following condition:

$$\sum_n \mathrm{tr}\big[F_n(\hat{G}_n)\big] < \infty.$$

Equation (6.9) is usually written in the following short form:

$$R(g) = \int_{\hat{G}} \mathrm{tr}\big[\lambda(g)\,\mathrm{d}F(\lambda)\big]. \tag{6.10}$$

We will often use the following simple statement.

Lemma 6.3 *Let λ be an irreducible representation of a topological group G in a complex Hilbert space H. Let $\mathbf{x} \in H$ be a common eigenvector of all operators $\lambda(g)$, $g \in G$. If λ is not trivial, then $\mathbf{x} = \mathbf{0}$.*

Proof Let $\lambda(g)\mathbf{x} = \mathbf{x}$, $g \in G$. Then the one-dimensional subspace of H generated by $\mathbf{x}$ is an invariant subspace of the representation λ. Because λ is irreducible, we have either $\mathbf{x} = \mathbf{0}$ or $\mathbf{x}$ generates all of H. In the latter case, assume that λ is not trivial. Then, there exists $g \in G$ with $\lambda(g) \neq 1$. But $\lambda(g)\mathbf{x} = \mathbf{x}$. It follows that $\mathbf{x} = \mathbf{0}$. $\qquad\square$

Let K be a compact subgroup of G. We say that $\lambda \in \hat{G}_K$ if the restriction of λ to K contains *at least* one trivial representation of the group K. K is called a *massive* subgroup of G if the restriction of any $\lambda \in \hat{G}_K$ to K contains *exactly* one trivial representation of K, see Vilenkin (1968).

Let $\{\mathbf{e}_j^{(\lambda)} : 1 \leq j \leq \dim \lambda\}$ be bases in the spaces $H^{(\lambda)}$ of the irreducible representations $\lambda \in \hat{G}$. Assume that K is massive, and let $\mathbf{e}_1^{(\lambda)}$ lie in the space of the trivial representation of K. The functions

$$\varphi_m^{(\lambda)}(g) := \big(\lambda(g)\mathbf{e}_1^{(\lambda)}, \mathbf{e}_m^{(\lambda)}\big)$$

depend only on the coset $t = gK$. The function $\varphi_m^{(\lambda)}(t)$ is called the *spherical function* corresponding to the above chosen basis. The function $\varphi_1^{(\lambda)}(t)$ is called the *zonal spherical function*. The zonal spherical function depends only on a point x in the set $X = K\backslash G/K$ of double cosets.

If G is a connected Lie group, then the zonal spherical function in the sense of Definition 6.2 is the zonal spherical function in the sense of the previous paragraph iff it is positive-definite.

The condition $\varphi(kt) = \varphi(t)$ for all $k \in K$ in Definition 6.2 may be formulated in the language of representation theory as follows. Let H be a linear span of the functions $\varphi(kt)$, $k \in K$. Formula

$$\bigl(U(k)f\bigr)(t) := f\bigl(k^{-1}(t)\bigr)$$

determines a representation U of the group K. The representation U corresponding to a zonal spherical function $\varphi(t)$ is trivial.

Let $\delta \in \hat{K}$. A complex-valued function $\varphi(t)$ on T of class C^∞ is called a *spherical function of type δ* if φ is an eigenfunction of each operator in $\mathbf{D}(T)$ and the representation U is a direct sum of finitely many copies of δ.

The functions $\varphi_m^{(\lambda)}(t)$ with $m > 1$ are called the *associated* spherical functions.

Let the restriction of λ to K contain at least one copy of the representation $\delta \in \hat{K}$, and let $e_m^{(\lambda)}$ belong to the space of the above copy. The associated spherical function $\varphi_m^{(\lambda)}(t)$ is the spherical function of type δ.

6.3.3 Unitary Representations of Commutative Topological Groups

In particular, let G be a *commutative* locally compact topological group. In this case G is of type I, all the irreducible unitary representations are one-dimensional (they are called *characters*), and (6.10) takes the form

$$R(g) = \int_{\hat{G}} \hat{g}(g)\,dF(\hat{g}), \tag{6.11}$$

where $\hat{g}(g) \in \mathbb{C}$ is the value of the character $\hat{g}$ on the element $g \in G$, and F is a real-valued finite measure on $(\hat{G}, \mathfrak{B}(\hat{G}))$. The set of characters $\hat{G}$ is again a commutative topological group with respect to multiplication of characters

$$(\hat{g}_1 + \hat{g}_2)(g) := \hat{g}_1(g)\hat{g}_2(g)$$

and the topology of uniform convergence on compact subsets. This group is called the *character group*.

For example, let $G = \mathbb{R}$. The character group $\hat{\mathbb{R}}$ can be identified with $\mathbb{R}$ by the following mapping:

$$\mathbb{R} \to \hat{\mathbb{R}}, \quad p \mapsto e^{ipt}, \quad p \in \hat{\mathbb{R}}, t \in \mathbb{R},$$

and (6.11) takes the form

$$R(t) = \int_{-\infty}^{\infty} e^{ipt} \, dF(p).$$

Similarly, for any positive integer n, we have

$$R(\mathbf{x}) = \int_{\mathbb{R}^n} e^{i(\mathbf{p},\mathbf{x})} \, dF(\mathbf{p}), \quad \mathbf{x} \in \mathbb{R}^n, \ \mathbf{p} \in \hat{\mathbb{R}}^n.$$

Let G be the group $\mathbb{Z}$ of all integers. The character group $\hat{\mathbb{Z}}$ can be identified with the *torus*

$$\mathbb{T} := \left\{ e^{i\varphi} : 0 \le \varphi < 2\pi \right\}$$

by the mapping

$$\mathbb{T} \to \hat{\mathbb{Z}}, \quad e^{i\varphi} \mapsto e^{in\varphi}, \quad n \in \mathbb{Z},$$

and (6.11) takes the form

$$R(n) = \int_{\mathbb{T}} e^{in\varphi} \, dF(\varphi).$$

Similarly, for any positive integer n, we have

$$R(\mathbf{n}) = \int_{\mathbb{T}^n} e^{i(\mathbf{n},\boldsymbol{\varphi})} \, dF(\boldsymbol{\varphi}), \quad \mathbf{n} \in \mathbb{Z}^n, \ \boldsymbol{\varphi} \in \mathbb{T}^n.$$

Finally, let $G = \mathbb{T}$. The character group $\hat{T}$ can be identified with $\mathbb{Z}$ by the mapping

$$\mathbb{Z} \to \hat{\mathbb{T}}, \quad n \mapsto e^{in\varphi}, \quad n \in \mathbb{Z}.$$

This follows from the *Pontryagin duality*: for any commutative locally compact group G the mapping

$$\hat{\hat{G}} \to G, \quad \hat{g}(g) \mapsto g,$$

is an isomorphism of topological groups $\hat{\hat{G}}$ and G, see Morris (1977). In this case, (6.11) takes the form

$$R(\varphi) = \sum_{n=-\infty}^{\infty} e^{in\varphi} F_n, \quad F_n \ge 0, \quad \sum_{n=-\infty}^{\infty} F_n < \infty.$$

Similarly, for any positive integer n, we have

$$R(\boldsymbol{\varphi}) = \sum_{\mathbf{n} \in \mathbb{Z}^n} e^{i(\mathbf{n},\boldsymbol{\varphi})} F_{\mathbf{n}}, \quad F_{\mathbf{n}} \ge 0, \quad \sum_{\mathbf{n} \in \mathbb{Z}^n} F_{\mathbf{n}} < \infty, \quad \boldsymbol{\varphi} \in \mathbb{T}^n.$$

Formula (6.11) is correct for some commutative topological groups that are not locally compact. For example, let $\{G_\alpha : \alpha \in A\}$ be a set of commutative locally compact groups, which is partially ordered by the relation "G_α is a subgroup of G_β". Assume that any two groups G_α and G_β have an upper bound G_γ with $\gamma \in A$, i.e., both G_α and G_β are subgroups of G_γ.

Introduce the following equivalence relation in the disjoint union of all G_α: the elements $g \in G_\alpha$ and $h \in G_\beta$ are equivalent iff G_α and G_β have such an upper bound G_γ that the image of g under inclusion $G_\alpha \subset G_\gamma$ is the same as the image of h under inclusion $G_\beta \subset G_\gamma$. Let $[g]$ be the equivalence class of an element $g \in G_\alpha$. In the set G of equivalence classes, introduce the group operation

$$[g] + [h] := [g + h], \quad g \in G_\alpha, \ h \in G_\beta.$$

On the right-hand side, $+$ denotes the group operation in any upper bound of G_α and G_β. Equip G with the weakest topology under which all inclusion mappings

$$G_\alpha \to G, \quad g \mapsto [g], \quad \alpha \in A, \ g \in G_\alpha$$

are continuous. This topology is called *inductive*, and the group G equipped with inductive topology is called the *inductive limit* of the system $\{G_\alpha : \alpha \in A\}$. It is denoted by $\mathrm{inj} \lim_{\alpha \in A} G_\alpha$.

For example, the inductive limit of the system $\{\mathbb{R}^n : n \geq 1\}$ is the group $\mathbb{R}_0^\infty$ of all finite sequences of real numbers, in which the sequence $\{\mathbf{x}_m : m \geq 1\}$ with

$$\mathbf{x}_m := (x_{m,1}, x_{m,2}, \ldots, x_{m,N(m)}, 0, 0, \ldots)$$

converges iff $\sup_m N(m) = N < \infty$ and the sequence of sections

$$\tilde{\mathbf{x}}_m := (x_{m,1}, x_{m,2}, \ldots, x_{m,N})$$

converges in $\mathbb{R}^N$.

When G_α is a subgroup of G_β, any character $\hat{g} \in \hat{G}_\beta$ may be restricted to G_α. In the Cartesian product of all $\hat{G}_\alpha$, consider the subgroup of all elements $\{\hat{g}_\alpha : \alpha \in A\}$ such that the restriction of the character $\hat{g}_\beta$ to G_α is equal to $\hat{g}_\alpha$ whenever G_α is a subgroup of G_β. This subgroup is called the *projective limit* of the system $\{\hat{G}_\alpha : \alpha \in A\}$ and is denoted by $\mathrm{proj} \lim_{\alpha \in A} \hat{G}_\alpha$. The *projective topology* on the projective limit is defined as the weakest topology for which the projections

$$\mathrm{proj} \lim_{\alpha \in A} \hat{G}_\alpha \to \hat{G}_\alpha, \quad (\hat{g}_\alpha) \mapsto \hat{g}_\alpha$$

are continuous. The projective limit $\mathrm{proj} \lim_{\alpha \in A} \hat{G}_\alpha$ and the corresponding inductive limit $\mathrm{inj} \lim_{\alpha \in A} G_\alpha$ are each the character group of other, see Samoĭlenko (1991).

For example, the projective limit of the system $\{\mathbb{R}^n : n \geq 1\}$ is the group $\mathbb{R}^\infty$ of all sequences of real numbers with the topology of coordinate-wise convergence. Moreover, $\mathbb{R}^\infty$ is the character group of $\mathbb{R}_0^\infty$.

A continuous positive-definite function on the injective limit $\operatorname{inj} \lim_{\alpha \in A} G_\alpha$ has the form

$$R(g) = \int_{\operatorname{proj} \lim_{\alpha \in A} \hat{G}_\alpha} \hat{g}(g) \, dF(\hat{g}),$$

while a continuous positive-definite function on the projective limit $\operatorname{proj} \lim_{\alpha \in A} G_\alpha$ has the form

$$R(g) = \int_{\operatorname{inj} \lim_{\alpha \in A} \hat{G}_\alpha} \hat{g}(g) \, dF(\hat{g}).$$

6.3.4 Unitary Representations of Compact Topological Groups

Let G be a compact topological group. Every irreducible unitary representation $\lambda \in \hat{G}$ is finite-dimensional.

There exists a unique probability measure μ on $(G, \mathfrak{B}(G))$ which is bi-invariant, i.e., for all $B \in \mathfrak{B}(G)$ and for all $g \in G$ we have $\mu(gB) = \mu(Bg) = \mu(B)$. This measure is called the *Haar measure*.

Let $\{e_j^{(\lambda)} : 1 \le j \le \dim \lambda\}$ be bases in the spaces $H^{(\lambda)}$ of the irreducible representations $\lambda \in \hat{G}$. The *Peter–Weyl theorem* states that the set

$$\sqrt{\dim \lambda} \big(\lambda(g) e_j^{(\lambda)}, e_k^{(\lambda)} \big), \quad \lambda \in \hat{G}, \ 1 \le j, k \le \dim \lambda$$

forms a basis for the space $L^2(G, d\mu)$.

Assume that K is massive in G, and let $e_1^{(\lambda)}$ lie in the space of the trivial representation of K. The generalised Peter–Weyl theorem states that the associated spherical functions

$$\varphi_m^{(\lambda)}(g) := \big(\lambda(g) e_1^{(\lambda)}, e_m^{(\lambda)} \big)$$

depend only on the coset $t = gK$, and the set

$$\sqrt{\dim \lambda} \varphi_m^{(\lambda)}(t), \quad \lambda \in \hat{G}_K, \ 1 \le m \le \dim \lambda$$

forms a basis for the space $L^2(T, dt)$, where dt is a unique G-invariant probability measure on the *homogeneous space* $T = G/K$. Moreover, the set

$$\sqrt{\dim \lambda} \varphi_1^{(\lambda)}(x), \quad \lambda \in \hat{G}_K$$

forms a basis for the space $L^2(X, d\nu(x))$, where $d\nu(x)$ is the unique G-bi-invariant probability measure on the space of double cosets $X = K \backslash G / K$.

Note that the zonal spherical function is uniquely defined, while the associated spherical functions depend on the choice of basis. One way to choose the basis in the space $H^{(\lambda)}$ is as follows. Let

$$G = G_1 \supset G_2 \supset \cdots \supset G_N = \{e\} \tag{6.12}$$

be a chain of subgroups of G satisfying the following *Gel'fand–Tsetlin condition*: for any k with $2 \leq k \leq N$, G_k is massive in G_{k-1}. Then, the space $H^{(\lambda)}$ is a direct sum of subspaces of the non-equivalent irreducible representations of G_2, the above subspaces are direct sums of subspaces of the non-equivalent irreducible representations of G_3, and so on. Because the chain is finite, the space $H^{(\lambda)}$ finally becomes a direct sum of its one-dimensional subspaces. The *Gel'fand–Tsetlin basis* consists of vectors of length 1 lying in the above subspaces. Each subspace is uniquely determined by the set $\lambda = (\lambda_1, \ldots, \lambda_{N-1})$, where each λ_j, $2 \leq j \leq n - 1$, is the irreducible component of the restriction of $\lambda_{j-1} \in \hat{G}_{j-1}$ to G_j.

Consider the chain

$$SO(n) \supset SO(n-1) \supset \cdots \supset SO(1). \tag{6.13}$$

There exists a unique group called the *spin group* and denoted by $\mathrm{Spin}(n)$ that satisfies the following conditions.

- $\mathrm{Spin}(n)$ is the *double cover* of the group $SO(n)$. In other words, there exists a continuous group homomorphism $\varrho: \mathrm{Spin}(n) \to SO(n)$ (the *covering homomorphism*) and for each $g \in SO(n)$ the inverse image $\varrho^{-1}(g)$ contains exactly two elements.
- The following sequence of group homomorphisms

$$1 \to \mathbb{Z}_2 \to \mathrm{Spin}(n) \xrightarrow{\varrho} SO(n) \to 1$$

is *exact*, i.e., the image of each homomorphism is equal to the kernel of the next. See any standard text in homological algebra, for example, Rotman (2009).

Each irreducible unitary representation of $\mathrm{Spin}(n)$ is completely determined by the *signature*—the set of integers or half-integers (half of an odd positive integer) $p_1, p_2, \ldots, p_{[n/2]}$ satisfying the following conditions: if $n = 2j + 1$, then $p_1 \geq p_2 \geq \cdots \geq p_j$; if $n = 2j$, then $p_1 \geq p_2 \geq \cdots \geq |p_j|$. If the signature consists of integers, then both elements of the inverse image $\varrho^{-1}(g)$ map to the same unitary operator. Therefore, the corresponding representation is a one-to-one representation of the group $SO(n)$. Otherwise the elements of the inverse image map to different unitary operators and the corresponding representation of the group $SO(n)$ is one-to-two.

The restriction of the irreducible unitary representation of the group $SO(2j + 1)$ with signature $(p_1, \ldots, p_j)$ to the subgroup $SO(2j)$ is equivalent to the direct sum of the irreducible unitary representations with signatures $(q_1, \ldots, q_j)$ satisfying the following conditions:

$$p_k \geq q_k \geq p_{k+1}, \quad 1 \leq k \leq j - 1,$$
$$p_j \geq q_j \geq -p_j. \tag{6.14}$$

The restriction of the irreducible unitary representation of the group $SO(2j)$ with signature $(p_1, \ldots, p_j)$ to the subgroup $SO(2j - 1)$ is equivalent to the direct sum of the irreducible unitary representations with signatures $(q_1, \ldots, q_{j-1})$ satisfying the following conditions:

$$p_k \geq q_k \geq |p_{k+1}|, \quad 1 \leq k \leq j - 1. \tag{6.15}$$

Therefore, each vector of the Gel'fand–Tsetlin basis is completely determined by the *Gel'fand–Tsetlin scheme*—a numerical table with $n - 1$ rows, where the jth row contains the signature of $\lambda_j \in \lambda$ and the numbers in the adjacent rows satisfy either (6.14) or (6.15).

The Gel'fand–Tsetlin basis is not unique. Each vector $\mathbf{e}_m$ of the basis may be multiplied by the *phase multiplier*, $\exp(\mathrm{i}\varphi_m)$. The choice of the phase multipliers is called the *phase choice*.

Put $G = \mathrm{SU}(2)$, the group of unitary 2×2 matrices with unit determinant. Any element $g \in \mathrm{SU}(2)$ is defined by the *Euler angles* (φ, θ, ψ) as

$$g = \begin{pmatrix} \cos(\theta/2)e^{\mathrm{i}(\varphi+\psi)/2} & \mathrm{i}\sin(\theta/2)e^{\mathrm{i}(\varphi-\psi)/2} \\ \mathrm{i}\sin(\theta/2)e^{\mathrm{i}(\psi-\varphi)/2} & \cos(\theta/2)e^{-\mathrm{i}(\varphi+\psi)/2} \end{pmatrix}.$$

Let K be the subgroup of all matrices of the form

$$\begin{pmatrix} e^{\mathrm{i}\varphi/2} & 0 \\ 0 & e^{-\mathrm{i}\varphi/2} \end{pmatrix}.$$

The group G is isomorphic to $\mathrm{Spin}(3)$. Therefore, the irreducible unitary representations of G are in one-to-one correspondence with nonnegative integer or half-integer numbers ℓ. It is possible to choose the Gel'fand–Tsetlin basis with respect to the chain $G \supset K \supset \{e\}$ (to fix the phase choice) in such a way that the matrix elements of the irreducible unitary representations λ_ℓ are so-called *Wigner D-functions*:

$$D_{mn}^{(\ell)}(\varphi, \theta, \psi) := e^{-\mathrm{i}m\varphi} d_{mn}^{(\ell)}(\theta) e^{-\mathrm{i}n\psi}, \tag{6.16}$$

where $d_{mn}^{(\ell)}(\theta)$ are *Wigner d-functions*:

$$d_{mn}^{(\ell)}(\theta) := (-1)^m \sqrt{\frac{(\ell+m)!(\ell-m)!}{(\ell+n)!(\ell-n)!}} \sin^{2\ell}(\theta/2)$$

$$\times \sum_{r=\max\{0,m+n\}}^{\min\{\ell+m,\ell+n\}} \binom{\ell+n}{r}\binom{\ell-n}{r-m-n}(-1)^{\ell-r+n} \cot^{2r-m-n}(\theta/2). \tag{6.17}$$

Let λ be an irreducible unitary representation of G in a Hilbert space H_1, and let μ be an irreducible unitary representation of G in a Hilbert space H_2. The group action

$$[\lambda \otimes \mu(g)](\mathbf{x} \otimes \mathbf{y}) := (\lambda(g)\mathbf{x}) \otimes (\mu(g)\mathbf{y})$$

extends by linearity to the unitary representation $\lambda \otimes \mu$ in the tensor product $H_1 \otimes H_2$. This representation is called the *tensor product* of the representations λ and μ.

Let Λ, M, and N be the Gel'fand–Tsetlin bases of the representation λ, μ, and $\nu \in \hat{G}$. There exist two natural bases in the space $H_1 \otimes H_2$. The first one, the *uncoupled basis*, is

$$\{\lambda \otimes \mu : \lambda \in \Lambda, \mu \in M\}.$$

The *coupled basis* is

$$\left\{\boldsymbol{\nu}_n : \nu \in \hat{G}, 1 \leq n \leq n(\lambda \otimes \mu, \nu)\right\},$$

where $n(\lambda \otimes \mu, \nu)$ is the multiplicity of ν in $\lambda \otimes \mu$. It follows that

$$\boldsymbol{\nu}_n = \sum_{\lambda \in \Lambda} \sum_{\mu \in M} (\boldsymbol{\nu}_n, \lambda \otimes \mu)_{H_1 \otimes H_2} \lambda \otimes \mu.$$

The numbers $(\boldsymbol{\nu}_n, \lambda \otimes \mu)_{H_1 \otimes H_2}$ are called the *Clebsch–Gordan coefficients* of the group G. They are denoted by

$$C(\lambda, \mu; \nu, n) := (\boldsymbol{\nu}_n, \lambda \otimes \mu)_{H_1 \otimes H_2}.$$

The Clebsch–Gordan coefficients are the elements of the unitary matrix C—the matrix of the intertwining operator between equivalent representations $\lambda \otimes \mu$ and $\sum_{\nu \in \hat{G}} \sum_{n=1}^{n(\lambda \otimes \mu, \nu)} \oplus \nu_n$. It follows that

$$\lambda(g) \otimes \mu(g) = C^* \left(\sum_{\nu \in \hat{G}} \sum_{n=1}^{n(\lambda \otimes \mu, \nu)} \oplus \nu_n \right) C.$$

Taking the matrix elements of both sides, we obtain

$$D^{(\lambda)}_{\lambda\lambda'}(g) D^{(\mu)}_{\mu\mu'}(g) = \sum_{\nu \in \hat{G}} \sum_{n=1}^{n(\lambda \otimes \mu, \nu)} \sum_{\nu, \nu' \in N} \overline{C(\lambda, \mu; \nu_n)} C(\lambda', \mu'; \nu'_n) D^{(\nu)}_{\nu\nu'}(g), \quad (6.18)$$

where $D^{(\lambda)}_{\lambda\lambda'}(g)$ and $D^{(\nu)}_{\nu\nu'}(g)$ are the matrix elements of the representations λ and μ in Gel'fand–Tsetlin bases.

6.3.5 The Fourier Transform

Let G be a local compact topological group of type I with Haar measure dg. The *left regular representation*

$$[U(g)f](h) := f(g^{-1}h), \quad f(h) \in L^2(G, dg)$$

is unitary. Therefore, it is equivalent to a direct integral of the irreducible unitary representations of the group G. The above integral has the form

$$U(g) = \int_{\hat{G}} [\lambda \otimes I_\lambda](g) \, d\mu(\lambda), \tag{6.19}$$

where I_λ is the trivial representation of G in the space H_λ of the irreducible unitary representation $\lambda \in \hat{G}$. The measure μ is called the *Plancherel measure*. Note that for the case of a commutative locally compact group, the Plancherel measure is proportional to the Haar measure on $\hat{G}$.

The direct integral (6.19) determines an isomorphism between the space $L^2(G)$ and the direct integral of the spaces $H_\lambda \otimes H_\lambda^*$ with respect to the Plancherel measure. For any function $f \in L^2(G) \cap L^1(G)$, the above isomorphism has the form

$$f \mapsto \lambda(f) = \int_G f(g)\lambda(g) \, dg$$

and is called the *Fourier transform* on the group G. In particular, if $G = \mathbb{R}$, we recover the classical Fourier transform.

For any function $f \in L^2(G) \cap L^1(G)$, and for μ-almost all $\lambda \in \hat{G}$, the operator $\lambda(f)$ is Hilbert–Schmidt. Moreover, the Fourier transform is an isometry, i.e., respects norms:

$$\int_G |f(g)|^2 \, dg = \int_{\hat{G}} \mathrm{tr}\big[\lambda(f)\lambda^*(f)\big] \, d\mu(\lambda).$$

This equation is called the *Plancherel formula*. There exists such a dense subset in $L^2(G) \cap L^1(G)$ that for any function f in this subset the following *inversion formula* holds:

$$f(g) = \int_{\hat{G}} \mathrm{tr}\big[\lambda(f)\lambda^*(g)\big] \, d\mu(\lambda).$$

6.3.6 Positive-Definite Functions on Homogeneous Spaces and Induced Representations

Let G be a locally compact separable topological group, and let K be a closed subgroup. Let G/K be the space of left cosets gK, $g \in G$. Let $p: G \to G/K$, $p(g) = gK$ be the standard projection from G to G/K. Define a quotient topology on G/K: a set $A \subseteq G/K$ is open in G/K iff the set $p^{-1}A$ is open in G. The mapping

$$G \times G/K \to G/K, \quad (g, g_0 K) \mapsto g g_0 K, \quad g, g_0 \in G$$

defines the transitive continuous action of the group G on the homogeneous space $T = G/K$. We denote this action by $g \cdot t$.

Let G be a locally compact topological group of type I with countable base, and let K be a compact subgroup. Let λ be an irreducible unitary representation of the group G in the Hilbert space $H^{(\lambda)}$. Let $\lambda \in \hat{G}_K$ and let $H_K^{(\lambda)}$ be the closed linear span of all one-dimensional subspaces of $H^{(\lambda)}$ where trivial representations act. Let $P_K^{(\lambda)}$ be the projector to $H_K^{(\lambda)}$. Let $dF_K(\lambda)$ be a measure on $(\hat{G}_K, \mathfrak{B}(\hat{G}))$ whose restriction to

$$\hat{G}_{K,n} := \{\lambda \in \hat{G}_K : \dim \lambda = n\}$$

takes values in the set of Hermitian nonnegative-definite trace-class operators on the standard space H_n and satisfying condition

$$\sum_{n \geq 1} \mathrm{tr}\left[F(\hat{G}_{K,n})\right] < \infty.$$

Let t_1 and t_2 be two points in T, and let g_1 and g_2 be arbitrary elements of left cosets modulo K defining the points t_1 and t_2. A positive-definite function $R(t_1, t_2)$ satisfying $R(g \cdot t_1, g \cdot t_2) = R(t_1, t_2)$ has the form

$$R(t_1, t_2) = \int_{\hat{G}_K} \mathrm{tr}\left[P_K^{(\lambda)} \lambda(g_2^{-1} g_1) P_K^{(\lambda)} \, dF_K(\lambda)\right]. \tag{6.20}$$

For a *massive* subgroup K, the expression $P_K^{(\lambda)} \lambda(g_2^{-1} g_1) P_K^{(\lambda)}$ becomes a zonal spherical function.

Matrix multiplication yields

$$\varphi_0^{(\lambda)}(t_1, t_2) = \sum_{m=1}^{\dim \lambda} \varphi_m^{(\lambda)}(t_1) \overline{\varphi_m^{(\lambda)}(t_2)}, \tag{6.21}$$

where $\varphi_m^{(\lambda)}(t_1) = \varphi_m^{(\lambda)}(t_1, o)$ are spherical functions, and $o = p(e)$.

There exists a unique equivalence class of measures on the measurable space $(T, \mathfrak{B}(T))$, satisfying the following condition: for any measure μ from the above-mentioned class, and for any $g \in G$ the measures $d\mu_g(t) = d\mu(g^{-1}t)$ and $d\mu(t)$ are equivalent. Any measure satisfying this condition is called a *quasi-invariant* measure on T.

Let W be a unitary representation of the (not necessarily compact) group K in a Hilbert space H. Introduce the following equivalence relation in the topological product $G \times H$. Two elements (g_1, u_1) and (g_2, u_2) are *equivalent* iff there exists an element $k \in K$ such that $(g_2, u_2) = (g_1 k, W(k^{-1}) u_1)$. The projection mapping maps any element $(g, u) \in G \times H$ to its equivalence class and defines the quotient topology on the set E_W of equivalence classes. Another projection map,

$$\pi: E_W \to X, \qquad \pi(g, u) = gK,$$

determines a vector bundle $\xi = (E_W, \pi, T)$.

Let $L^2(E_W)$ be the space of all sections f of the vector bundle ξ satisfying the following conditions:

1. For any $u \in H$, the function $x \mapsto (f(x), u)_H$ is μ-measurable.
2. $\int_T \|f(t)\|_H^2 \, d\mu(t) < \infty$.

The space $L^2(E_W)$ can be also understood as a space of "twisted" functions on the base space. If W is the trivial representation of the group K in the Hilbert space $\mathbb{C}$, then $E_W = T \times \mathbb{C}$, and $L^2(E_W) = L^2(T, d\mu(t))$, the space of ordinary square-integrable functions. The vector bundle ξ is called the *homogeneous, or equivariant, vector bundle*. This is because the left action of G on T induces an associate left action of G on E_W by $(g_0, u) \mapsto (gg_0, u)$. Moreover, this left action also generates a unitary representation U of the group G in the space $L^2(E_W)$ by

$$(U(g)f)(x) := \left(\frac{d\mu_g(x)}{d\mu(x)} \right)^{1/2} f(g^{-1}x), \quad g \in G, \ f \in L^2(E_W), \ x \in X, \quad (6.22)$$

where $d\mu_g(x)/d\mu(x)$ denotes the Radon–Nikodym derivative of the shifted measure μ_g with respect to the measure μ. This representation is called the representation *induced* from the representation V.

Frobenius reciprocity states that the multiplicity of $V \in \hat{G}$ in U is equal to the multiplicity of W in V. See any standard text in the theory of group representations, for example, Fulton and Harris (1991). The representation induced from the direct sum $W_1 \oplus W_2 \oplus \cdots \oplus W_N$ is the direct sum of representations induced from W_1, W_2, ..., W_N.

6.3.7 Irreducible Unitary Representations of Low Dimensional Noncompact Lie Groups by Examples

Example 6.4 The group $G = \mathrm{SL}(2, \mathbb{R})$ has the series of representations $\lambda^{\sigma, \varepsilon}$, $\sigma \in \mathbb{C}$, $\varepsilon \in \{0, 1/2\}$ in the Hilbert space $L^2(S^1, (2\pi)^{-1} \, d\varphi)$ with orthonormal basis $\{e^{-in\varphi} : n \in \mathbb{Z}\}$. The representations $\lambda^{-1/2 + i\varrho, 0}$, $\varrho \geq 0$ are unitary and irreducible. They are called the representations of the *first principal series*. The representations $\lambda^{-1/2 + i\varrho, 1/2}$, $\varrho > 0$ are unitary and irreducible. They are called the representations of the *second principal series*. The representations $\lambda^{\varrho, 0}$, $-1 < \varrho < -1/2$ are equivalent to unitary and irreducible. These representations become unitary in the orthonormal basis

$$\left(\frac{\Gamma(-n - \varrho - \varepsilon)}{\Gamma(\varrho - n - \varepsilon + 1)} \right)^{1/2} e^{-in\varphi}, \quad n \in \mathbb{Z}. \quad (6.23)$$

They are called the representations of the *supplementary series*. If $\sigma - \varepsilon$ is a negative integer, then the space of the representation $\lambda^{\sigma, \varepsilon}$ contains two invariant subspaces H_σ^+ and H_σ^-. The restrictions $\lambda^{\sigma, \varepsilon, +}$ (resp. $\lambda^{\sigma, \varepsilon, -}$) of the representation $\lambda^{\sigma, \varepsilon}$ to the space H_σ^+ (resp. H_σ^-) are equivalent to unitary irreducible representations. The representations $\lambda^{\sigma, \varepsilon, +}$ with $(\sigma, \varepsilon) \neq (-1/2, 1/2)$ are called the representations of the *first discrete series*. The representations $\lambda^{\sigma, \varepsilon, -}$ with $(\sigma, \varepsilon) \neq (-1/2, 1/2)$ are called the representations of the *second discrete series*. The representations $\lambda^{-1/2, 1/2, +}$, and $\lambda^{-1/2,1/2,-}$ are the *limiting points* of the discrete series. The representations of the first discrete series and the representation $\lambda^{-1/2,1/2,+}$ become unitary in the orthonormal basis (6.23) with $n \leq \sigma - \varepsilon$, while those of the second discrete series and $\lambda^{-1/2,1/2,-}$ become unitary in the orthonormal basis

$$\left(\frac{\Gamma(n - \varrho + \varepsilon)}{\Gamma(\varrho + n + \varepsilon + 1)} \right)^{1/2} e^{-in\varphi}, \quad n \geq -\sigma - \varepsilon.$$

Finally, the space of the representation $\lambda^{-1,0}$ contains the one-dimensional invariant subspace H^0, where the *trivial representation* acts.

The restriction of the representations of the first principal and supplementary series to the compact subgroup $K = SO(2)$ of the group G is equivalent to the direct sum of the irreducible unitary representations of K with even signatures $p \in \mathbb{Z}$. The restriction of the representations of the second principal series to K is equivalent to the direct sum of the above representations with odd signatures $p \in \mathbb{Z}$. The restriction of the representations $\lambda^{\sigma,\varepsilon,+}$, $\sigma - \varepsilon \leq -1$, to K is equivalent to the direct sum of the above representations with signatures $\varepsilon - \sigma + 2p$, $p \geq 0$, while the restriction of the representations $\lambda^{\sigma,\varepsilon,-}$, $\sigma - \varepsilon \leq -1$ to K is equivalent to the direct sum of the above representations with signatures $\sigma + \varepsilon - 2p$, $p \geq 0$.

The *Cartan decomposition* states that every matrix $g \in SL(2, \mathbb{R})$ has the unique representation $g = u_\varphi g_\tau u_\psi$ with

$$u_\varphi = \begin{pmatrix} e^{i\varphi/2} & 0 \\ 0 & e^{-i\varphi/2} \end{pmatrix}, \quad g_\tau = \begin{pmatrix} \cosh(\tau/2) & \sinh(\tau/2) \\ \sinh(\tau/2) & \cosh(\tau/2) \end{pmatrix}.$$

The parameters φ, ψ, and τ are called the *Euler angles*.

Introduce the following functions: if $m, n \in \mathbb{Z}$ with $m \leq n$, then

$$f_{mn}^{\sigma,\varepsilon}(\varphi, \tau, \psi) := e^{-i[(m+\varepsilon)\varphi + (n+\varepsilon)\psi]} \frac{\Gamma(\sigma + n + \varepsilon + 1)}{\Gamma(\sigma + m + \varepsilon + 1)(n - m)!}$$
$$\times \left(\sinh(\tau/2) \right)^{n-m} \left(\cosh(\tau/2) \right)^{2\sigma - n + m}$$
$$\times {}_2F_1\left(-\sigma - m - \varepsilon, n + \varepsilon - \sigma; n - m + 1; \tanh^2(\tau/2) \right),$$

while if $n < m$, then

$$f_{mn}^{\sigma,\varepsilon}(\varphi, \tau, \psi) := e^{-i[(m+\varepsilon)\varphi + (n+\varepsilon)\psi]} \frac{\Gamma(\sigma + m + \varepsilon + 1)}{\Gamma(\sigma + n + \varepsilon + 1)(m - n)!}$$
$$\times \left(\sinh(\tau/2) \right)^{m-n} \left(\cosh(\tau/2) \right)^{2\sigma - m + n}$$
$$\times {}_2F_1\left(-\sigma - n - \varepsilon, m + \varepsilon - \sigma; m - n + 1; \tanh^2(\tau/2) \right),$$

where ${}_2F_1$ is the hypergeometric function (Subsection 6.4.2). The matrix elements of the representations of the principal series have the following form:

$$\lambda_{mn}^{-1/2+i\varrho,\varepsilon}(\varphi, \tau, \psi) = f_{mn}^{-1/2+i\varrho,\varepsilon}(\varphi, \tau, \psi), \quad m, n \in \mathbb{Z}. \tag{6.24}$$

The matrix elements of the representations of the supplementary series have the form

$$\lambda^{\varrho,0}_{mn}(\varphi, \tau, \psi) = \left(\frac{\Gamma(\varrho - m + 1)\Gamma(-\varrho - n)}{\Gamma(-\varrho - m)\Gamma(\varrho - n + 1)} \right)^{1/2} f^{\varrho,0}_{mn}(\varphi, \tau, \psi), \quad m, n \in \mathbb{Z}.$$
(6.25)

The matrix elements of the representations of the first discrete series and of the representation $\lambda^{-1/2,1/2,+}$ have the form

$$\lambda^{\sigma,\varepsilon,+}_{mn}(\varphi, \tau, \psi) = \left(\frac{\Gamma(\sigma - m - \varepsilon + 1)\Gamma(-\sigma - n - \varepsilon)}{\Gamma(-\sigma - m - \varepsilon)\Gamma(\sigma - n - \varepsilon + 1)} \right)^{1/2} f^{\sigma,\varepsilon}_{mn}(\varphi, \tau, \psi) \quad (6.26)$$

with $m, n \leq \sigma - \varepsilon$. The matrix elements of the representations of the second discrete series and of the representation $\lambda^{-1/2,1/2,-}$ have the form

$$\lambda^{\sigma,\varepsilon,-}_{mn}(\varphi, \tau, \psi) = \left(\frac{\Gamma(-\sigma - m - \varepsilon)\Gamma(\sigma - n - \varepsilon + 1)}{\Gamma(-\sigma - n - \varepsilon)\Gamma(\sigma - m - \varepsilon + 1)} \right)^{1/2} f^{\sigma,\varepsilon}_{mn}(\varphi, \tau, \psi) \quad (6.27)$$

with $m, n \geq -\sigma - \varepsilon$. Finally, $\lambda^{0,0}_{00}(\varphi, \tau, \psi) = 1$ (trivial representation).

Example 6.5 The group $G = \mathrm{SL}(2, \mathbb{C})$ has the series of representations $\lambda^{\sigma,n}$, $\sigma \in \mathbb{C}$, $n \in \mathbb{Z}$ in the space of the square-integrable sections of the homogeneous vector bundle ξ_n over the sphere S^2 with the basis of spin-weighted spherical harmonics $_nY_{\ell m}(\theta, \varphi)$ (Subsection 5.2.2). The representations $T^{i\varrho-2,n}$ with $\varrho \geq 0$ are unitary and irreducible in the orthonormal basis $\sqrt{4\pi} \cdot {}_nY_{\ell m}(\theta, \varphi)$, $\ell \geq 0$, $-\ell \leq m \leq \ell$. They are called the representations of the *principal series*. The representations $\lambda^{-2-\varrho,0}$ with $-2 < \varrho < 0$ are equivalent to unitary and irreducible. These representations become unitary in the basis

$$\left(\frac{\Gamma(\ell + 2 + \varrho)}{\Gamma(\ell - \varrho)} \right)^{1/2} \sqrt{4\pi} \cdot {}_nY_{\ell m}(\theta, \varphi), \quad \ell \geq 0, \ -\ell \leq m \leq \ell.$$

They are called the representations of the *supplementary series*. Finally, the space of the representation $\lambda^{0,0}$ contains the one-dimensional invariant subspace H^0, where the *trivial representation* acts.

The restriction of the representations $\lambda^{\sigma,n}$ of both the principal and supplementary series to the compact subgroup $K = \mathrm{SU}(2)$ is equivalent to the direct sum of the irreducible unitary representations π^ℓ, where n and 2ℓ are simultaneously either even or odd numbers, and $2\ell \geq |n|$.

The Cartan decomposition states that every matrix $g \in G$ can be represented as

$$g = k(\varphi_1, \theta_1, 0) \begin{pmatrix} e^{\theta/2} & 0 \\ 0 & e^{-\theta/2} \end{pmatrix} k(\varphi_2, \theta_2, \psi_2),$$

where $k(\varphi, \theta, \psi)$ is the element of K with Euler angles (φ, θ, ψ), and $\theta \in \mathbb{R}$. The middle matrix belongs to the inverse image of the *hyperbolic rotation* in the plane

(x_2, x_3):

$$x_2' := x_2 \cosh\theta + x_3 \sinh\theta,$$

$$x_3' := x_2 \sinh\theta + x_3 \cosh\theta$$

under the covering homomorphism. The parameters φ_1, θ_1, θ, φ_2, θ_2, and ψ_2 are called the *Euler angles*.

By Cartan decomposition, the matrix elements of the irreducible unitary representations of the group G have the form

$$\lambda_{\ell_1 m_1, l_2 m_2}^{\sigma,n}(\varphi_1, \theta_1, \theta, \varphi_2, \theta_2, \psi_2) = \sum_{n_1=\ell_1}^{\ell_1} \sum_{n_2=\ell_2}^{\ell_2} D_{m_1 n_1}^{\ell_1}(\varphi_1, \theta_1, 0) f_{\ell_1 n_1, \ell_2 n_2}^{\sigma,n}(\theta)$$

$$\times D_{n_2 m_2}^{\ell_2}(\varphi_2, \theta_2, \psi_2),$$

where D are Wigner D-functions, while $f_{\ell_1 n_1, \ell_2 n_2}^{\sigma,n}(\theta)$ are the matrix elements of the hyperbolic rotation. It is known from Ström (1967) that $f_{\ell_1 n_1, \ell_2 n_2}^{\sigma,n}(\theta) = 0$ if $n_1 \neq n_2$. Therefore, the last formula takes the form

$$\lambda_{\ell_1 m_1, l_2 m_2}^{\sigma,n}(\varphi_1, \theta_1, \theta, \varphi_2, \theta_2, \psi_2) = \sum_{p=-\min\{\ell_1,\ell_2\}}^{\min\{\ell_1,\ell_2\}} D_{m_1 p}^{\ell_1}(\varphi_1, \theta_1, 0) f_{\ell_1 p, \ell_2 p}^{\sigma,n}(\theta)$$

$$\times D_{pm_2}^{\ell_2}(\varphi_2, \theta_2, \psi_2).$$

The matrix elements of the hyperbolic rotation were calculated by Ström (1967, 1968). Denote

$$D_{\ell_1 p, \ell_2 p}^{\sigma,n}(\theta) := \sum_{d=\max\{0,-n/2-p\}}^{\min\{\ell_1-n/2,\ell_1-p\}} \sum_{d'=\max\{0,-n/2-p\}}^{\min\{\ell_2-n/2,\ell_2-p\}} B_{\ell_1 \ell_2 p d d'}^{\sigma,n} e^{-\lambda(2d'+p+n/2-\sigma/2)}$$

$$\times {}_2F_1\left(\ell_2 - \sigma/2, n/2 + p + d + d' + 1; \ell_1 + \ell_2 + 2; 1 - e^{-2\theta}\right),$$

where

$$B_{\ell_1 \ell_2 p d d'}^{\sigma,n} := \overline{a_{\ell_1}^{\sigma,n}} a_{\ell_2}^{\sigma,n} \frac{(-1)^{\ell_1+\ell_2-2p+d+d'}}{(\ell_1 + \ell_2 + 1)!}$$

$$\times \left[(2\ell_1 + 1)(\ell_1 - n/2)!(\ell_1 + n/2)!(\ell_1 - p)!(\ell_1 + p)!\right]^{1/2}$$

$$\times \left[(2\ell_2 + 1)(\ell_2 - n/2)!(\ell_2 + n/2)!(\ell_2 - p)!(\ell_2 + p)!\right]^{1/2}$$

$$\times \frac{(n/2 + p + d + d')!}{d!(\ell_1 - p - d)!(\ell_1 - n/2 - d)!(p + n/2 + d)!}$$

$$\times \frac{(\ell_1 + \ell_2 - n/2 - p - d - d')!}{d'!(\ell_2 - p - d')!(\ell_2 - n/2 - d')!(p + n/2 + d')!}$$

and

$$a_\ell^{\sigma,n} := \prod_{s=|n/2|}^{\ell} \frac{-2s + \sigma + 2}{(4s^2 - (\sigma + 2)^2)^{1/2}}.$$

The matrix elements of the principal series are

$$f_{\ell_1 p, \ell_2 p}^{i\varrho-2,n}(\theta) = D_{\ell_1 p, \ell_2 p}^{i\varrho-2,n}(\theta), \tag{6.28}$$

while the matrix elements of the supplementary series are

$$f_{\ell_1 p, \ell_2 p}^{-2-\varrho,n}(\theta) = \left(\frac{\Gamma(\ell_1 + 2 + \varrho)\Gamma(\ell_2 - \varrho)}{\Gamma(\ell_1 - \varrho)\Gamma(\ell_2 + 2 + \varrho)} \right)^{1/2} D_{\ell_1 p, \ell_2 p}^{-2-\varrho,n}(\theta). \tag{6.29}$$

6.3.8 The Spherical Fourier Transform

Assume the notation of Sect. 6.2. Let G be a semisimple Lie group with finite centre, let K be a maximal compact subgroup, let $T = G/K$, and let p be the standard projection from G to T. Let $\mathfrak{a}^\mathbb{C}$ be the complexification of the real vector space $\mathfrak{a}$, and let $(\mathfrak{a}^\mathbb{C})^*$ be the complex linear dual space to $\mathfrak{a}^\mathbb{C}$. The *Harish-Chandra zonal spherical function* is

$$\varphi_\lambda(g) := \int_K a(gk)^{i\lambda - \varrho} dk, \quad \lambda \in (\mathfrak{a}^\mathbb{C})^*,$$

where $a(g) \in A$ is defined by the Iwasawa decomposition $g = k(g)a(g)n(g)$ with $k(g) \in K$ and $n(g) \in N$. We have $\varphi_\lambda = \varphi_\nu$ iff there exists $w \in W$ such that $\mu = w\lambda$.

Let $T_0 = K \backslash G/K$ be the space of double cosets, and let dt be the G-bi-invariant measure on T_0. The *spherical Fourier transform* of a function $f \in L^1(T_0, dt)$ is

$$\tilde{f}(\lambda) := \int_{T_0} f(t)\overline{\varphi_\lambda(t)}\, dt.$$

The inversion formula is

$$f(t) = d^{-1} \int_{\mathfrak{a}^*} \tilde{f}(\lambda)\varphi_\lambda(t)|c(\lambda)|^{-2}\, d\lambda,$$

where d is the order of the Weyl group, and $c(\lambda) := \pi(i\lambda)/\pi(\varrho)$. The function $\pi(\lambda)$ is the product of beta functions (Subsection 6.4.1)

$$\pi(\lambda) := \prod_{\alpha \in \Sigma_+} B\big(m_\alpha/2, m_{2\alpha}/4 + 2\langle\lambda, \alpha\rangle/\langle\alpha, \alpha\rangle\big).$$

Let G have real rank one. Then there exists $\gamma \in \mathfrak{a}^*$ such that either $\Sigma = \{\gamma, -\gamma\}$ or $\Sigma = \{\gamma, -\gamma, 2\gamma, -2\gamma\}$. Let $H \in \mathfrak{a}$ be such that $\alpha(H) = 1$, and let $H_s = sH$,

$a_s = \exp(H_s)$. Let r be the length of the geodesic line connecting o with $p(a_s)$. The mapping $(r, kM) \to ka_s \cdot o$ is a diffeomorphism of $(0, +\infty) \times (K/M)$ on $T' = T \setminus \{o\}$ (geodesic polar coordinates). All the possible cases are shown in lines 2–5 of Table 2.2 (Chap. 2). The numbers α and β are

$$\alpha = (m_\gamma + 2m_{2\gamma})/2, \qquad \beta = (m_{2\gamma} - 1)/2,$$

where $m_{2\gamma} = 0$ if $2\gamma \notin \Sigma$. The Harish-Chandra zonal spherical function is

$$\varphi_\lambda(a_t) = \varphi_\lambda^{(\alpha,\beta)}(t),$$

where $\varphi_\lambda^{(\alpha,\beta)}(t)$ is the *Jacobi function*:

$$\varphi_\lambda^{(\alpha,\beta)}(t) := {}_2F_1\left(\frac{\alpha + \beta + 1 - i\lambda}{2}, \frac{\alpha + \beta + 1 + i\lambda}{2}; \alpha + 1; -\sinh^2(t)\right).$$

Put $r := a_t$. Normalise Haar measure dg on G in such a way that for any integrable bi-invariant function f we have

$$\int_G f(g)\, dg = \int_0^\infty f(r)\Delta(r)\, dr,$$

where

$$\Delta(r) := (2\sinh r)^{2\alpha+1}(2\cosh r)^{2\beta+1}.$$

It follows that the spherical Fourier transform of f equals the *Jacobi transform* of order (α, β) of the function $r \mapsto f(r)$:

$$\tilde{f}(\lambda) := \int_0^\infty f(r)\varphi_\lambda^{(\alpha,\beta)}(r)\Delta(r)\, dr.$$

The Plancherel measure is

$$d\mu(\lambda) = \frac{1}{2\pi}|c(\lambda)|^{-2}\, d\lambda$$

with

$$c(\lambda) := \frac{2^{\varrho-i\lambda}\Gamma(\alpha+1)\Gamma(i\lambda)}{\Gamma((i\lambda+\varrho)/2)\Gamma((i\lambda+\alpha-\beta+1)/2)}$$

and $\varrho = \alpha + \beta + 1$. The inversion formula is

$$f(r) = \frac{1}{2\pi}\int_0^\infty \tilde{f}(\lambda)\varphi_\lambda^{(\alpha,\beta)}(r)|c(\lambda)|^{-2}\, d\lambda. \tag{6.30}$$

If the values of α and β differ from those shown in lines 2–5 of Table 2.2, the Jacobi transform has no group-theoretic meaning. The inversion formula, however, remains true if $\alpha > -1$, $\alpha + \beta + 1 \geq 0$, and $\alpha - \beta + 1 \geq 0$.

The group ISO(n) from the first line of Table 2.2 is not semisimple. However, put $A^+ := (0, \infty)$, $o := \mathbf{0} \in T = \mathbb{R}^n$. The mapping $(kM, r) \to kr \cdot o$ is a diffeomorphism of $(K/M) \times A^+$ on $T' = \mathbb{R}^n \setminus \{\mathbf{0}\}$ (spherical coordinates). The Harish-Chandra zonal spherical function is

$$\varphi_\lambda(r) = 2^{(n-2)/2} \Gamma(n/2) \frac{J_{(n-2)/2}(\lambda r)}{(\lambda r)^{(n-2)/2}}.$$

Let $n = 2$. The Haar measure dg on ISO(2) can be normalised in such a way that for any integrable bi-invariant function f we have

$$\int_G f(g) \, dg = \int_0^\infty f(r) r \, dr.$$

The spherical Fourier transform is the *Hankel transform* of order 0 (see Subsection 6.4.2):

$$\tilde{f}(\lambda) := \int_0^\infty f(r) J_0(\lambda r) r \, dr.$$

6.3.9 Paley–Wiener Theorems

The above theorems under consideration give the intrinsic characterisation of the image of various linear spaces under the Fourier transform. For example, let $G = \mathrm{SL}(2, \mathbb{R})$, let dg be the Haar measure (3.1), and let $d\mu$ be the corresponding Plancherel measure. Consider the group $G' = \mathrm{SU}(1, 1)$, which is isomorphic to G. Let K' be the subgroup of the group G' containing the diagonal matrices with diagonal elements $e^{i\varphi}$ and $e^{-i\varphi}$. Let $\Delta = d^2/d\varphi^2$ be the Laplace operator on the group K'. Let $B_{\sigma,\varepsilon}$ be the intertwining operator between equivalent representations $\lambda^{\sigma,\varepsilon}$ and $\lambda^{-\sigma-1,\varepsilon}$ which do not belong to discrete series or its limiting points. Let $B_{\sigma,\varepsilon,+}$ be the intertwining operator between the equivalent representations $\lambda^{\sigma,\varepsilon,+}$ and $\lambda^{-\sigma-1,\varepsilon,+}$ of the first discrete series or its limiting point in the space H_σ^+ and zero operator on $L^2(K') \ominus H_\sigma^+$. Let $B_{\sigma,\varepsilon,-}$ be the intertwining operator between the equivalent representations $\lambda^{\sigma,\varepsilon,-}$ and $\lambda^{-\sigma-1,\varepsilon,-}$ of the second discrete series or its limiting point in the space H_σ^- and zero operator on $L^2(K') \ominus H_\sigma^-$.

A function $T(\sigma, \varepsilon)$, $\sigma \in \mathbb{C}$, $\varepsilon \in \{0, 1/2\}$ taking values in the space of Hilbert–Schmidt operators in $L^2(K')$ is said to belong to the *Paley–Wiener space* on the group G' if it satisfies the following conditions.

- For any nonnegative integers n_1 and n_2, the operator $\Delta^{n_1} T(\sigma, \varepsilon) \Delta^{n_2}$ is a Hilbert–Schmidt operator in the space $L^2(K')$.
- The function $\sigma \mapsto T(\sigma, \varepsilon)$ is a rapidly decreasing entire function of finite exponential type, i.e., for any nonnegative integer n, there exist two nonnegative real

numbers $C_n = C_n(T)$ and $a = a(T)$ such that

$$\left\| \sigma^n T(\sigma, \varepsilon) \right\|_2 \leq C_n e^{a|\operatorname{Im}\sigma|},$$

where $\| \cdot \|_2$ denotes the Hilbert–Schmidt norm.

- The *symmetry relations* hold true, i.e., $B_{\sigma,\varepsilon} T(\sigma, \varepsilon) = T(-\sigma - 1, \varepsilon) B_{\sigma,\varepsilon}$ if $\lambda^{\sigma,\varepsilon}$ is irreducible; $B_{\sigma,\varepsilon,\pm} T(\sigma, \varepsilon) = T(-\sigma - 1, \varepsilon) B_{\sigma,\varepsilon,\pm}$ otherwise.

Theorem 6.6 *The Fourier transform on the group G' is an isomorphism between the space $C_0^\infty(G')$ of infinitely differentiable compactly supported functions on G' and the Paley–Wiener space.*

6.4 Special Functions

In this section, we remind some definitions and formulae from the theory of special functions.

6.4.1 The Gamma Function and the Beta Function

The *gamma function* of a complex variable z with $\operatorname{Re} z > 0$ is defined by

$$\Gamma(z) := \int_0^\infty t^{z-1} e^{-t} \, dt.$$

By partial integration, we obtain

$$\Gamma(z - 1) = \frac{\Gamma(z)}{z - 1}. \tag{6.31}$$

This formula is used to extend Γ to an analytic function of $z \in \mathbb{C} \setminus \mathbb{Z}_-$, where $\mathbb{Z}_- = \{0, -1, \ldots, -n, \ldots\}$. The points $z \in \mathbb{Z}_-$ are the poles. By induction, we obtain from (6.31)

$$\frac{\Gamma(z - m)}{\Gamma(z)} = \prod_{k=1}^{m} \frac{1}{z - k}. \tag{6.32}$$

The *duplication formula* states that

$$\Gamma(2z) = \frac{2^{2z-1}}{\sqrt{\pi}} \Gamma(z) \Gamma(z + 1/2), \tag{6.33}$$

while *Euler's reflection formula* states

$$\Gamma(z) \Gamma(1 - z) = \frac{\pi}{\sin(\pi z)}. \tag{6.34}$$

The *mirror symmetry* states that

$$\Gamma(\bar{z}) = \overline{\Gamma(z)}, \tag{6.35}$$

while the *absolute value formula* states that

$$|\Gamma(iy)| = \sqrt{\frac{\pi}{y \sinh(\pi y)}}. \tag{6.36}$$

The *beta function* is defined by

$$B(a, b) := \frac{\Gamma(a)\Gamma(b)}{\Gamma(a+b)}.$$

6.4.2 The Fox H-function

Let m, n, p, and q be four nonnegative integers with $0 \le m \le q$ and $0 \le n \le p$. Let $a_1 \ldots, a_p$, $b_1, \ldots, b_q$ be points in the complex plane, and let $A_1, \ldots, A_p$, $B_1, \ldots, B_q$ be positive real numbers such that no pole of $\Gamma(b_j + B_j s)$, $j = 1, 2, \ldots, m$ coincides with any pole of $\Gamma(1 - a_k - A_k s)$, $k = 1, 2, \ldots, n$. Then there exists an infinite contour $\mathscr{L}$ that separates the poles of $\Gamma(1 - a_k - A_k s)$ at $s = A_k^{-1}(1 - a_k + j)$, $j \in \mathbb{Z}_+$ from the poles of $\Gamma(B_i s + b_i)$ at $s = B_i^{-1}(-b_i - \ell)$, $\ell \in \mathbb{Z}_+$. The *Fox H-function* is defined by

$$
\begin{aligned}
H_{p,q}^{m,n} &\left(z \left| \begin{matrix} (a_1, A_1), \ldots, (a_p, A_p) \\ (b_1, B_1), \ldots, (b_q, B_q) \end{matrix} \right. \right) \\
&:= \frac{1}{2\pi i} \int_{\mathscr{L}} \frac{\Gamma(B_1 s + b_1) \cdots \Gamma(B_m s + b_m)}{\Gamma(A_{n+1} s + a_{n+1}) \cdots \Gamma(A_p s + a_p)} z^{-s} \\
&\quad \times \frac{\Gamma(1 - a_1 - A_1 s) \cdots \Gamma(1 - a_n - A_n s)}{\Gamma(1 - b_{m+1} - B_{m+1} s) \cdots \Gamma(1 - b_q - B_q s)} \, ds.
\end{aligned}
$$

A special value of the Fox H-function, when $A_j = B_k = 1$, is called the *Meijer G-function*.

In the case of $p = q = m$, $n = 0$, and real argument $x \ge 1$, there are no poles at all inside the above contour, therefore

$$G_{p,p}^{p,0} \left(x \left| \begin{matrix} a_1, \ldots, a_m \\ b_1, \ldots, b_p \end{matrix} \right. \right) = 0, \quad x \ge 1. \tag{6.37}$$

The classical Meijer integral from two G-functions is:

$$\int_0^\infty \tau^{\alpha-1} G_{u,v}^{s,t} \left(w\tau \left| \begin{matrix} c_1, \ldots, c_t, c_{t+1}, \ldots, c_u \\ d_1, \ldots, d_s, d_{s+1}, \ldots, d_v \end{matrix} \right. \right)$$

$$\times G_{p,q}^{m,n} \left(z\tau \left| \begin{matrix} a_1, \ldots, a_n, a_{n+1}, \ldots, a_p \\ b_1, \ldots, b_m, b_{m+1}, \ldots, b_q \end{matrix} \right. \right) d\tau$$

$$= w^{-\alpha} G_{v+p,u+q}^{m+t,n+s} \left(\frac{z}{w} \left| \begin{matrix} a_1, \ldots, a_n, 1-\alpha-d_1, \ldots, 1-\alpha-d_v, a_{n+1}, \ldots, a_p \\ b_1, \ldots, b_m, 1-\alpha-c_1, \ldots, 1-\alpha-c_u, b_{m+1}, \ldots, b_q \end{matrix} \right. \right).$$

$$(6.38)$$

The symmetry relation states that

$$G_{p,q}^{m,n} \left(z \left| \begin{matrix} a_1, \ldots, a_p \\ b_1, \ldots, b_q \end{matrix} \right. \right) = G_{q,p}^{n,m} \left(\frac{1}{z} \left| \begin{matrix} 1-b_1, \ldots, 1-b_q \\ 1-a_1, \ldots, 1-a_p \end{matrix} \right. \right). \tag{6.39}$$

The argument transformation states that

$$G_{p,q}^{m,n} \left(z \left| \begin{matrix} \alpha+a_1, \ldots, \alpha+a_p \\ \alpha+b_1, \ldots, \alpha+b_q \end{matrix} \right. \right) = z^\alpha G_{p,q}^{m,n} \left(z \left| \begin{matrix} a_1, \ldots, a_p \\ b_1, \ldots, b_q \end{matrix} \right. \right). \tag{6.40}$$

Many elementary and special functions are specialised values of the Meijer G-function. For example,

$$G_{1,1}^{0,1} \left(x \left| \begin{matrix} a \\ b \end{matrix} \right. \right) = \frac{x^b}{\Gamma(a-b)} (\max\{0, x-1\})^{a-b-1}. \tag{6.41}$$

The *Jacobi polynomials* $P_n^{(\alpha,\beta)}(z)$ appear as

$$P_n^{(\alpha,\beta)}(z) = -\frac{1}{\pi} \lim_{m\to n} \frac{\sin(\pi m)\Gamma(\alpha+m+1)}{\Gamma(\alpha+\beta+m+1)} G_{2,2}^{1,2} \left(z \left| \begin{matrix} m+1, -\alpha-\beta-m \\ 0, -\alpha \end{matrix} \right. \right).$$

The value of the Jacobi polynomial at point 1 is

$$P_n^{(\alpha,\beta)}(1) = \frac{\Gamma(\alpha+n+1)}{n!\Gamma(\alpha+1)}. \tag{6.42}$$

The derivative of the Jacobi polynomial is

$$\frac{d}{dx} P_n^{(\alpha,\beta)}(x) = \frac{\alpha+\beta+n+1}{2} P_{n-1}^{(\alpha+1,\beta+1)}(x). \tag{6.43}$$

The *Gegenbauer polynomials* $C_\ell^\lambda(x)$ appear as

$$G_{2,2}^{0,2} \left(z \left| \begin{matrix} a, c \\ b, b+1/2 \end{matrix} \right. \right) = \frac{\Gamma(2b-2c+2)\vartheta(|z|-1)}{\Gamma(a-2b+c-1/2)(2(a-2b+c-1))_{2b-2c+1}}$$

$$\times z^b (z-1)^{a-2b+c-3/2} C_{2b-2c+1}^{a-2b+c-1}(\sqrt{z}),$$

where $(a)_n := \Gamma(a+n)/\Gamma(a)$ is the Pochhammer symbol, and

$$\theta(x) := \begin{cases} 1, & x \geq 0, \\ 0, & x < 0 \end{cases}$$

is the unit step function. The value of the Gegenbauer polynomial at point 1 is

$$C_\ell^\lambda(1) = \frac{\Gamma(2\lambda + \ell)}{\ell! \Gamma(2\lambda)}. \tag{6.44}$$

The *Legendre polynomials* $P_n(x)$ are just Gegenbauer polynomials with $\lambda = 0$.

The representation of the Meijer G-function through the *generalised hypergeometric functions* has the form

$$G_{p,q}^{m,n}\left(z \left| \begin{matrix} a_1, \ldots, a_p \\ b_1, \ldots, b_q \end{matrix} \right.\right)$$

$$= \sum_{k=1}^{m} \frac{\prod_{j \in \{1,2,\ldots,m\} \setminus \{k\}} \Gamma(b_j - b_k) \prod_{j=1}^{m} \Gamma(1 - a_j + b_k)}{\prod_{j=n+1}^{p} \Gamma(a_j - b_k) \prod_{j=m+1}^{q} \Gamma(1 - b_j + b_k)} z^{b_k}$$

$$\times {}_pF_{q-1}\left(\begin{matrix} 1 - a_1 + b_k, \ldots, 1 - a_p + b_k; (-1)^{p-m-n} z \\ 1 + a_1 - a_k, \ldots, 1 + a_{k-1} - a_k, 1 + a_{k+1} - a_k, \ldots, 1 + a_q - a_k \end{matrix} \right),$$

$$\tag{6.45}$$

where the generalised hypergeometric function ${}_pF_{q-1}$ is defined by the following *hypergeometric series*:

$$ {}_pF_{q-1}\left(\begin{matrix} a_1, \ldots, a_p; z \\ b_1, \ldots, b_{q-1} \end{matrix} \right) = \sum_{k=0}^{\infty} \frac{(a_1)_k \cdots (a_p)_k}{(b_1)_k \cdots (b_{q-1})_k} \frac{z^k}{k!}.$$

In particular, the following formula follows:

$$ {}_1F_2(a_1; b_1, b_2; z) = \frac{\Gamma(b_1)\Gamma(b_2)}{\Gamma(a_1)} G_{1,3}^{1,1}\left(-z \left| \begin{matrix} 1 - a_1 \\ 0, 1 - b_1, 1 - b_2 \end{matrix} \right.\right). \tag{6.46}$$

The following formula holds:

$$(\sqrt{z} + 1)^{-2a} {}_2F_1\left(a, b; 2b; 4\sqrt{z}/(\sqrt{z}+1)^2\right) = \frac{\Gamma(b+1/2)\Gamma(b-a+1/2)}{\Gamma(a)}$$

$$\times G_{2,2}^{1,1}\left(z \left| \begin{matrix} 1 - a, b - a + 1/2 \\ 0, 1/2 - b \end{matrix} \right.\right). \tag{6.47}$$

For the generalised hypergeometric function, the following formula holds:

$$ {}_pF_q(0, a_2, \ldots, a_p; b_1, \ldots, b_q; z) = 1. \tag{6.48}$$

The following formula holds for the function $_3F_2$:

$$_3F_2(1, b, c; 2, e; z) = \frac{e-1}{(b-1)(c-1)z}\left[_2F_1(b-1, c-1; e-1; z) - 1\right]. \quad (6.49)$$

The argument-simplification formula states:

$$_2F_1\left(a, b; 2b; 4z/(z+1)^2\right) = (z+1)^{2a}\,_2F_1\left(a, a-b+1/2; b+1/2; z^2\right). \quad (6.50)$$

The *Bessel functions* appear as

$$J_\nu(z) = G_{0,2}^{1,0}\left(\frac{z^2}{4}\left|\begin{matrix}\cdot\\\nu/2, -\nu/2\end{matrix}\right.\right). \quad (6.51)$$

The *parity relation* reads

$$J_{-n}(z) = (-1)^n J_n(z), \quad n \in \mathbb{Z}. \quad (6.52)$$

The Bessel function has the following series representation:

$$J_\nu(z) = \sum_{m=0}^{\infty} \frac{(-1)^m z^{2m+\nu}}{2^m m! \Gamma(\nu+m+1)}. \quad (6.53)$$

The integral transform

$$\tilde{f}(y) := \int_0^\infty f(x) J_\nu(xy) x\,dx$$

is called the *Hankel transform* of order ν. The inversion formula for Hankel transform (Vilenkin, 1968) has the form

$$f(y) = \int_0^\infty \tilde{f}(y) J_\nu(xy) y\,dy. \quad (6.54)$$

The following formula holds:

$$_1F_2(1; 2, c; z) = I_{c-2}(2\sqrt{z})\Gamma(c)z^{-c/2} + \frac{1-c}{z}. \quad (6.55)$$

Here I denotes the *modified Bessel function*

$$I_\nu(z) := e^{-\nu\pi i/2} J_\nu\left(e^{\pi i/2}z\right).$$

By Prudnikov et al. (1990, formulae 8.4.27.1 and 8.4.27.3), the *Lommel functions* are

$$s_{\mu,\nu}(2\sqrt{x}) := AG_{1,3}^{1,1}\left(x\left|\begin{matrix}(\mu+1)/2\\(\mu+1)/2, \nu/2, -\nu/2\end{matrix}\right.\right)$$

and

$$S_{\mu,\nu}(2\sqrt{x}) := B\,G_{1,3}^{3,1}\left(x\left|\begin{matrix}(\mu+1)/2\\(\mu+1)/2,\,\nu/2,\,-\nu/2\end{matrix}\right.\right),$$

where

$$A := 2^{\mu-1}\Gamma\left(\frac{\mu-\nu+1}{2}\right)\Gamma\left(\frac{\mu+\nu+1}{2}\right),$$

$$B := \frac{2^{\mu-1}}{\Gamma((1-\mu-\nu)/2)\Gamma((1-\mu+\nu)/2)}.$$

6.4.3 Harmonics

Let G be a compact topological group, and let K be a massive subgroup. Let U be a unitary representation of the group G induced by the trivial representation of the group K. Let μ be a fixed G-invariant measure on Borel sets of the homogeneous space $T = G/K$. The representation U acts on the Hilbert space $L^2(T, d\mu)$. By Frobenius reciprocity,

$$L^2(T, d\mu) = \sum_{\lambda \in \hat{G}_K} \oplus H^{(\lambda)},$$

where $H^{(\lambda)}$ is the space of the representation λ.

For example, let $N \geq 2$ be an integer, let $G = \mathrm{SO}(N)$, and let $K = \mathrm{SO}(N-1)$. The homogeneous space T is the sphere S^{N-1}. Formulae (6.14) and (6.15) show that $\lambda \in \hat{G}_K$ iff the signature of λ has the form $(\ell, 0, \ldots, 0)$, where ℓ is a nonnegative integer. Denote this representation by λ_ℓ.

Let Δ be the Laplace operator in the space $\mathbb{R}^N$. A polynomial $f \colon \mathbb{R}^n \to \mathbb{C}$ is called *harmonic* if $\Delta f = 0$. Let H_ℓ be the space of restrictions of all homogeneous harmonic polynomials of degree ℓ to the sphere S^{N-1}. The representation λ_ℓ acts on H_ℓ by

$$\left(\lambda_\ell(g)f\right)(t) = f\left(g^{-1}t\right), \quad g \in \mathrm{SO}(N),\ t \in S^{N-1}.$$

Denote by $h(S^{N-1}, \ell)$ the dimension of the representation λ_ℓ. We have

$$h\left(S^{N-1}, \ell\right) = \frac{(2\ell+N-2)(\ell+N-3)!}{(N-2)!\ell!}. \tag{6.56}$$

Let $\varphi_m^{(\ell)}$ be the spherical functions corresponding to the Gel'fand–Tsetlin basis of the chain (6.13). The functions

$$Y_{\ell m}(t) := \sqrt{\dim \lambda_\ell}\,\varphi_m^{(\ell)}(t), \quad t \in S^{N-1},$$

form a basis in the space $L^2(S^{N-1}, d\mu)$ of square-integrable functions with respect to the $\mathrm{SO}(N)$-invariant probability measure $d\mu$ on the sphere S^{N-1}. It is customary to divide these functions by

$$\sqrt{\frac{2\pi^{N/2}}{\Gamma(N/2)}}$$

to obtain a basis in the space of square-integrable functions with respect to Lebesgue measure on S^{N-1}. The functions

$$Y_{\ell m}(t) := \sqrt{\frac{h(S^{N-1}, \ell)\Gamma(N/2)}{2\pi^{N/2}}} \varphi_m^{(\ell)}(t) \tag{6.57}$$

are called *spherical harmonics*.

It is possible to give an analytic formula for spherical harmonics. Let m_0, m_1, $\ldots$, m_{N-2} be integers satisfying the condition

$$\ell = m_0 \geq m_1 \geq \cdots \geq m_{N-2} \geq 0.$$

Let $t = (t_1, t_2, \ldots, t_N)$ be a point in the space $\mathbb{R}^N$. Let

$$r_k := \sqrt{t_{k+1}^2 + t_{k+2}^2 + \cdots + t_N^2},$$

where $k = 0, 1, \ldots, N - 2$. Consider the functions

$$H(m_k, \pm, t) := \left(\frac{t_{N-1} + it_N}{r_{N-2}}\right)^{\pm m_{N-2}} r_{N-2}^{m_{N-2}} \prod_{k=0}^{N-3} r_k^{m_k - m_{k+1}}$$

$$\times C_{m_k - m_{k+1}}^{m_{k+1} + (N-k-2)/2}\left(\frac{t_{k+1}}{r_k}\right),$$

and denote

$$Y(m_k, \pm, t) := r_0^{-m} H(m_k, \pm, t).$$

The functions $Y(m_k, \pm, t)$ are orthogonal in the Hilbert space $L^2(S^{N-1})$ of the square-integrable functions on the unit sphere S^{N-1}, and the square of the length of the vector $Y(m_k, \pm, t)$ is

$$L(m_k)$$

$$= 2\pi \prod_{k=1}^{N-2} \frac{\pi 2^{k-2m_k-N+2}\Gamma(m_{k-1} + m_k + N - 1 - k)}{(m_{k-1} + (N - 1 - k)/2)(m_{k-1} - m_k)![\Gamma(m_k + (N - 1 - k)/2)]^2}.$$

The functions $Y(m_k, \pm, t)/\sqrt{L(m_k)}$ are complex-valued spherical harmonics.

Every space H_ℓ contains a spherical harmonic corresponding to the zonal spherical function. It is called the *zonal spherical harmonic*. The corresponding choice of parameters is $m_1 = \cdots = m_{N-2} = 0$. The zonal spherical harmonic depends on the point in the double coset $X = K\backslash G/K$, which is the meridian $0 \leq \theta \leq \pi$ and has the form

$$\frac{C_\ell^{(N-3)/2}(\cos\theta)}{C_\ell^{(N-3)/2}(1)}.$$

The bi-invariant probability measure on X has the form

$$d\nu = \frac{(N-1)!}{2^{N-1}\Gamma^2(N/2)}\sin^{N-1}\theta\,d\theta,$$

and the functions $\sqrt{h(S^{N-1},\ell)}\dfrac{C_\ell^{(N-3)/2}(\cos\theta)}{C_\ell^{(N-3)/2}(1)}$ form an orthonormal basis in the Hilbert space $L^2(X, d\nu)$.

Let $m = m(m_k, \pm)$ be the number of the symbol $(m_0, m_1, \ldots, m_{N-2}, \pm)$ in the lexicographic ordering. The *real-valued spherical harmonics*, $S_\ell^m(t)$, can be defined as

$$S_\ell^m(t) := \begin{cases} Y(m_k, +, t)/\sqrt{L(m_k)}, & m_{N-2} = 0, \\ \sqrt{2}\,\mathrm{Re}\,Y(m_k, +, t)/\sqrt{L(m_k)}, & m_{N-2} > 0, m = m(m_k, +), \\ -\sqrt{2}\,\mathrm{Im}\,Y(m_k, -, t)/\sqrt{L(m_k)}, & m_{N-2} > 0, m = m(m_k, -). \end{cases}$$

The *addition theorem for spherical harmonics*, see e.g. Erdélyi et al. (1953b, Chap. XI, Sect. 4, Theorem 4) reads

$$\frac{C_m^{(N-2)/2}(\cos\varphi)}{C_m^{(N-2)/2}(1)} = \frac{2\pi^{N/2}}{\Gamma(N/2)h(S^{N-1},\ell)}\sum_{m=1}^{h(S^{N-1},\ell)} S_\ell^m\left(\frac{\mathbf{x}}{\|\mathbf{x}\|}\right)S_\ell^m\left(\frac{\mathbf{y}}{\|\mathbf{y}\|}\right). \tag{6.58}$$

The above choice of the groups G and K corresponds to the first line of Table 2.1. The remaining lines can be analysed similarly. The group K is massive in G; the representations $\lambda \in \hat{G}_K$ may be enumerated by nonnegative integers ℓ. The dimension of the representation λ_ℓ is given by (2.37). All functions lying in the space $\mathcal{H}_\ell$ of the above representation are eigenfunctions of the Laplace–Beltrami operator on $T = G/K$, therefore we call them *harmonics*. The harmonics are given by (2.38).

The representation λ_ℓ in (2.38) is equivalent to λ_ℓ^+. For if not, then the sequence $\{h(T, \ell) : \ell \geq 0\}$ would contain at least two equal terms, namely $\dim\lambda_\ell$ and $\dim\lambda_\ell^+$ for some ℓ. Equation (2.37) shows that this is not the case. Therefore, it is possible to choose a basis $\{e_m : 1 \leq m \leq h(T, \ell)\}$ in such a way that the matrix elements of the representation λ_ℓ become real-valued. Therefore, the harmonics are also real-valued. We denote the real-valued harmonics by $S_\ell^m(t)$.

The space of double cosets $K\backslash G/K$ is one half of a geodesic line, and the zonal harmonic is given by (2.35).

The analysis of Table 2.2 is more challenging. For clarity, change our notation. We consider a compact group K from the third column and a massive subgroup M from the fourth column. For all five lines, the space $T = G/K$ is the sphere S^{N-1}, where N is the number from the last column. Since K in its action on S^{N-1} is contained in the group $SO(N)$, we have that $\mathcal{H}_m$, the set of restrictions of all homogeneous harmonic polynomials of degree m to S^{N-1}, is invariant under the action of K. Moreover, the chain $SO(N) \supset K$ satisfies the Gel'fand–Tsetlin condition.

Let $\alpha > -1$ and $\beta > -1$ be two real numbers, let k and ℓ be integers with $k \geq \ell \geq 0$. Put

$$R_{k,\ell}^{(\alpha,\beta)}(x,y) := \frac{P_\ell^{\alpha,\beta+k-\ell+1/2}(2y-1)}{P_\ell^{\alpha,\beta+k-\ell+1/2}(1)} y^{(k-\ell)/2} \frac{P_{k-\ell}^{\beta,\beta}(x/\sqrt{y})}{P_{k-\ell}^{\beta,\beta}(1)}$$

and define the measure $dv_{\alpha,\beta}(x,y)$ as

$$dv_{\alpha,\beta}(x,y) := \frac{\Gamma(\alpha+\beta+5/2)}{\Gamma(\alpha+1)\Gamma(\beta+1)\Gamma(1/2)}(1-y)^\alpha (y-x^2)^\beta \, dxdy.$$

The first two lines of Table 2.2 have been already analysed. In the case of the third line, consider the sphere S^{2n-1} as the unit sphere in the space $\mathbb{C}^n$. Let $\mathscr{P}_m$ be the space of all homogeneous polynomials of degree m in the variables $z_1 = x_1 + x_2 i$, $z_2 = x_3 + x_4 i, \ldots, z_n = x_{2n-1} + x_{2n} i$. Let p and q be two nonnegative integers. Denote

$$\mathscr{P}_{p,q} := \left\{ f \in \mathscr{P}_{p+q} : f(\alpha z_1, \ldots, \alpha z_n) = \alpha^p \overline{\alpha}^q f(z_1, \ldots, z_n), \alpha \in \mathbb{C} \right\}.$$

Let $\mathscr{H}_{p,q} = \mathscr{H}_{p+q} \cap \mathscr{P}_{p,q}$. Then

$$\mathscr{H}_m = \sum_{p=0}^{m} \oplus \mathscr{H}_{p,m-p}$$

and each space $\mathscr{H}_{p,m-p}$ carries the irreducible unitary representation of the group K. The representations in the spaces $\mathscr{H}_{p,m-p}$ and $\mathscr{H}_{m-p,p}$ are mutually dual and their dimensions are given by (2.44) with $\alpha = n - 1$ and $\beta = 0$. The set $X = M\backslash K/M$ has the following parameterisation:

$$X = \left\{ (r\cos\varphi, r\sin\varphi) : r \in [0,1], \varphi \in [0,2\pi] \right\}.$$

The bi-invariant probability measure on X has the form

$$dv(r,\varphi) = dv_{n-2,-1/2}(r\cos\varphi, r^2),$$

and the real part of the zonal spherical function has the form

$$R_{p,m-p}^{n-2,-1/2}(r\cos\varphi, r^2).$$

In the case of the fourth line the set $X = M\backslash K/M$ has the following parameterisation:

$$X = \left\{ (r\cos\varphi, r\sin\varphi) : r \in [0,1], \varphi \in [0,\pi] \right\}.$$

The bi-invariant probability measure on X has the form

$$d\nu(r, \varphi) = d\nu_{2n-3,1/2}(r \cos \varphi, r^2).$$

Moreover,

$$\mathscr{H}_p = \sum_q \oplus \mathscr{H}_{p,q},$$

where $p \geq q \geq 0$ are integers with $p - q$ even. The space $\mathscr{H}_{p,q}$ carries the irreducible unitary representation of the group $K = \mathrm{Sp}(n) \times \mathrm{Sp}(1)$ which is the tensor product of representations with signatures $((p + q)/2, (p - q)/2, \ldots, (p - q)/2)$ and (q). The dimension of the above representation is given by (2.44) with $\alpha = 2n - 1$ and $\beta = 1$. The zonal spherical function has the form

$$R^{2n-1,1/2}_{(p+q)/2,(p-q)/2}(r \cos \varphi, r^2).$$

The case of the fifth line differs from the previous case in the following way. The bi-invariant probability measure on X has the form

$$d\nu(r, \varphi) = d\nu_{3,5/2}(r \cos \varphi, r^2).$$

The space $\mathscr{H}_{p,q}$ carries the irreducible unitary representation of the group $K = \mathrm{Spin}(9)$ with signature $((p + q)/2, q, q, q)/2$. The dimension of the above representation is given by (2.44) with $\alpha = 7$ and $\beta = 3$. The zonal spherical function has the form

$$R^{3,5/2}_{(p+q)/2,(p-q)/2}(r \cos \varphi, r^2).$$

What happens when a unitary representation U of the group G is induced by a *not necessarily trivial* representation W of a massive subgroup K in the Hilbert space H? The representation U acts on the Hilbert space $L^2(E_W)$ of "twisted" functions (Subsection 6.3.6). The functions (2.50) form a basis in the above space. In particular, when $G = \mathrm{SO}(3)$, $K = \mathrm{SO}(2)$, and $W(\varphi) = e^{is\varphi}$, the functions (2.50) become the *spin-weighted spherical harmonics*.

6.5 Rigged Hilbert Spaces

Let H_0 be a complex Hilbert space with inner product $(\cdot,\cdot)_0$ and norm $\|\cdot\|_0$. Let H_+ be a dense linear subset of H_0. Assume that H_+ is itself a Hilbert space with respect to another inner product $(\cdot,\cdot)_+$ and the corresponding norm $\|\cdot\|_+$ satisfies the inequality

$$\|\varphi\|_0 \leq c\|\varphi\|_+, \quad \varphi \in H_+.$$

For any $f \in H_0$, let $\ell_f : H_+ \to \mathbb{C}$ be the following function:

$$\ell_f(\varphi) := (f, \varphi)_0, \quad \varphi \in H_+.$$

Then ℓ_f is an anti-linear continuous functional with zero kernel on H_+. Introduce another norm in H_0 by

$$\|f\|_- := \sup_{\varphi \in H_+ \setminus \{0\}} \frac{|(f, \varphi)_0|}{\|\varphi\|_+}.$$

Let H_- be the completion of H_0 with respect to this norm.

Let $O \colon H_+ \to H_0$ be the embedding operator. Let $I = O^*$. The bilinear form

$$(f, g)_- := (If, Ig)_+$$

is the inner product on H_- consistent with the norm $\| \cdot \|_-$.

The just-constructed rigging of the Hilbert space H_0 by the spaces H_+ and H_- is called a *rigged Hilbert space*. It is denoted by $H_- \supset H \supset H_+$.

A rigged Hilbert space can also be constructed using the pair H_0 and H_- with $\|f\|_- \leq c\|f\|_0$ for all $f \in H_0$. Indeed, the continuous bilinear form $(f, g)_-$ on $H_0 \times H_0$ has the representation

$$(f, g)_- = (Kf, g)_0, \quad f, g \in H_0,$$

where K is a continuous positive-definite linear operator in H_0. Then, the range of K is dense in H_0, because H_0 is dense in H_-. If $Kf = 0$, then for all $g \in H_0$ we have $(f, g)_- = (Kf, g)_0 = 0$, i.e., $f = 0$. Therefore, there exists the inverse operator K^{-1} mapping the range of K on H_0. Let H_+ be the completion of the range of K with respect to the inner product

$$(\varphi, \psi)_+ := \left(K^{-1}\varphi, \psi \right)_0.$$

The constructed chain $H_- \supset H \supset H_+$ is a rigged Hilbert space.

Example 6.7 Let R be a set, let $\mathfrak{R}$ be a σ-field of subsets of R, and let μ be a measure on $(R, \mathfrak{R})$. Put $H_0 := L^2(R, d\mu)$. Let $K(x, y) \colon R \times R \to \mathbb{C}$ be a bounded $\mathfrak{R} \times \mathfrak{R}$-measurable function. It is called a *positive-definite kernel* if for all $f \in H_0$ we have

$$0 \leq \int_R \int_R K(x, y) f(x) \overline{f(y)} \, d\mu(x) \, d\mu(y) < \infty.$$

Put

$$(f, g)_- := \int_R \int_R K(x, y) f(x) \overline{g(y)} \, d\mu(x) \, d\mu(y).$$

If there exist $f \in H_0 \setminus \{0\}$ with $(f, f)_- = 0$, consider the quotient space of H_0 with respect to the linear subspace of all such f and denote it again by H_0. Denote by H_- the completion of the space H_0 with respect to the above inner product. The chain $H_- \supset H \supset H_+$ may be constructed as above.

6.6 Abelian and Tauberian Theorems

Let X be a set, let $\mathfrak{X}$ be a σ-field of subsets of X, and let μ be a measure on $\mathfrak{X}$. Let Y be another set, and let $K(x, y) \colon X \times Y \to \mathbb{C}$ be such a function, that for any $y_0 \in Y$ the function $K(x, y_0) \colon X \to \mathbb{C}$ is measurable. Let $f \colon X \to \mathbb{C}$ be another measurable function. Put

$$\hat{f}(y) := \int_X K(x, y) f(x) \, d\mu(x), \tag{6.59}$$

where the integral in the right-hand side may converge absolutely, conditionally, or in some generalised sense.

An *Abelian theorem* is a theorem that states the following: "*if $f(x)$ has some kind of asymptotic behaviour, then $\hat{f}(y)$ has another kind of asymptotic behaviour*". A *Tauberian theorem* is a theorem that states the following: "*if $\hat{f}(y)$ has another kind of asymptotic behaviour **and** $f(x)$ **satisfies a Tauberian condition**, then $f(x)$ has some kind of asymptotic behaviour*".

We are interested in Abelian and Tauberian theorems in the following situation. Let X be the set of all positive integers, let $\mathfrak{X}$ be the σ-field of all subsets of the set X. Let Y be the interval $[-1, 1]$. Let α and β be two real numbers with $\alpha \geq \beta \geq -1/2$. Let $P_m^{(\alpha,\beta)}(x)$ be Jacobi polynomials. Define

$$K(\ell, y) := R_\ell^{(\alpha,\beta)}(y) = \frac{P_\ell^{(\alpha,\beta)}(y)}{P_\ell^{(\alpha,\beta)}(1)}.$$

Let $\mu^{(\alpha,\beta)}$ be the following probability measure on the interval $[0, \pi]$:

$$d\mu^{(\alpha,\beta)}(\theta) := \frac{\Gamma(\alpha+\beta+2)}{\Gamma(\alpha+1)\Gamma(\beta+1)} \sin^{2\alpha+1}(\theta/2) \cos^{2\beta+1}(\theta/2) \, d\theta.$$

Then we have

$$\int_0^\pi R_{\ell_1}^{(\alpha,\beta)}(\cos\theta) R_{\ell_2}^{(\alpha,\beta)}(\cos\theta) \, d\mu^{(\alpha,\beta)}(\theta) = \frac{\delta_{\ell_1\ell_2}}{w_{\ell_1}^{(\alpha,\beta)}},$$

where

$$w_\ell^{(\alpha,\beta)} := \frac{\Gamma(\beta+1)(2\ell+\alpha+\beta+1)\Gamma(\ell+\alpha+\beta+1)\Gamma(\ell+\alpha+1)}{\Gamma(\alpha+1)\Gamma(\alpha+\beta+2)\ell!\Gamma(\ell+\beta+1)}. \tag{6.60}$$

Here $(2\ell+\alpha+\beta+1)\Gamma(\ell+\alpha+\beta+1)$ is to be interpreted as 1 if $\ell = 0$ and $\alpha = \beta = -1/2$.

Let $f(\ell) = c_\ell$ be a sequence of real numbers. Put $\mu(\ell) := \ell^{-2\alpha-2} w_\ell^{(\alpha,\beta)}$. The integral in the right-hand side of (6.59) is the following formal series:

$$f(\theta) := \sum_{\ell=1}^\infty \frac{c_\ell w_\ell^{(\alpha,\beta)}}{\ell^{2\alpha+2}} R_\ell^{(\alpha,\beta)}(\cos\theta). \tag{6.61}$$

Put $c_\ell := \ell^\sigma$, $\sigma > 0$. If $0 < \sigma < \alpha + 1/2$, then the series in the right-hand side of (6.61) is absolutely convergent. If $\alpha + 1/2 \le \sigma \le \alpha + 3/2$, then the above series is conditionally convergent. Consider the following function:

$$f_r(\theta) := \sum_{\ell=1}^{\infty} \frac{r^\ell \ell^\sigma w_\ell^{(\alpha,\beta)}}{\ell^{2\alpha+2}} R_\ell^{(\alpha,\beta)}(\cos\theta), \quad 0 < r < 1.$$

If $\alpha + 3/2 \le \sigma < 2\alpha + 2$, then $f_r(\theta)$ converges as $r \uparrow 1$ to a function $f(\theta) \in L^1([0,\pi], \mathrm{d}\mu^{(\alpha,\beta)})$. In other words, the series (6.61) is *Abel summable*.

To formulate results, we need a definition. A measurable function $L : [x_0, \infty) \to \mathbb{R}$ is called *slowly varying at infinity* if for all $a > 0$,

$$\lim_{x \to \infty} \frac{L(ax)}{L(x)} = 1.$$

In what follows, L denotes a slowly varying function.

The Abelian theorem under absolute convergence is as follows.

Theorem 6.8 *If $0 < \sigma < \alpha + 1/2$ and $c_\ell \sim \ell^\sigma L(\ell)$ as $\ell \to \infty$, then the series (6.61) is absolutely convergent and*

$$f(\theta) \sim \frac{2^\sigma \Gamma(\beta+1)\Gamma(\sigma/2)}{\Gamma(\alpha+\beta+2)\Gamma(\alpha+1-\sigma/2)} \frac{L(\theta^{-1})}{\theta^\sigma} \tag{6.62}$$

as $\theta \downarrow 0$.

In what follows we write $a_\ell \sim b_\ell$ as $\ell \to \infty$ as short for

$$\lim_{\ell \to \infty} \frac{a_\ell}{b_\ell} = 1.$$

Similarly, we write $f(\theta) \sim g(\theta)$ as $\theta \downarrow 0$ for

$$\lim_{\theta \downarrow 0} \frac{f(\theta)}{g(\theta)} = 1.$$

The Tauberian theorem under absolute convergence is as follows.

Theorem 6.9 *Let $f(\theta)$ be the sum of the absolutely convergent Jacobi series (6.61) satisfying the asymptotic relation (6.62). Assume that the coefficients c_ℓ satisfy any of the following Tauberian conditions:*

1. *There exists a function $\omega(\lambda) : (1, +\infty) \to \mathbb{R}$ such that*

$$\limsup_{\ell \to \infty} \max_{\ell \le m \le \lambda \ell} \frac{|(\ell/m)^\sigma c_m - c_\ell|}{\ell^\sigma L(\ell)} \le \omega(\lambda), \quad \lambda > 1,$$

$$\lim_{\lambda \downarrow 1} \omega(\lambda) = 1.$$

2. *There exists a function* $\omega(\lambda): (1, +\infty) \to [0, \infty)$ *such that*

$$\liminf_{\ell \to \infty} \min_{\ell \leq m \leq \lambda \ell} \left(\frac{\pm c_m}{m^\sigma L(m)} - \frac{\pm c_\ell}{\ell^\sigma L(\ell)} \right) \geq -\omega(\lambda), \quad \lambda > 1,$$

$$\lim_{\lambda \downarrow 1} \omega(\lambda) = 1.$$

3. $c_\ell \geq 0$ *for sufficiently large* ℓ *and there exists a function* $\omega(\lambda): (1, +\infty) \to [0, \infty)$ *such that*

$$\liminf_{\ell \to \infty} \min_{\ell \leq m \leq \lambda \ell} (c_m - c_\ell) \geq -\omega(\lambda), \quad \lambda > 1,$$

$$\lim_{\lambda \downarrow 1} \omega(\lambda) = 1.$$

Then $c_\ell \sim \ell^\sigma L(\ell)$ *as* $\ell \to \infty$.

In order to formulate the Abelian theorem under conditional convergence, we need a definition.

Definition 6.10 A positive function $f(x)$ is *quasi-monotone* if it is of bounded variation on compact subsets of $[0, \infty)$ and there exists $\eta > 0$ such that

$$\int_0^x y^\eta |df(y)| = O(x^\eta f(x)), \quad x \to \infty.$$

To check that a function is quasi-monotone, one can use the Quasi-Monotonicity Theorem by Bojanic and Karamata (1963):

Theorem 6.11 *Let* f *be slowly varying and of bounded variation on compact subsets of* $[0, \infty)$. *Then* f *is quasi-monotone iff there exist two positive nondecreasing functions* f_1 *and* f_2 *with*

$$f_i(2x) = O(f_i(x)), \quad i = 1, 2$$

as $x \to \infty$ *such that*

$$f(x) = \frac{f_1(x)}{f_2(x)}, \quad x \geq 0.$$

Theorem 6.12 *Assume that there exist a real number* $\sigma \in [\alpha + 1/2, \alpha + 3/2)$ *and a slowly varying and quasi-monotone function* $L(x)$ *such that*

$$c_\ell = \ell^\sigma L(\ell).$$

Then the series (6.61) *is conditionally convergent and its sum satisfies* (6.62).

The Tauberian theorem under conditional convergence is as follows.

Theorem 6.13 *Let $f(\theta)$ be the sum of the conditionally convergent Jacobi series (6.61) satisfying the asymptotic relation (6.62). Assume that the coefficients c_ℓ satisfy the following Tauberian condition: $c_\ell \geq 0$ for sufficiently large ℓ and there exists a function $\omega(\lambda): (1, +\infty) \to [0, \infty)$ such that*

$$\liminf_{\ell \to \infty} \min_{\ell \leq m \leq \lambda \ell} (c_m - c_\ell) \geq -\omega(\lambda), \quad \lambda > 1,$$

$$\lim_{\lambda \downarrow 1} \omega(\lambda) = 1.$$

Then $c_\ell \sim \ell^\sigma L(\ell)$ as $\ell \to \infty$.

In order to formulate the Abelian theorem under Abel summability, consider the following formal series:

$$f_r(\theta) := \sum_{\ell=1}^{\infty} r^\ell \ell^{\sigma - 2\alpha - 2} L(\ell) w_\ell^{(\alpha, \beta)} R_\ell^{(\alpha, \beta)}(\cos \theta).$$

If σ lies in the domain of Abel summability, then $\alpha + 3/2 \leq \sigma < 2\alpha + 2$. There exists a unique positive integer m with $\alpha + m + 1/2 \leq \sigma < \alpha + m + 3/2$. Denote $L_0(\ell) := L(\ell)$ and $L_k(\ell) := L_{k-1}(\ell + 1) - L_{k-1}(\ell)$, $k \geq 1$.

Theorem 6.14 *Assume that $\alpha + 3/2 \leq \sigma < 2\alpha + 2$, that the sequence $\ell^k L_k(\ell)$ is slowly varying and there exist functions $\omega_k(\lambda): (1, +\infty) \to [0, \infty)$ such that*

$$\limsup_{\ell \to \infty} \max_{\ell \leq m \leq \lambda \ell} \left| (\ell/m)^\sigma m^k L_k(m) - \ell^k L_k(\ell) \right| \leq \omega(\lambda), \quad \lambda > 1,$$

$$\lim_{\lambda \downarrow 1} \omega(\lambda) = 1,$$

for $0 \leq k \leq m$. Then $f_r(\theta)$ converges as $r \uparrow 1$ to a function $f(\theta)$ satisfying (6.62).

The Tauberian theorem under Abel summability is as follows.

Theorem 6.15 *Let f be an integrable function with respect to the measure $\mu^{(\alpha, \beta)}$ satisfying (6.62) with $\alpha + 3/2 < \sigma < 2\alpha + 2$. Then*

$$\int_{-1}^{1} f(x) R_\ell^{(\alpha, \beta)}(x) \, d\mu^{(\alpha, \beta)}(x) \sim \ell^{\sigma - 2\alpha - 2} L(\ell)$$

as $\ell \to \infty$.

Consider the limiting case of $\sigma = 0$. The Abelian theorem is as follows.

Theorem 6.16 *Consider the series*

$$f(\theta) := \sum_{\ell=1}^{\infty} L(\ell) w_\ell^{(\alpha, \beta)} \ell^{-2\alpha - 2} R_\ell^{(\alpha, \beta)}(\cos \theta).$$

If the series $\sum_{\ell=1}^{\infty} L(\ell)/\ell$ diverges, then

$$f(\theta) \sim \frac{2\Gamma(\beta+1)}{\Gamma(\alpha+\beta+2)\Gamma(\alpha+1)} \int_0^{\theta^{-1}} \frac{L(u)\,\mathrm{d}u}{u}$$

as $\theta \downarrow 0$.

6.7 Bibliographical Remarks

Manifolds are classical objects of studies in differential geometry. There exist two basic approaches to manifolds. Our treatment is based on the classical book by Helgason (2001). Yet another approach is based on the result by Whitney (1936): any manifold of dimension m and of class C^r with $r \geq 1$ is homeomorphic with a manifold lying in $\mathbb{R}^{2m+1}$. Therefore, we do not need to use any charts, atlases and so on. This approach may be found, for example, in Adler and Taylor (2007).

Vector bundles are another object of investigations in differential geometry. We mention the classical books by Kobayashi and Nomizu (1996a,b).

Differential operators on manifolds are considered in classical books by Helgason (2000, 2008).

There is a vast literature in *Lie groups*, *Lie algebras*, and *spherical functions*. We base our considerations on Helgason (2001), but would like to mention also Adams (1969, 1996), Chevalley (1946, 1951, 1955), Gangolli and Varadarajan (1988), Gorbatsevich et al. (1997), Jacobson (1979), Knapp (2001, 2002), and Varadarajan (1984, 1999).

Many books are devoted to *group actions*. We mention three classical books by Bredon (1972), Montgomery and Zippin (1974), and Pontryagin (1966).

The precise link between the theory of positive-definite functions and the theory of group representations is described in Naĭmark (1972). The theory of group representations is considered in many excellent books. We mention Borel (1991), Fulton and Harris (1991), Goodman and Wallach (2009), Howe and Tan (1992), Humphreys (1978), Lang (1985), Onishchik and Vinberg (1990), Procesi (2007), Rossmann (2002), Vogan (1987), Wallach (1988, 1992), and Zhelobenko (1973, 2006). Applications of the theory of group representations to probability and statistics were considered in books by Diaconis (1988) and Hannan (1965).

The Gel'fand–Tsetlin basis for the chains satisfying the Gel'fand–Tsetlin condition was proposed by Gel'fand and Tsetlin (1950a,b). This basis may be used for constructing harmonics on the spaces S^n, $\mathbb{R}P^n$, and $\mathbb{C}P^n$ of Table 2.1. In the case of the space $\mathbb{H}P^n$, the group $\mathrm{Sp}(n) \times \mathrm{Sp}(1)$ is *not* massive in $\mathrm{Sp}(n+1)$. The multiplicities of representations for this case have been calculated by Lepowsky (1971). The remaining case of the projective plane over octonions has been partly solved by Lepowsky (1971), Camporesi (2005), and Mashimo (2006).

The *Clebsch–Gordan coefficients* for the groups SO(3) and SU(2) appeared in quantum mechanics, see Varshalovich et al. (1988). Their values depend on phase convention. The Clebsch–Gordan coefficients for the group SU(n) appear in the construction of unifying theories. Their numerical calculation was performed in Alex et al. (2011).

Theorem 2.18 and Examples 2.21, 6.4, and 6.5 show that it is very important to have a complete description of the *unitary dual* $\hat{G}$ of a given topological group G of type I, i.e., the set of all irreducible unitary representations of G. The description of the unitary dual of the group $\mathrm{SL}(2, \mathbb{C})$ was given by Gel'fand and Naĭmark (1946), while that of the group $\mathrm{SL}(2, \mathbb{R})$ was found by Bargmann (1947). Note that our parameterisation of the unitary dual differs from that of Kostant (1975).

An interesting still unsolved problem is the description of the unitary dual of real reductive Lie groups, i.e. such Lie groups for which the corresponding Lie algebra is a direct sum of a commutative and a semisimple Lie algebra. Any semisimple Lie algebra is a sum of simple Lie algebras. Complex simple Lie algebras were classified by Killing (1888, 1889a,b, 1890). He found four infinite series of classical complex simple Lie algebras and five exceptional algebras. Real simple Lie algebras were classified by Cartan (1914). He found 17 infinite series and 22 exceptional algebras. For any simple Lie algebra $\mathfrak{g}$, there exists a unique simply connected group G that has $\mathfrak{g}$ as its Lie algebra. The other groups with this property are quotient groups of G with respect to the subgroups of the centre of G.

It is known that any irreducible unitary representation λ of a real reductive group G is *admissible*, i.e., the restriction of λ to the maximal compact subgroup K of the group G contains only finitely many copies of any irreducible unitary representation of K. In 1973, R.P. Langlands published a preprint (reprinted in Langlands, 1989), where he classified all admissible representations of real reductive Lie groups. It is easy to determine which of the above representations admit a nontrivial invariant sesquilinear (linear in one argument and anti-linear in the other) form. It remains to find when the above form is positive-definite. This was done by Barbasch (1989) for the case of complex classical Lie groups. In the general case, computers are widely use for calculations. One of the intermediate steps in these calculations has been described by Vogan (2007).

The Plancherel theorem for the case of the group $G = \mathbb{R}$ was proved by Plancherel (1910). Berry (1929) generalised the Plancherel theorem to the case of $G = \mathbb{R}^n$, while Weil (1940) proved the above theorem for commutative locally compact groups. An extension to the case of a separable unimodular group is due to Segal (1950). The explicit determination of the Plancherel measure for the case of $G = \mathrm{SL}(2, \mathbb{C})$ was performed by Gel'fand and Naĭmark (1947), for the case of $G = \mathrm{SL}(n, \mathbb{C})$ by Gel'fand and Naĭmark (1948), for the case of a complex semisimple Lie group by Harish-Chandra (1951), for the case of $G = \mathrm{SL}(2, \mathbb{R})$ by Harish-Chandra (1952), and for a wide class of groups including all semisimple Lie groups by Harish-Chandra (1975, 1976a,b).

Induced representations were introduced by Frobenius (1898). Wigner (1939) used the method of induced representation to construct the irreducible unitary representations of the inhomogeneous Lorentz group. Later on, this method was used by Gel'fand and Naĭmark (1946) and Bargmann (1947) for the case of the groups $\mathrm{SL}(2, \mathbb{C})$ and $\mathrm{SL}(2, \mathbb{R})$. Gel'fand and Naĭmark (1950) used the above method to construct almost all (with respect to the Plancherel measure) irreducible unitary representations of complex classical groups. Bruhat (1956) constructed irreducible unitary representations of real Lie groups. For an introduction to induced representations, see Barut and Rączka (1986).

The Plancherel formula for spherical Fourier transform was proved by Mautner (1951). The Plancherel measure was determined by Harish-Chandra (1966).

Paley–Wiener theorems for the Fourier transform on $\mathbb{R}$ appeared in Paley and Wiener (1934), hence the name. The Paley–Wiener theorem on the group $SL(2, \mathbb{R})$ (Theorem 6.6) was proved by Ehrenpreis and Mautner (1955, 1957, 1959).

There is nothing "special" about *special functions*. The first examples of special functions appeared in the early 18th century as solutions to differential equations of mathematical physics and results of integration. The most complete account of the theory of special functions by the middle of the past century is contained in the publications of the Bateman project, Erdélyi et al. (1953a,b, 1954a,b, 1955).

At first glance, the theory of special functions is a chaotic collection of formulae. One possible way to unify the subject is as follows. Many special functions appear when one solves a partial differential equation by the method of separation of variables in a special coordinate system. When the variables are separated, the coordinate space splits into manifolds. In many interesting cases these manifolds are homogeneous spaces of some Lie group G, while the solutions of the partial differential equations under consideration are eigenfunctions of G-invariant differential operators. Moreover, G acts on the space of all eigenfunctions that correspond to the same eigenvalue by linear transformation. Thus, the above space carries a representation of the group G. The representation theory gives a short way to establish various properties of special functions.

Separation of variables is discussed by Miller (1968, 1977). A treatment of the theory of special functions from the point of view of representation theory is given by Vilenkin (1968) and Vilenkin and Klimyk (1991, 1993, 1994, 1995).

It is clear that limit formulae, integration, and summation for both elementary and special functions are very important for spectral theory of random fields. Modern treatments of these questions are contained in Prudnikov et al. (1986, 1988, 1990, 1992a,b) and Brychkov (2008).

The general theory of spherical harmonics on compact homogeneous spaces was developed by Cartan (1929), Weyl (1934), and Weil (1940). The case of the sphere S^{N-1} is described in detail in Erdélyi et al. (1953a).

Let H_m be the space of restrictions of all homogeneous harmonic polynomials in N variables of a degree m to the sphere S^{N-1}. The representation λ_m of the group $SO(N)$ acts on H_m. The restriction of λ_m to the group K (Table 2.2, column 3) is a direct sum of irreducible unitary representations of K. The description of the spaces $H_{p,q}$ of the above representations (Subsection 6.4.3) is taken from Johnson and Wallach (1977) for the case of hyperbolic spaces over complex numbers and quaternions, and from Johnson (1976) for the case of the hyperbolic plane over octonions.

Rigged Hilbert spaces were introduced by Gel'fand and Kostyučenko (1955), hence their other name, *Gel'fand triples*.

Let $a_0, a_1, \ldots, a_n, \ldots$ be a sequence of complex numbers. Abel (1826) proved the *Abel continuity theorem*: if $\sum_{n=0}^{\infty} a_n = s$, then

$$\lim_{r \uparrow 1} \sum_{n=0}^{\infty} a_n r^n = s.$$

The example of the sequence $a_n = (-1)^n$ shows that the converse does not hold. Tauber (1897) proved that under Tauberian condition $\lim_{n \to \infty} n a_n = 0$ the converse theorem is true. Hardy (1910) asked whether the above Tauberian condition can be replaced by the weaker condition that $a_n = O(n^{-1})$. Littlewood (1911) proved that this is indeed the case. The names *Abelian and Tauberian theorems* were proposed by Hardy and Littlewood (1913). An excellent survey of Tauberian theory has been written by Wiener (1930). Abelian and Tauberian theorems for Jacobi series were proved by Bingham (1978). See also the historical survey by Korevaar (2004).

Abelian and Tauberian theorems are widely used in probability. See the books by Bingham et al. (1987) and Seneta (1976) and papers by Leonenko and Olenko (1991, 1992, 1993), Olenko (1996, 2006, 2007a,b), and Olenko and Klykavka (2006a,b).

References

N.H. Abel. Untersuchungen über die Reihe $1 + \frac{m}{1}x + \frac{m(m-1)}{1\cdot2}x^2 + \frac{m(m-1)(m-2)}{1\cdot2\cdot3}x^3 + \cdots$. *J. Reine Angew. Math.*, 1:311–339, 1826.

J.F. Adams. *Lectures on Lie groups*. Benjamin, New York, 1969.

J.F. Adams. *Lectures on exceptional Lie groups*. Chicago Lect. Math. University of Chicago Press, Chicago, 1996.

R.J. Adler and J.E. Taylor. *Random fields and geometry*. Springer Monogr. Math. Springer, New York, 2007.

A. Alex, M. Kalus, A. Huckleberry, and J. von Delft. A numerical algorithm for the explicit calculation of SU(n) and SL$(n, \mathbb{C})$ Clebsch–Gordan coefficients. *J. Math. Phys.*, 52:21 p., 2011. doi:10.1063/1.3521562.

D. Barbasch. The unitary dual for complex classical Lie groups. *Invent. Math.*, 96(1):103–176, 1989.

V. Bargmann. Irreducible unitary representations of the Lorentz group. *Ann. of Math. (2)*, 48(3):568–640, 1947. URL http://www.jstor.org/stable/1969129.

A.O. Barut and R. Rączka. *Theory of group representations and applications*. World Scientific, Singapore, second edition, 1986.

A.C. Berry. The Fourier transform theorem. *J. Math. Phys.*, 8:106–118, 1929.

N.H. Bingham. Tauberian theorems for Jacobi series. *Proc. Lond. Math. Soc. (3)*, 36:285–309, 1978.

N.H. Bingham, C.M. Goldie, and J.L. Teugels. *Regular variation*, volume 27 of *Encycl. Math. Appl.* Cambridge University Press, Cambridge, 1987.

R. Bojanic and J. Karamata. On slowly varying functions and asymptotic relations. Technical Report 432, Math. Research Center, 1963.

A. Borel. *Linear algebraic groups*, volume 126 of *Graduate Texts in Math.* Springer, New York, second edition, 1991.

G.E. Bredon. *Introduction to compact transformation groups*, volume 46 of *Pure Appl. Math.* Academic Press, New York, 1972.

F. Bruhat. Sur les représentations induites des groupes de Lie. *Bull. Soc. Math. Fr.*, 84:97–205, 1956. URL http://www.numdam.org/item?id=BSMF_1956__84__97_0.

Yu.A. Brychkov. *Handbook of special functions. Derivatives, integrals, series and other formulas.* CRC Press, Boca Raton, 2008.

R. Camporesi. The Helgason Fourier transform for homogeneous vector bundles over compact Riemannian symmetric spaces—the local theory. *J. Funct. Anal.*, 220:97–117, 2005.

E. Cartan. Les groupes réels simples finis et continus. *Ann. Sci. École Norm. Supérieure*, 31:263–355, 1914.

E. Cartan. Sur la détermination d'un système orthogonal complet dans un espace de Riemann symmétrique clos. *Rend. Circ. Mat. Palermo*, 53:217–252, 1929.

C. Chevalley. *Theory of Lie Groups. I.*, volume 8 of *Princeton Math. Ser.* Princeton University Press, Princeton, 1946.

C. Chevalley. *Théorie des groupes de Lie. Tome II. Groupes algébriques*, volume 1152 of *Actualités Sci. Ind.* Hermann & Cie, Paris, 1951.

C. Chevalley. *Théorie des groupes de Lie. Tome III. Théorèmes généraux sur les algèbres de Lie*, volume 1226 of *Actualités Sci. Ind.* Hermann & Cie, Paris, 1955.

P. Diaconis. *Group representations in probability and statistics*, volume 11 of *IMS Lect. Notes Monogr. Ser.* Institute of Mathematical Statistics, Hayward, 1988.

L. Ehrenpreis and F.I. Mautner. Some properties of the Fourier transform on semi-simple Lie groups. I. *Ann. of Math. (2)*, 61:406–439, 1955. URL http://www.jstor.org/stable/1969808.

L. Ehrenpreis and F.I. Mautner. Some properties of the Fourier transform on semi-simple Lie groups. II. *Trans. Am. Math. Soc.*, 84:1–55, 1957. URL http://www.jstor.org/stable/1992890.

L. Ehrenpreis and F.I. Mautner. Some properties of the Fourier transform on semi-simple Lie groups. III. *Trans. Am. Math. Soc.*, 90:431–484, 1959.

A. Erdélyi, W. Magnus, F. Oberhettinger, and F.G. Tricomi. *Higher transcendental functions, volume I.* McGraw-Hill, New York, 1953a.

A. Erdélyi, W. Magnus, F. Oberhettinger, and F.G. Tricomi. *Higher transcendental functions, volume II.* McGraw-Hill, New York, 1953b.

A. Erdélyi, W. Magnus, F. Oberhettinger, and F.G. Tricomi. *Tables of integral transforms, volume I.* McGraw-Hill, New York, 1954a.

A. Erdélyi, W. Magnus, F. Oberhettinger, and F.G. Tricomi. *Tables of integral transforms, volume II.* McGraw-Hill, New York, 1954b.

A. Erdélyi, W. Magnus, F. Oberhettinger, and F.G. Tricomi. *Higher transcendental functions, volume III.* McGraw-Hill, New York, 1955.

G. Frobenius. Über Relationen zwischen den Charakteren einer Gruppe und denen ihrer Untergruppen. *Sitzber. Preuss. Akad. Wiss.*, pages 501–515, 1898.

W. Fulton and J. Harris. *Representation theory. A first course*, volume 129 of *Graduate Texts in Math.* Springer, New York, 1991.

R. Gangolli and V.S. Varadarajan. *Harmonic analysis of spherical functions on real reductive groups*, volume 101 of *Ergeb. Mat. Grenzgeb.* Springer, Berlin, 1988.

I.M. Gel'fand and A.G. Kostyučenko. Expansion in eigenfunctions of differential and other operators. *Dokl. Akad. Nauk SSSR (N.S.)*, 103:349–352, 1955. In Russian.

I.M. Gel'fand and M.A. Naĭmark. Unitary representations of the Lorentz group. *Acad. Sci. USSR. J. Phys.*, 10:93–94, 1946.

I.M. Gel'fand and M.A. Naĭmark. Unitary representations of the Lorentz group. *Izv. Akad. Nauk SSSR, Ser. Mat.*, 11:411–504, 1947. In Russian.

I.M. Gel'fand and M.A. Naĭmark. The analogue of Plancherel's formula for the complex unimodular group. *Dokl. Akad. Nauk SSSR (N.S.)*, 63:609–612, 1948. In Russian.

I.M. Gel'fand and M.A. Naĭmark. *Unitary representations of the classical groups*, volume 36 of *Trudy Mat. Inst. Steklov.* Izdat. Acad. Nauk SSSR, Moscow–Leningrad, 1950. In Russian.

I.M. Gel'fand and M.L. Tsetlin. Finite-dimensional representations of the group of unimodular matrices. *Dokl. Akad. Nauk SSSR (N.S.)*, 71:825–828, 1950a. In Russian.

I.M. Gel'fand and M.L. Tsetlin. Finite-dimensional representations of groups of orthogonal matrices. *Dokl. Akad. Nauk SSSR (N.S.)*, 71:1017–1020, 1950b. In Russian.

R. Goodman and N.R. Wallach. *Symmetry, representations, and invariants*, volume 255 of *Graduate Texts in Math.* Springer, Dordrecht, 2009.

V.V. Gorbatsevich, A.L. Onishchik, and E.B. Vinberg. *Foundations of Lie theory and Lie transformation groups.* Springer, Berlin, 1997.

E.J. Hannan. *Group representations and applied probability*, volume 3 of *Methuen's Suppl. Rev. Ser. Appl. Probab.* Methuen, London, 1965.

G.H. Hardy. Theorems related to the summability and convergence of slowly oscillating series. *Proc. Lond. Math. Soc. (2)*, 8:301–320, 1910.

G.H. Hardy and J.E. Littlewood. Tauberian theorems concerning series of positive terms. *Messenger of Math.*, 42:191–192, 1913.

Harish-Chandra. Plancherel formula for complex semi-simple Lie groups. *Proc. Natl. Acad. Sci. USA*, 37:813–818, 1951.

Harish-Chandra. Plancherel formula for the 2×2 real unimodular group. *Proc. Natl. Acad. Sci. USA*, 38:337–342, 1952.

Harish-Chandra. Discrete series for semisimple Lie groups. II. Explicit determination of the characters. *Acta Math.*, 116:1–111, 1966.

Harish-Chandra. Harmonic analysis on real reductive groups. I. The theory of the constant term. *J. Funct. Anal.*, 19:104–204, 1975.

Harish-Chandra. Harmonic analysis on real reductive groups. II. Wavepackets in the Schwartz space. *Invent. Math.*, 36:1–55, 1976a.

Harish-Chandra. Harmonic analysis on real reductive groups. III. The Maass–Selberg relations and the Plancherel formula. *Ann. of Math. (2)*, 104(1):117–201, 1976b.

S. Helgason. *Groups and geometric analysis. Integral geometry, invariant differential operators, and spherical functions*, volume 83 of *Math. Surv. Monogr.* Am. Math. Soc., Providence, 2000. Corrected reprint of the 1984 original.

S. Helgason. *Differential geometry, Lie groups, and symmetric spaces*, volume 34 of *Graduate Studies in Math.* Am. Math. Soc., Providence, 2001. Corrected reprint of the 1978 original.

S. Helgason. *Geometric analysis on symmetric spaces*, volume 39 of *Math. Surv. Monogr.* Am. Math. Soc., Providence, second edition, 2008.

R. Howe and E.-C. Tan. *Nonabelian harmonic analysis. Applications of* $SL(2, \mathbb{R})$. Universitext. Springer, New York, 1992.

J.E. Humphreys. *Introduction to Lie algebras and representation theory*, volume 9 of *Graduate Texts in Math.* Springer, New York, 1978.

N. Jacobson. *Lie algebras.* Dover, New York, 1979.

K.D. Johnson. Composition series and intertwining operators for the spherical principal series. II. *Trans. Am. Math. Soc.*, 215:269–293, 1976. URL http://www.jstor.org/stable/1999726.

K.D. Johnson and N.R. Wallach. Composition series and intertwining operators for the spherical principal series. I. *Trans. Am. Math. Soc.*, 229:137–173, 1977.

W. Killing. Die Zusammensetzung der stetigen endlichen Transformationsgruppen. I. *Math. Ann.*, 31:252–290, 1888.

W. Killing. Die Zusammensetzung der stetigen endlichen Transformationsgruppen. II. *Math. Ann.*, 33:1–48, 1889a.

W. Killing. Die Zusammensetzung der stetigen endlichen Transformationsgruppen. III. *Math. Ann.*, 34:57–122, 1889b.

W. Killing. Die Zusammensetzung der stetigen endlichen Transformationsgruppen. IV. *Math. Ann.*, 36:161–189, 1890.

A.W. Knapp. *Representation theory of semisimple groups. An overview based on examples.* Princeton University Press, Princeton, 2001.

A.W. Knapp. *Lie groups beyond an introduction*, volume 140 of *Prog. Math.* Birkhäuser Boston, Boston, second edition, 2002.

S. Kobayashi and K. Nomizu. *Foundations of differential geometry I.* Wiley Classics Libr. Wiley, New York, 1996a. Reprint of the 1963 original.

S. Kobayashi and K. Nomizu. *Foundations of differential geometry II.* Wiley Classics Libr. Wiley, New York, 1996b. Reprint of the 1969 original.

J. Korevaar. *Tauberian theory. A century of developments*, volume 329 of *Grundlehren Mat. Wiss.* Springer, Berlin, 2004.

B. Kostant. On the existence and irreducibility of certain series of representations. In I. Gel'fand, editor, *Lie groups and their representations*, pages 231–329. Halsted, New York, 1975.

S. Lang. SL(2, $\mathbb{R}$), volume 105 of *Graduate Texts in Math.* Springer, New York, 1985.

R.P. Langlands. On the classification of irreducible representations of real algebraic groups. In P. Sally and D. Vogan, editors, *Representation theory and harmonic analysis on semisimple Lie groups*, volume 31 of *Math. Surv. Monogr.*, pages 101–170. Am. Math. Soc., Providence, 1989.

N. Leonenko and A.Ja. Olenko. Tauberian and Abelian theorems for Hankel-type transformations, and limit theorems for functionals of random fields. *Dokl. Akad. Nauk Ukrain. SSR*, 9:61–64, 1991. In Russian.

N. Leonenko and A.Ja. Olenko. Tauberian and Abelian theorems for the correlation function of a homogeneous isotropic random field. *Ukr. Math. J.*, 43(12):1539–1548, 1992.

N. Leonenko and A.Ja. Olenko. Tauberian theorems for correlation functions and limit theorems for spherical averages of random fields. *Random Oper. Stoch. Equ.*, 1(1):57–67, 1993.

J. Lepowsky. Multiplicity formulas for certain semisimple Lie groups. *Bull. Am. Math. Soc.*, 77:601–605, 1971.

J.E. Littlewood. The converse of Abel's theorem on power series. *Proc. Lond. Math. Soc. (2)*, 9:434–448, 1911.

K. Mashimo. On the branching theorem of the pair (F$_4$, Spin$_9$). *Tsukuba J. Math.*, 30(1):31–47, 2006.

F.I. Mautner. Fourier analysis and symmetric spaces. *Proc. Natl. Acad. Sci. USA*, 37:529–533, 1951.

W. Miller. *Lie theory and special functions*, volume 43 of *Math. Sci. Eng.* Academic Press, New York, 1968.

W. Miller. *Symmetry and separation of variables*, volume 4 of *Encycl. Math. Appl.* Addison–Wesley, Reading, 1977.

D. Montgomery and L. Zippin. *Topological transformation groups.* Krieger, Nuntington, 1974.

S.A. Morris. *Pontryagin duality and the structure of locally compact abelian groups.* London Math. Soc. Lect. Note Ser. Cambridge University Press, Cambridge, 1977.

M.A. Naĭmark. *Normed algebras.* Wolters–Noordhoff Ser. Monogr. Textb. Pure Appl. Math. Wolters–Noordhoff, Groningen, third edition, 1972.

A.Ja. Olenko. Tauberian and Abelian theorems for random fields with strong dependence. *Ukr. Math. J.*, 48(3):412–427, 1996.

A.Ja. Olenko. A Tauberian theorem for fields with the OR spectrum. I. *Theory Probab. Math. Stat.*, 72:113–123, 2006.

A.Ja. Olenko. A Tauberian theorem for fields with the OR spectrum. II. *Theory Stoch. Process.*, 13(1–2):194–204, 2007a.

A.Ja. Olenko. Some properties of weight functions in Tauberian theorems. II. *Theory Stoch. Process.*, 13(1–2):194–204, 2007b.

A.Ja. Olenko and B. Klykavka. Some properties of weight functions in Tauberian theorems. I. *Theory Stoch. Process.*, 3–4(3–4):123–136, 2006a.

A.Ja. Olenko and B. Klykavka. A Tauberian theorem for fields on a plane. *Dopov. Nats. Akad. Nauk Ukr. Mat. Prirodozn. Tekh. Nauki*, 6:19–25, 2006b. In Ukrainian.

A.L. Onishchik and E.B. Vinberg. *Lie groups and algebraic groups.* Springer Ser. Sov. Math. Springer, Berlin, 1990.

R.E.A.C. Paley and N. Wiener. *Fourier transforms in the complex domain*, volume 19 of *Am. Math. Soc. Colloquium Publ.* Am. Math. Soc., Providence, 1934.

M. Plancherel. Contribution à l'étude de la représentation d'une fonction arbitraire par les intégrales définies. *Rend. Circ. Mat. Palermo*, 30:289–335, 1910.

L.S. Pontryagin. *Topological groups.* Gordon & Breach, New York, 1966.

C. Procesi. *Lie groups. An approach through invariants and representations.* Universitext. Springer, New York, 2007.

A.P. Prudnikov, Yu.A. Brychkov, and O.I. Marichev. *Integrals and series. Vol. 1. Elementary functions*. Gordon & Breach, New York, 1986.

A.P. Prudnikov, Yu.A. Brychkov, and O.I. Marichev. *Integrals and series. Vol. 2. Special functions*. Gordon & Breach, New York, second edition, 1988.

A.P. Prudnikov, Yu.A. Brychkov, and O.I. Marichev. *Integrals and series. Vol. 3. More special functions*. Gordon & Breach, New York, 1990.

A.P. Prudnikov, Yu.A. Brychkov, and O.I. Marichev. *Integrals and series. Vol. 4. Direct Laplace transforms*. Gordon & Breach, New York, 1992a.

A.P. Prudnikov, Yu.A. Brychkov, and O.I. Marichev. *Integrals and series. Vol. 5. Inverse Laplace transforms*. Gordon & Breach, New York, 1992b.

W. Rossmann. *Lie groups. An introduction through linear groups*. Oxford Graduate Texts in Math. Oxford University Press, Oxford, 2002.

J.J. Rotman. *An introduction to homological algebra*. Universitext. Springer, New York, second edition, 2009.

Yu.S. Samoĭlenko. *Spectral theory of families of selfadjoint operators*, volume 57 of *Math. Appl. (Soviet Ser.)*. Kluwer Academic, Dordrecht, 1991. Translated from the Russian by E.V. Tisjachnij.

I.E. Segal. An extension of Plancherel's formula to separable unimodular groups. *Ann. of Math. (2)*, 52:272–292, 1950. URL http://www.jstor.org/stable/1969470.

E. Seneta. *Regularly varying functions*, volume 508 of *Lect. Notes Math*. Springer, Berlin, 1976.

S. Ström. A note on the matrix element of a unitary representation of the homogeneous Lorentz group. *Ark. Fys.*, 33:465–469, 1967.

S. Ström. Matrix elements of the supplementary series of unitary representations of SL(2, ℂ). *Ark. Fys.*, 38:373–381, 1968.

A. Tauber. Ein Satz aus der Theorie der unendlichen Reihen. *Monatsh. Math. U. Phys.*, 8:273–277, 1897.

V.S. Varadarajan. *Lie groups, Lie algebras, and their representations*, volume 102 of *Graduate Texts in Math*. Springer, New York, 1984.

V.S. Varadarajan. *An introduction to harmonic analysis on semisimple Lie groups*, volume 16 of *Cambridge Stud. Adv. Math*. Cambridge University Press, Cambridge, 1999.

D.A. Varshalovich, A.N. Moskalev, and V.K. Khersonskiĭ. *Quantum theory of angular momentum. Irreducible tensors, spherical harmonics, vector coupling coefficients, 3nj symbols*. World Scientific, Teaneck, 1988.

N.Ya. Vilenkin. *Special functions and the theory of group representations*, volume 22 of *Transl. Math. Monogr*. Am. Math. Soc., Providence, 1968. Translated from the Russian by V.N. Singh.

N.Ya. Vilenkin and A.U. Klimyk. *Representation of Lie groups and special functions. Vol. 1. Simplest Lie groups, special functions and integral transforms*, volume 72 of *Math. Appl. (Sov. Ser.)*. Kluwer Academic, Dordrecht, 1991.

N.Ya. Vilenkin and A.U. Klimyk. *Representation of Lie groups and special functions. Vol. 3. Classical and quantum groups and special functions*, volume 75 of *Math. Appl. (Sov. Ser.)*. Kluwer Academic, Dordrecht, 1993.

N.Ya. Vilenkin and A.U. Klimyk. *Representation of Lie groups and special functions. Vol. 2. Class I representations, special functions, and integral transforms*, volume 74 of *Math. Appl. (Sov. Ser.)*. Kluwer Academic, Dordrecht, 1994.

N.Ya. Vilenkin and A.U. Klimyk. *Representation of Lie groups and special functions. Recent advances*, volume 316 of *Math. Appl*. Kluwer Academic, Dordrecht, 1995.

D.A. Vogan. *Unitary representations of reductive Lie groups*, volume 118 of *Ann. Math. Stud*. Princeton University Press, Princeton, 1987.

D.A. Vogan. The character table for E_8. *Not. Am. Math. Soc.*, 54(9):1122–1134, 2007.

N.R. Wallach. *Real reductive groups. I*, volume 132 of *Pure Appl. Math*. Academic Press, Boston, 1988.

N.R. Wallach. *Real reductive groups. II*, volume 132-II of *Pure Appl. Math*. Academic Press, Boston, 1992.

A. Weil. *L'intégration dans les groupes topologiques et ses applications.* Number 869 in *Actual. Sci. Ind.* Hermann, Paris, 1940.

H. Weyl. Harmonics on homogeneous manifolds. *Ann. of Math. (2),* 35(3):486–499, 1934. URL http://www.jstor.org/stable/1968746.

H. Whitney. Differentiable manifolds. *Ann. of Math. (2),* 37(3):645–680, 1936. URL http://www.jstor.org/stable/1968482.

N. Wiener. Tauberian theorems. *Ann. of Math. (2),* 33(1):1–100, 1930.

E. Wigner. On unitary representations of the inhomogeneous Lorentz group. *Ann. of Math. (2),* 40(1):149–204, 1939. URL http://www.jstor.org/stable/1968551.

D.P. Zhelobenko. *Compact Lie groups and their representations,* volume 40 of *Transl. Math. Monogr.* Am. Math. Soc., Providence, 1973.

D.P. Zhelobenko. *Principal structures and methods of representation theory,* volume 228 of *Transl. Math. Monogr.* Am. Math. Soc., Providence, 2006.

Index

A. Malyarenko, *Invariant Random Fields on Spaces with a Group Action*,
Probability and Its Applications, DOI 10.1007/978-3-642-33406-1,
© Springer-Verlag Berlin Heidelberg 2013